Graph Partitioning and Graph Clustering

CONTEMPORARY MATHEMATICS

588

Graph Partitioning and Graph Clustering

10th DIMACS Implementation Challenge Workshop
February 13–14, 2012
Georgia Institute of Technology
Atlanta, GA

David A. Bader
Henning Meyerhenke
Peter Sanders
Dorothea Wagner
Editors

American Mathematical Society
Center for Discrete Mathematics
and Theoretical Computer Science

American Mathematical Society
Providence, Rhode Island

2010 *Mathematics Subject Classification.* Primary 05C85, 68W05, 05C82, 68W10, 68R05, 68R10, 05C05, 05C65.

Library of Congress Cataloging-in-Publication Data

Cataloging-in-Publication Data has been applied for by the AMS. See http://www.loc.gov/publish/cip/.

Contemporary Mathematics ISSN: 0271-4132 (print); ISSN: 1098-3627 (online)

∞ The paper used in this book is acid-free and falls within the guidelines
established to ensure permanence and durability.
Visit the AMS home page at `http://www.ams.org/`

10 9 8 7 6 5 4 3 2 1 13 12 11 10 09 08

Contents

Preface

This collection is related to the Workshop of the 10th DIMACS Implementation Challenge, which took place in Atlanta, Georgia (USA) on February 13-14, 2012. The purpose of DIMACS Implementation Challenges[1] is to assess the practical performance of algorithms in a respective problem domain. These challenges are scientific competitions in areas of interest where worst case and probabilistic analysis yield unrealistic results. Where analysis fails, experimentation can provide insights into realistic algorithm performance and thereby help to bridge the gap between theory and practice. For this purpose common benchmark instances, mostly from real applications, are established. By evaluating different implementations on these instances, the challenges create a reproducible picture of the state of the art in the area under consideration. This helps to foster an effective technology transfer within the research areas of algorithms, data structures, and implementation techniques as well as a transfer back to the original applications.

The topics of the previous nine challenges are as follows (in chronological order): *Network Flows and Matching* (1990-91), *Maximum Clique, Graph Coloring and Satisfiability* (1992-93), *Parallel Algorithms for Combinatorial Problems* (1993-94), *Fragment Assembly and Genome Rearrangements* (1994-95), *Priority Queues, Dictionaries, and Multi-Dimensional Point Sets* (1995-96), *Near Neighbor Searches* (1998-99), *Semidefinite and Related Optimization Problems* (1999-2000), *The Traveling Salesman Problem* (2000-01), and *Shortest Path Problems* (2005-06).

1. Introducing the 10th Challenge – Graph Partitioning and Graph Clustering

The 10th challenge considered the two related problems of partitioning and clustering graphs. Both are ubiquitous subtasks in many application areas. Generally speaking, techniques for graph partitioning and graph clustering aim at the identification of vertex subsets with many internal and few external edges. To name only a few, problems addressed by graph partitioning and graph clustering algorithms are:

- What are the communities within an (online) social network?
- How do I speed up a numerical simulation by mapping it efficiently onto a parallel computer?
- How must components be organized on a computer chip such that they can communicate efficiently with each other?
- What are the segments of a digital image?
- Which functions are certain genes (most likely) responsible for?

[1] `http://dimacs.rutgers.edu/Challenges/`

For a more detailed treatment of applications and solution techniques, the interested reader is referred to the surveys of Fortunato[2], Schaeffer[3], and Schloegel et al.[4] on the different topics.

Within the algorithms community, techniques for solving the problems above have been developed at least since the early 1970s—while some of the applications are newer. Improving known and developing new solution techniques are aspects of ongoing research.

The primary goal of this challenge was to create a reproducible picture of the state of the art in the area of graph partitioning and graph clustering algorithms. To this end, a standard set of benchmark instances was identified. Then participants were invited to submit solutions to different challenge problems. This way different algorithms and implementations were tested against the benchmark instances. Thereby future researchers are enabled to identify techniques that are most effective for a respective partitioning or clustering problem—by using our benchmark set and by comparing their results to the challenge results.

2. Key Results

The main results of the 10th DIMACS Implementation Challenge include:

- Extension of a file format used by several graph partitioning and graph clustering libraries for graphs, their geometry, and partitions. Formats are described on the challenge website.[5]
- Collection and online archival[5] of a common testbed of input instances and generators (including their description) from different categories for evaluating graph partitioning and graph clustering algorithms. For the actual challenge, a core subset of the testbed was chosen.
- Definition of a new combination of measures to assess the quality of a clustering.
- Definition of a measure to assess the work an implemention performs in a parallel setting. This measure is used to normalize sequential and parallel implementations to a common base line.
- Experimental evaluation of state-of-the-art implementations of graph partitioning and graph clustering codes on the core input families.
- A nondiscriminatory way to assign scores to solvers that takes both running time and solution quality into account.
- Discussion of directions for further research in the areas of graph partitioning and graph clustering.
- The paper *Benchmarks for Network Analysis*, which was invited as a contribution to the *Encyclopedia of Social Network Analysis and Mining.*

The primary location of information regarding the 10th DIMACS Implementation Challenge is the website `http://www.cc.gatech.edu/dimacs10/`.

---

[2]Santo Fortunato, Community detection in graphs, Physics Reports 486 (2010), no. 3–5, 75–174.

[3]Satu E. Schaeffer, Graph clustering, Computer Science Review 1 (2007), no. 1, 27–64.

[4]K. Schloegel, G. Karypis, and V. Kumar, Graph partitioning for high-performance scientific simulations, Sourcebook of parallel computing (Jack Dongarra, Ian Foster, Geoffrey Fox, William Gropp, Ken Kennedy, Linda Torczon, and Andy White, eds.) Morgan Kaufmann Publishers, 2003, pp. 491–541.

[5]`http://www.cc.gatech.edu/dimacs10/downloads.shtml`

3. Challenge Description

3.1. Data Sets. The collection of benchmark inputs of the 10th DIMACS Implementation Challenge includes both synthetic and real-world data. All graphs are undirected. Formerly directed instances were symmetrized by making every directed edge undirected. While this procedure necessarily loses information in a number of real-world applications, it appeared to be necessary since most existing software libraries can handle undirected graphs only. Directed graphs (or unsymmetric matrices) are left for further work.

Synthetic graphs in the collection include random graphs (Erdős-Rényi, R-MAT, random geometric graphs using the unit disk model), Delaunay triangulations, and graphs that mimic meshes from dynamic numerical simulations. Real-world inputs consist of co-author and citation networks, road networks, numerical simulation meshes, web graphs, social networks, computational task graphs, and graphs from adapting voting districts (redistricting).

For the actual challenge two subsets were chosen, one for graph partitioning and one for graph clustering. The first one (for graph partitioning) contained 18 graphs, which had to be partitioned into 5 different numbers of parts each, yielding 90 problem instances. The second one (for graph clustering) contained 30 graphs. Due to the choice of objective functions for graph clustering, no restriction on the number of parts or their size was necessary in this category.

3.2. Categories. One of the main goals of the challenge was to compare different techniques and algorithmic approaches. Therefore participants were invited to join different challenge competitions aimed at assessing the performance and solution quality of different implementations. Let $G = (V, E, \omega)$ be an undirected graph with edge weight function ω.

3.2.1. *Graph Partitioning.* Here the task was to compute a partition Π of the vertex set V into k parts of size at most $(1+\epsilon)\lceil\frac{|V|}{k}\rceil$. The two objective functions used to assess the partitioning quality were edge cut (EC, total number of edges with endpoints in different parts) and maximum communication volume (CV). CV sums for each part p and each vertex v therein the number of parts adjacent to v but different from p. The final result is the maximum over each part.

For each instance result (EC and CV results were counted as one instance *each*), the solvers with the first six ranks received a descending number of points (10, 6, 4, 3, 2, 1), a scoring system borrowed from former Formula 1 rules.

Three groups submitted solutions to the graph partitioning competition. Only one of the submitted solvers is a graph partitioner by nature, the other two are actually hypergraph partitioners. Both hypergraph partitioners use multilevel recursive bisection. While their quality, in particular for the communication volume, is generally not bad, the vast majority of best ranked solutions (139 out of 170) are held by the graph partitioner KaPa.

3.2.2. *Graph Clustering.* The clustering challenge was divided into two separate competitions with different optimization criteria. For the first competition the objective modularity had to be optimized. Modularity has been a very popular measure in the last years, in particular in the field of community detection. It follows the intra-cluster-density vs. inter-cluster-sparsity paradigm. However, some criticism has emerged recently.[6] Also, solvers performing implicit optimization based on the intra-cluster-density vs. inter-cluster-sparsity paradigm were supposed to have a fair chance, too. That is why we developed a second competition with a mix of four other clustering objectives. The rationale was that the combination of these measures would lead to meaningful clusters and avoid pathological cases of single measures. The exact definition of the objective functions can be found at the challenge website.[7]

The modularity competition saw the largest number of entries, with 15 solvers from eight groups. Two solvers led the field, CGGCi_RG and VNS. Of the two, CGGCi_RG scored the most points and obtained the highest number of best ranked solutions. The four solvers entering the mix clustering competition were submitted by two groups (two each). Three solvers headed the top of the ranking, with a slight advantage for the two COMMUNITY-EL implementations.

3.2.3. *Pareto Challenges.* For all quality competitions there was one corresponding Pareto challenge. The rationale of the Pareto challenges was to take the work into account an algorithm requires to compute a solution. Hence, the two dimensions considered here were *quality* and *work*. Work was normalized with respect to the machine performance, measured by a graph-based benchmark. To this end, we used the shortest path benchmark produced for the 9th DIMACS Implementation Challenge. Participants were asked to run this sequential benchmark on their machine. Both the performance obtained in the shortest path benchmark and the number of processing cores (raised to the power of 0.9) used for the 10th DIMACS Implementation Challenge were taken into account for normalizing the amount of work invested for obtaining the solution.

For each challenge instance result, each submitted solver received a Pareto dominance count, which expresses by how many other algorithms it was Pareto-dominated in terms of work *and* running time; then algorithms were ranked by this number (lower count = better) and received points according to the Formula 1 scoring scheme described above.

Several groups submitted solutions from more than one solver to the respective Pareto challenges, making use of the fact that here a lower solution quality might be compensated by a better running time and vice versa. Still, the Pareto challenges were won in all cases by the same groups that also won the respective quality competitions. We attribute this double success (i) to the superior quality which could not be dominated in many cases and (ii) to the Formula 1 scoring scheme, which might have given an advantage to groups who submitted solutions from several solvers. More information on the challenge results are available online.[8]

3.3. URL to Resources. The main website of the 10th DIMACS Implementation Challenge can be found at its permanent location `http://www.cc.gatech.edu/dimacs10/`. The following subdirectories contain:

- `archive/data/`: Testbed instances archived for long-term access.
- `talks/`: Slides of the talks presented at the workshop.
- `papers/`: Papers on which the workshop talks are based.

[6] Andrea Lancichinetti and Santo Fortunato, Limits of modularity maximization in community detection, Phys. Rev. E 84 (2011), 066122.

[7] `http://www.cc.gatech.edu/dimacs10/data/dimacs10-rules.pdf`

[8] `http://www.cc.gatech.edu/dimacs10/talks/orga-talk-dimacs-results.pdf`

- `results/`: Partitions submitted as part of the challenge as well as code for their evaluation and the resulting data

All respective files can be found and downloaded by following links from the homepage. Researchers are particularly encouraged to download and use the graphs we compiled and archived.

4. Contributions to this Collection

In this section we give a short overview of the papers that were selected for this collection. All of them were presented at the Workshop of the 10th DIMACS Implementation Challenge and contributed to the success of the event. Not all solvers described in these papers actually entered the challenge. Also, not all solvers that entered the challenge are part of this collection.

4.1. Graph Partitioning. The winner in terms of graph partitioning quality was KaPa, by Sanders and Schulz, described in their paper *High Quality Graph Partitioning.* KaPa combines the solutions of several related solvers developed by the same authors. It is a set of algorithms which use a combination of strategies. Among these strategies are network flows, evolutionary algorithms, edge ratings for approximate maximum weighted matchings in the multilevel process, repetitive improvement cycles, and problem-specific local search techniques based on the Fiduccia-Mattheyses (FM) heuristic.

Abusing a Hypergraph Partitioner for Unweighted Graph Partitioning, by Fagginger Auer and Bisseling, describes Mondriaan, a package for matrix and hypergraph partitioning, and its (ab)use for graph partitioning. While Mondriaan usually computes worse edge cuts than state-of-the-art graph partitioners, the solutions are generally acceptable.

In *Parallel Partitioning with Zoltan: Is Hypergraph Partitioning Worth It?*, Rajamanickam and Boman describe a partitioner which is very powerful in that it is designed for scalable parallelism on large asymmetric hypergraphs.

Çatalyürek, Deveci, Kaya, and Uçar present in *UMPa: A Multi-objective, multi-level partitioner* a system doing recursive multi-objective hypergraph bipartitioning that takes the bottleneck communication volume as primary objective function into account but also looks for solutions with small total communication.

The related task of repartitioning dynamic graphs is addressed by Meyerhenke in *Shape Optimizing Load Balancing for MPI-Parallel Adaptive Numerical Simulations.* Diffusive methods are employed to determine both *how many* elements have to migrate between processors as well as *which* elements are chosen for migration. The properties of the diffusive processes usually lead to nicely shaped partitions.

In *Graph Partitioning for Scalable Distributed Graph Computations*, by Buluc and Madduri, the authors develop a method for partitioning large-scale sparse graphs with skewed degree distribution. The approach aims to partition the graph into balanced parts with low edge cuts, a challenge for these types of graphs, so that they can be used on distributed-memory systems where communication is often a major bottleneck in running time. The authors derive upper bounds on the communication costs incurred for a two-dimensional partitioning during breadth-first search. The performance results using the large-scale DIMACS challenge graphs shows that reducing work and communication imbalance among partitions is more important than minimizing the total edge cut.

4.2. Graph Clustering. *Using Graph Partitioning for Efficient Network Modularity Optimization*, by Djidjev and Onus, describes how to formulate modularity maximization in graph clustering as a minimum cut problem in a complete weighted graph. In general, the according graph contains also negative weights. However, the resulting minimum cut problem can be attacked by applying modifications of existing powerful codes for graph partitioning.

The solver VNS, by Aloise, Caporossi, Hansen, Liberti, and Perron, performs *Modularity Maximization in Networks by Variable Neighborhood Search*, a metaheuristic and variant of local search. A local search or improving heuristic consists of defining a neighborhood of a solution, choosing an initial solution x, and then moving to the best neighbor x' of x if the objective function value is improved. If no such neighbor exists, the heuristic stops, otherwise it is iterated. VNS improves this simple technique to escape from local optima. To this end, it applies the idea of neighborhood change. By increasing the neighborhood distance iteratively, even "mountain tops" surrounding local optima can be escaped.

The algorithm family K-COMMUNITY, developed by Verma and Butenko in *Network Clustering via Clique Relaxations: A Community Based Approach*, are based on the relaxation concept of a *generalized community*. Instead of requiring a community to be a perfect clique, a generalized k-community is defined as a connected subgraph such that the incident vertices of every edge have at least k common neighbors within the subgraph. The algorithm family computes clusters by finding k-communities for large (variable) k and placing them in different clusters.

Identifying Base Clusters for Maximizing Modularity, by Srinivasan, Chakraborty, and Bhowmick, introduces the concept of identifying base clusters as a preprocessing step for agglomerative modularity maximization methods. Base clusters are groups of vertices that are always assigned to the same community, independent of the modularity maximization algorithm employed or the order in which the vertices are processed. In a computational study on two agglomerative modularity maximization methods, the CNM method introduced by Clauset et al. and the Louvain method by Blondel et al., the effect of using base clusters as a preprocessing is shown.

Complete Hierarchical Cut-Clustering: A Case Study on Expansion and Modularity, by Hamann, Hartmann, and Wagner, studies the behavior of the cut-clustering algorithm of Flake et al., a clustering approach which is based on minimum s-t-cuts. The algorithm uses a parameter that provides a quality guarantee on the clusterings in terms of expansion. This is particularly interesting since expansion is a measure which is already NP-hard to compute. While Flake et al. examine their algorithm with respect to the semantic meaning of the clusters, Hamann et al. systematically analyze the quality of the clusterings beyond the guaranteed bounds with respect to the approved measures expansion and modularity.

In *A Partitioning-based divisive clustering technique for maximizing the modularity*, by Çatalyürek, Kaya, Langguth and Uçar, the authors present a new, divisive algorithm for computing high modularity clusterings. The approach is based upon recursive bipartitions using graph partitioning subroutines, and steps for refining the obtained clusters. The study includes an experimental evaluation. On a variety of problem instances from the literature, this new method performs well, and in a number of cases, finds the best known modularity scores on these test graphs.

An Ensemble Learning Strategy for Graph Clustering, by Ovelgönne and Geyer-Schulz, describes the heuristic CGGCi_RG, whose main idea is to combine several weak classifiers into a strong classifier. From the maximal overlap of clusterings computed by weak classifiers, the algorithm searches for a solution with high quality. This way difficult choices are deferred after easy decisions have been fixed, which leads to a high quality due to a better control of the search space traversal. It turns out that the quality of the initial clusterings is of minor importance for the quality of the final result given enough iterations.

While graph partitioning is rooted in the parallel computing community, the picture appears to be different for graph clustering as only two clustering papers employ significant parallelism. The agglomerative algorithm in *Parallel Community Detection for Massive Graphs*, by Riedy, Meyerhenke, Ediger, and Bader, starts out with each vertex as its own cluster. In each following iteration, beneficial cluster merges improving the objective function value are identified and performed in parallel by means of weighted matchings. The implementation is capable of clustering graphs with a few billion edges in less than 10 minutes on a standard Intel-based server.

The second paper that uses considerable parallelism to accelerate the solution process is *Graph Coarsening and Clustering on the GPU*, by Fagginger Auer and Bisseling. This paper also uses an agglomerative approach with matchings. It alleviates the problem of small matchings due to star subgraphs by merging siblings, i. e., neighbors of neighbors that do not share an edge. High performance is achieved by careful algorithm design, optimizing the interplay of the CPU and the employed graphics hardware.

5. Directions for Further Research

In the field of graph partitioning, important directions for further research mentioned at the workshop are the widespread handling of directed graphs (or unsymmetric matrices in case of matrix partitioning) and an improved consideration of the objective function maximum communication volume. One possible approach—also presented at the workshop—is to use hypergraphs instead of graphs. But this seems to come at the price of worse performance and/or worse edge cut quality. For the related problem of repartitioning with migration minimization, highly scalable tools with a good solution quality are sought.

An active graph clustering research area is the development of objective functions whose optimization leads to realistic and meaningful clusterings. While modularity has been very popular over recent years, current studies show that its deficiencies can be severe and hard to avoid. The analysis of massive graphs for clustering purposes is still in its infancy. Only two submissions for the graph clustering challenge made use of significant parallelism. And only one of them was able to process the largest graph in the challenge core benchmark, a web graph with 3.3 billion edges. Considering the size of today's online social networks and WWW (to name a few), there is a need to scale the analysis algorithms to larger input sizes.

High quality graph partitioning

Peter Sanders and Christian Schulz

ABSTRACT. We present an overview over our graph partitioners KaFFPa (Karlsruhe Fast Flow Partitioner) and KaFFPaE (KaFFPa Evolutionary). KaFFPa is a multilevel graph partitioning algorithm which on the one hand uses novel local improvement algorithms based on max-flow and min-cut computations and more localized FM searches and on the other hand uses more sophisticated global search strategies transferred from multi-grid linear solvers. KaFFPaE is a distributed evolutionary algorithm to solve the Graph Partitioning Problem. KaFFPaE uses KaFFPa and provides new effective crossover and mutation operators. By combining these with a scalable communication protocol we obtain a system that is able to improve the best known partitioning results for many inputs.

1. Introduction

Problems of *graph partitioning* arise in various areas of computer science, engineering, and related fields. For example in route planning, community detection in social networks and high performance computing. In many of these applications large graphs need to be partitioned such that there are few edges between blocks (the elements of the partition). For example, when you process a graph in parallel on k processors you often want to partition the graph into k blocks of about equal size so that there is as little interaction as possible between the blocks. In this paper we focus on a version of the problem that constrains the maximum block size to $(1+\epsilon)$ times the average block size and tries to minimize the total cut size, i.e., the number of edges that run between blocks. It is well known that this problem is NP-complete [**5**] and that there is no approximation algorithm with a constant ratio factor for general graphs [**5**]. Therefore mostly heuristic algorithms are used in practice. A successful heuristic for partitioning large graphs is the *multilevel graph partitioning* (MGP) approach depicted in Figure 1 where the graph is recursively *contracted* to achieve smaller graphs which should reflect the same structure as the input graph. After applying an *initial partitioning* algorithm to the smallest graph, the contraction is undone and, at each level, a *local refinement* method is used to improve the partitioning induced by the coarser level.

Although several successful multilevel partitioners have been developed in the last 13 years, we had the impression that certain aspects of the method are not well understood. We therefore have built our own graph partitioner KaPPa [**13**]

2010 *Mathematics Subject Classification.* Primary 68W40, 68W10, 90C27, 05C70.
Partially supported by DFG SA 933/10-1.

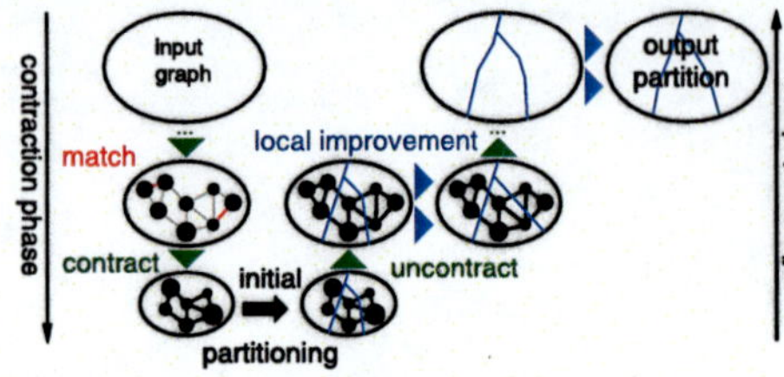

FIGURE 1. Multilevel graph partitioning.

(Karlsruhe Parallel Partitioner) with focus on scalable parallelization. Somewhat astonishingly, we also obtained improved partitioning quality through rather simple methods. This motivated us to make a fresh start putting all aspects of MGP on trial. This paper gives an overview over our most recent work, KaFFPa [**22**] and KaFFPaE [**21**]. KaFFPa is a classical matching based graph partitioning algorithm with focus on local improvement methods and overall search strategies. It is a system that can be configured to either achieve the best known partitions for many standard benchmark instances or to be the fastest available system for large graphs while still improving partitioning quality compared to the previous fastest system.

KaFFPaE is a technique which integrates an evolutionary search algorithm with our multilevel graph partitioner KaFFPa and its scalable parallelization. It uses novel mutation and combine operators which in contrast to previous evolutionary methods that use a graph partitioner [**8, 23**] do not need random perturbations of edge weights. The combine operators enable us to combine individuals of different kinds (see Section 5 for more details). Due to the parallelization our system is able to compute partitions that have quality comparable or better than previous entries in Walshaw's well known partitioning benchmark *within a few minutes* for graphs of moderate size. Previous methods of Soper et. al [**23**] required runtimes of up to one week for graphs of that size. We therefore believe that in contrast to previous methods, our method is very valuable in the area of high performance computing.

The paper is organized as follows. We begin in Section 2 by introducing basic concepts which is followed by related work in Section 3. In Section 4 we present the techniques used in the multilevel graph partitioner KaFFPa. We continue describing the main components of our evolutionary algorithm KaFFPaE in Section 5. A summary of extensive experiments to evaluate the performance of the algorithm is presented in Section 6. We have implemented these techniques in the graph partitioner KaFFPaE (Karlsruhe Fast Flow Partitioner Evolutionary) which is written in C++. Experiments reported in Section 6 indicate that KaFFPaE is able to compute partitions of very high quality and scales well to large networks and machines.

2. Preliminaries

2.1. Basic concepts. Consider an undirected graph $G = (V, E, c, \omega)$ with edge weights $\omega : E \to \mathbb{R}_{>0}$, node weights $c : V \to \mathbb{R}_{\geq 0}$, $n = |V|$, and $m = |E|$. We extend c and ω to sets, i.e., $c(V') := \sum_{v \in V'} c(v)$ and $\omega(E') := \sum_{e \in E'} \omega(e)$. $\Gamma(v) := \{u : \{v, u\} \in E\}$ denotes the neighbors of v. We are looking for *blocks* of nodes $V_1,\ldots,V_k$ that partition V, i.e., $V_1 \cup \cdots \cup V_k = V$ and $V_i \cap V_j = \emptyset$ for $i \neq j$. The *balancing constraint* demands that $\forall i \in \{1..k\} : c(V_i) \leq L_{\max} :=$

$(1+\epsilon)c(V)/k + \max_{v\in V} c(v)$ for some parameter ϵ. The last term in this equation arises because each node is atomic and therefore a deviation of the heaviest node has to be allowed. The objective is to minimize the total *cut* $\sum_{i<j} w(E_{ij})$ where $E_{ij} := \{\{u,v\} \in E : u \in V_i, v \in V_j\}$. A clustering is also a partition of the nodes, however k is usually not given in advance and the balance constraint is removed. A vertex $v \in V_i$ that has a neighbor $w \in V_j, i \neq j$, is a boundary vertex. An abstract view of the partitioned graph is the so called *quotient graph*, where vertices represent blocks and edges are induced by connectivity between blocks. Given two clusterings $\mathcal{C}_1$ and $\mathcal{C}_2$ the *overlay clustering* is the clustering where each block corresponds to a connected component of the graph $G_{\mathcal{E}} = (V, E\backslash\mathcal{E})$ where $\mathcal{E}$ is the union of the cut edges of $\mathcal{C}_1$ and $\mathcal{C}_2$, i.e. all edges that run between blocks in $\mathcal{C}_1$ or $\mathcal{C}_2$. We will need the of overlay clustering to define a combine operation on partitions in Section 5. By default, our initial inputs will have unit edge and node weights. However, even those will be translated into weighted problems in the course of the algorithm.

A matching $M \subseteq E$ is a set of edges that do not share any common nodes, i.e., the graph (V, M) has maximum degree one. *Contracting* an edge $\{u,v\}$ means to replace the nodes u and v by a new node x connected to the former neighbors of u and v. We set $c(x) = c(u) + c(v)$ so the weight of a node at each level is the number of nodes it is representing in the original graph. If replacing edges of the form $\{u,w\}$,$\{v,w\}$ would generate two parallel edges $\{x,w\}$, we insert a single edge with $\omega(\{x,w\}) = \omega(\{u,w\}) + \omega(\{v,w\})$. *Uncontracting* an edge e undoes its contraction. In order to avoid tedious notation, G will denote the current state of the graph before and after a (un)contraction unless we explicitly want to refer to different states of the graph. The *multilevel approach* to graph partitioning consists of three main phases. In the *contraction* (coarsening) phase, we iteratively identify matchings $M \subseteq E$ and contract the edges in M. Contraction should quickly reduce the size of the input and each computed level should reflect the structure of the input network. Contraction is stopped when the graph is small enough to be directly partitioned using some expensive other algorithm. In the *refinement* (or uncoarsening) phase, the matchings are iteratively uncontracted. After uncontracting a matching, a refinement algorithm moves nodes between blocks in order to improve the cut size or balance.

3. Related Work

There has been a huge amount of research on graph partitioning so that we refer the reader to [**26**] for more material on multilevel graph partitioning and to [**15**] for more material on genetic approaches for graph partitioning. All general purpose methods that are able to obtain good partitions for large real world graphs are based on the multilevel principle outlined in Section 2. Well known software packages based on this approach include, Jostle [**26**], Metis [**14**], and Scotch [**20**]. KaSPar [**19**] is a graph partitioner based on the central idea to (un)contract only a single edge between two levels. KaPPa [**13**] is a "classical" matching based MGP algorithm designed for scalable parallel execution. MQI [**16**] and Improve [**2**] are flow-based methods for improving graph cuts when cut quality is measured by quotient-style metrics such as *expansion* or *conductance*. This approach is only feasible for $k = 2$. Improve uses several minimum cut computations to improve the *quotient cut* score of a proposed partition. Soper et al. [**23**] provided the first algorithm that combined an evolutionary search algorithm with a multilevel graph partitioner. Here crossover and mutation operators have been used to compute edge biases, which yield hints for the underlying multilevel graph partitioner. Benlic et al. [4] provided a multilevel memetic algorithm for balanced graph partitioning. This approach is able to compute many entries in Walshaw's Benchmark Archive [23] for the case $\epsilon = 0$. Very recently an algorithm called PUNCH [8] has been introduced. This approach is not based on the multilevel principle. However, it creates a coarse version of the graph based on the notion of natural cuts. Natural cuts are relatively sparse cuts close to denser areas. They are discovered by finding minimum cuts between carefully chosen regions of the graph. They introduced an evolutionary algorithm which is similar to Soper et al. [23], i.e. using a combine operator that computes edge biases yielding hints for the underlying graph partitioner. Experiments indicate that the algorithm computes very good partitions for road networks. For instances without a natural structure natural cuts are not very helpful.

4. Karlsruhe Fast Flow Partitioner

The aim of this section is to provide an overview over the techniques used in KaFFPa which is used by KaFFPaE as a base case partitioner. KaFFPa [22] is a classical matching based multilevel graph partitioner. Recall that a multilevel graph partitioner basically has three phases: coarsening, initial partitioning and uncoarsening.

Coarsening. KaFFPa makes contraction more systematic by separating two issues: A *rating function* indicates how much sense it makes to contract an edge based on *local* information. A *matching* algorithm tries to maximize the sum of the ratings of the contracted edges looking at the *global* structure of the graph. While the rating function allows a flexible characterization of what a "good" contracted graph is, the simple, standard definition of the matching problem allows to reuse previously developed algorithms for weighted matching. Matchings are contracted until the graph is "small enough". In [13] we have observed that the rating function $\text{expansion}^{*2}(\{u,v\}) := \frac{\omega(\{u,v\})^2}{c(u)c(v)}$ works best among other edge rating functions, so that this rating function is also used in KaFFPa.

KaFFPa employs the *Global Path Algorithm (GPA)* as a matching algorithm. It was proposed in [17] as a synthesis of the Greedy algorithm and the Path Growing Algorithm [10]. This algorithm achieves a half-approximation in the worst case, but empirically, GPA gives considerably better results than Sorted Heavy Edge Matching and Greedy (for more details see [13]). GPA scans the edges in order of decreasing weight but rather than immediately building a matching, it first constructs a collection of paths and even cycles. Afterwards, optimal solutions are computed for each of these paths and cycles using dynamic programming.

Initial Partitioning. The contraction is stopped when the number of remaining nodes is below the threshold $\max(60k, n/(60k))$. The graph is then small enough to be partitioned by some initial partitioning algorithm. KaFFPa employs Scotch as an initial partitioner since it empirically performs better than Metis.

Uncoarsening. Recall that the refinement phase iteratively uncontracts the matchings contracted during the contraction phase. After a matching is uncontracted, local search based refinement algorithms move nodes between block boundaries in order to reduce the cut while maintaining the balancing constraint. Local improvement algorithms are usually variants of the FM-algorithm [12]. Our variant

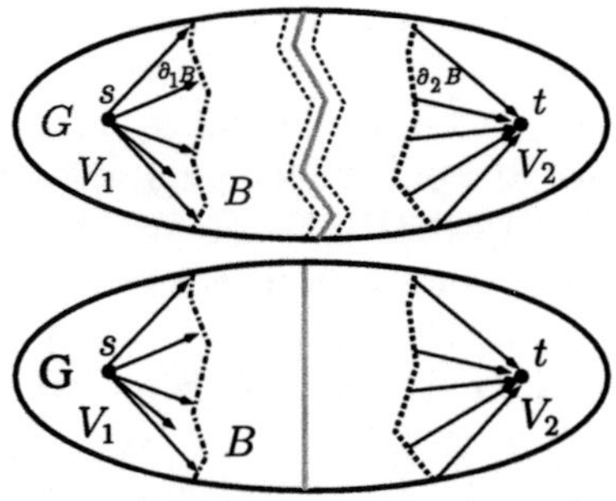

FIGURE 2. The construction of a feasible flow problem G' is shown on the top and an improved cut within the balance constraint in G is shown on the bottom.

of the algorithm is organized in rounds. In each round, a priority queue P is used which is initialized with all vertices that are incident to more than one block, in a random order. The priority is based on the gain $g(v) = \max_P g_P(v)$ where $g_P(v)$ is the decrease in edge cut when moving v to block P. Ties are broken randomly if there is more than one block that yields the maximum gain when moving v to it. Local search then repeatedly looks for the highest gain node v. Each node is moved at most once within a round. After a node is moved its unmoved neighbors become eligible, i.e. its unmoved neighbors are inserted into the priority queue. When a stopping criterion is reached all movements to the best found cut that occurred within the balance constraint are undone. This process is repeated several times until no improvement is found.

Max-Flow Min-Cut Local Improvement. During the uncoarsening phase KaFFPa additionally uses more advanced refinement algorithms. The first method is based on max-flow min-cut computations between pairs of blocks, i.e., a method to improve a given bipartition. Roughly speaking, this improvement method is applied between all pairs of blocks that share a non-empty boundary. The algorithm basically constructs a flow problem by growing an area around the given boundary vertices of a pair of blocks such that each min cut in this area yields a feasible bipartition of the original graph *within* the balance constraint. We explain how flows can be employed to improve a partition of *two blocks* V_1, V_2 without violating the balance constraint. That yields a local improvement algorithm. First we introduce a few notations. Given a set of nodes $B \subset V$ we define its *border* $\partial B := \{u \in B \mid \exists (u,v) \in E : v \notin B\}$. The set $\partial_1 B := \partial B \cap V_1$ is called *left border* of B and the set $\partial_2 B := \partial B \cap V_2$ is called *right border* of B. A B *induced subgraph* G' is the node induced subgraph $G[B]$ plus two nodes s, t that are connected to the border of B. More precisely s is connected to all left border nodes $\partial_1 B$ and all right border nodes $\partial_2 B$ are connected to t. All of these new edges get the edge weight ∞. Note that the additional edges are directed. G' has the *cut property* if each (s,t)-min-cut induces a cut within the balance constraint in G.

The basic idea is to construct a B induced subgraph G' having the cut property. Each min-cut will then yield a feasible improved cut within the balance constraint in G. By performing two Breadth First Searches (BFS) we can find such a set B. Each node touched during these searches belongs to B. The first BFS is done in the subgraph of G induced by V_1. It is initialized with the boundary nodes of V_1. As soon as the weight of the area found by this BFS would exceed $(1+\epsilon)c(V)/2-c(V_1)$, we stop the BFS. The second BFS is done for V_2 in an analogous fashion. The constructed subgraph G' has the cut property since the worst case new weight of V_2 is lower or equal to $c(V_2)+(1+\epsilon)c(V)/2-c(V_2) = (1+\epsilon)c(V)/2$. Indeed the same holds for the worst case new weight of V_1. There are multiple ways to improve this method, i.e. iteratively applying the method, searching in larger areas for feasible cuts and applying most balanced minimum cut heuristics. For more details we refer the reader to [**22**].

Multi-try FM. The second novel method for improving a given partition is called multi-try FM. This local improvement method moves nodes between blocks in order to decrease the cut. Previous k-way methods were initialized with *all* boundary nodes, i.e., all boundary nodes are eligible for movement at the beginning. Our method is repeatedly initialized with a *single* boundary node, thus achieving a more localized search. More details about k-way methods can be found in [**22**]. Multi-try FM is organized in rounds. In each round we put *all* boundary nodes of the current block pair into a todo list T. Subsequently, we begin a k-way local search starting with a *single* random node v of T if it is still a boundary node. Note that the difference to the global k-way search is in the initialisation of the search. The local search is only started from v if it was not touched by a previous localized k-way search in this round. Either way, the node is removed from the todo list. A localized k-way search is not allowed to move a node that has been touched in a previous run. This assures that at most n nodes are touched during a round of the algorithm. The algorithm uses the adaptive stopping criterion from KaSPar [**19**].

Global Search. KaFFPa extended the concept of *iterated multilevel algorithms* which was introduced by [**24**]. The main idea is to iterate the coarsening and uncoarsening phase. Once the graph is partitioned, edges that are between two blocks are not contracted. An F-cycle works as follows: on *each* level we perform at most *two recursive calls* using different random seeds during contraction and local search. A second recursive call is only made the second time that the algorithm reaches a particular level. Figure 3 illustrates a F-cycle. As soon as the graph is partitioned, edges that are between blocks are not contracted. This ensures nondecreasing quality of the partition since our refinement algorithms guarantee no worsening and break ties randomly. These so called *global search strategies* are more effective than plain restarts of the algorithm. *Extending this idea* will yield the combine and mutation operators described in Section 5.

5. KaFFPa Evolutionary

We now describe the techniques used in KaFFPaE. The general idea behind evolutionary algorithms (EA) is to use mechanisms which are highly inspired by biological evolution such as selection, mutation, recombination and survival of the fittest. An EA starts with a population of individuals (in our case partitions of the graph) and evolves the population into different populations over several rounds. In each round, the EA uses a selection rule based on the fitness of the individuals (in our case the edge cut) of the population to select good individuals and combine them to obtain improved offspring. Note that we can use the cut as a fitness function since our partitioner almost always generates partitions that are within the given balance constraint. Our algorithm generates only one offspring per generation.

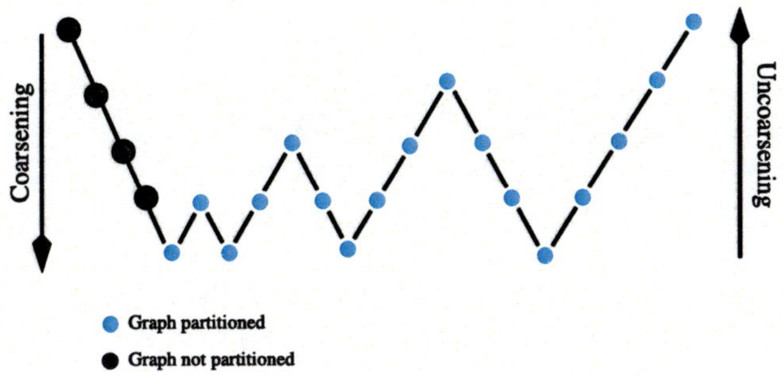

FIGURE 3. An F-cycle for the graph partitioning problem.

Such an evolutionary algorithm is called *steady-state* [7]. A typical structure of an evolutionary algorithm is depicted in Algorithm 1.

For an evolutionary algorithm it is of major importance to keep the diversity in the population high, i.e. the individuals should not become too similar, in order to avoid a premature convergence of the algorithm. In classical evolutionary algorithms, this is done using a mutation operator. It is also important to have operators that introduce unexplored search space to the population. Through a new kind of crossover and mutation operators, introduced in Section 5.1, we introduce more elaborate diversification strategies which allow us to search the search space more effectively.

Algorithm 1 A classic general steady-state evolutionary algorithm.

procedure *steady-state-EA*
  create initial population P
  while stopping criterion not fulfilled
    select parents p_1, p_2 from P
    combine p_1 with p_2 to create offspring o
    mutate offspring o
    evict individual in population using o
  return the fittest individual that occurred

5.1. Combine Operators. We now describe the general combine operator framework. This is followed by three instantiations of this framework. In contrast to previous methods that use a multilevel framework our combine operators do not need perturbations of edge weights since we integrate the operators into our partitioner and do not use it as a complete black box. Furthermore all of our combine operators assure that the offspring has a partition quality *at least as good as the best of both parents*. Roughly speaking, the combine operator framework combines an individual/partition $\mathcal{P} = V_1^{\mathcal{P}}, ..., V_k^{\mathcal{P}}$ (which has to fulfill a balance constraint) with a clustering $\mathcal{C} = V_1^{\mathcal{C}}, ..., V_{k'}^{\mathcal{C}}$. Note that the clustering does not necessarily has to fulfill a balance constraint and k' is not necessarily given in advance. All instantiations of this framework use a different kind of clustering or partition. The partition and the clustering are both used as input for our multi-level graph partitioner KaFFPa in the following sense. Let $\mathcal{E}$ be the set of edges that are cut edges, i.e. edges that run between two blocks, in $\mathcal{P}$ *or* $\mathcal{C}$. All edges in $\mathcal{E}$ are blocked during

FIGURE 4. At the far left, a graph G with two partitions, the dark and the light line, is shown. Cut edges are not eligible for the matching algorithm. Contraction is done until no matchable edge is left. The best of the two given partitions is used as initial partition.

the coarsening phase, i.e. they *are not contracted* during the coarsening phase. In other words these edges are not eligible for the matching algorithm used during the coarsening phase and therefore are not part of any matching computed. An illustration of this can be found in Figure 4.

The stopping criterion for the multi-level partitioner is modified such that it stops when no contractable edge is left. Note that the coarsest graph is now exactly the same as the quotient graph $\mathcal{Q}'$ of the overlay clustering of $\mathcal{P}$ and $\mathcal{C}$ of G (see Figure 5). Hence vertices of the coarsest graph correspond to the connected components of $G_{\mathcal{E}} = (V, E\backslash\mathcal{E})$ and the weight of the edges between vertices corresponds to the sum of the edge weights running between those connected components in G. As soon as the coarsening phase is stopped, we apply the partition $\mathcal{P}$ to the coarsest graph and use this as initial partitioning. This is possible since we did not contract any cut edge of $\mathcal{P}$. Note that due to the specialized coarsening phase and this specialized initial partitioning we obtain a high quality initial solution on a very coarse graph which is usually not discovered by conventional partitioning algorithms. Since our refinement algorithms guarantee no worsening of the input partition and use random tie breaking we can assure nondecreasing partition quality. Note that the refinement algorithms can effectively exchange good parts of the solution on the coarse levels by moving only a few vertices. Figure 5 gives an example.

When the offspring is generated we have to decide which solution should be evicted from the current population. We evict the solution that is *most similar* to the offspring among those individuals in the population that have a cut worse or equal than the offspring itself. The difference of two individuals is defined as the size of the symmetric difference between their sets of cut edges. This ensures some diversity in the population and hence makes the evolutionary algorithm more effective.

5.1.1. *Classical Combine using Tournament Selection.* This instantiation of the combine framework corresponds to a classical evolutionary combine operator C_1. That means it takes two individuals P_1, P_2 of the population and performs the combine step described above. In this case $\mathcal{P}$ corresponds to the partition having the smaller cut and $\mathcal{C}$ corresponds to the partition having the larger cut. Random tie breaking is used if both parents have the same cut. The selection process is based on the tournament selection rule [**18**], i.e. P_1 is the fittest out of two random individuals R_1, R_2 from the population. The same is done to select P_2. Note that in contrast to previous methods the generated offspring will have a cut smaller or equal to the cut of $\mathcal{P}$. Due to the fact that our multi-level algorithms are randomized,

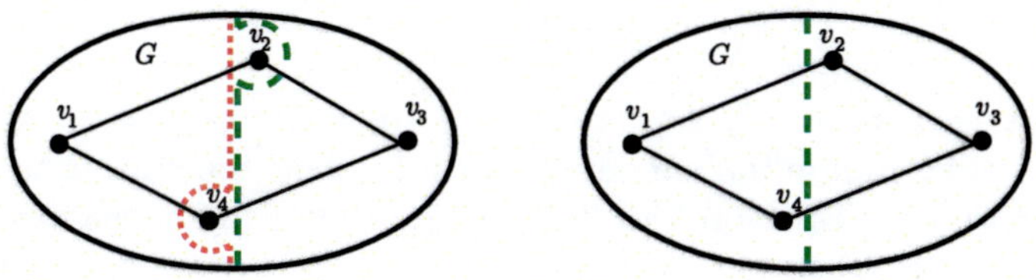

FIGURE 5. A graph G and two bipartitions; the dotted and the dashed line (left). Curved lines represent a large cut. The four vertices correspond to the coarsest graph in the multilevel procedure. Local search algorithms can effectively exchange v_2 or v_4 to obtain the better partition depicted on the right hand side (dashed line).

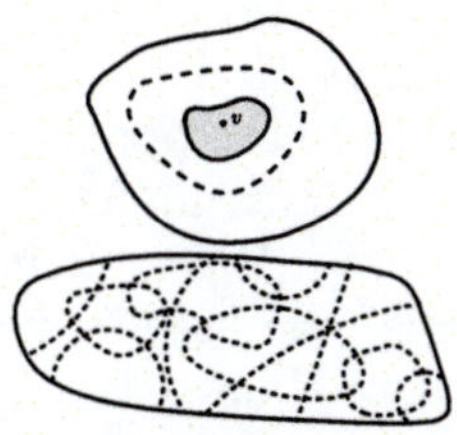

FIGURE 6. On top we see the computation natural cuts. A BFS Tree starting at v is grown. The core is gray. The dashed line is the natural cut. It is the minimum cut between the contracted versions of the core and the ring (solid line). Several natural cuts are detected (bottom).

a combine operation performed twice using the same parents can yield different offspring.

5.1.2. *Cross Combine / (Transduction).* In this instantiation of the combine framework C_2, the clustering $\mathcal{C}$ corresponds to a partition of G. But instead of choosing an individual from the population we create a new individual in the following way. We choose k' uniformly at random in $[k/4, 4k]$ and ϵ' uniformly at random in $[\epsilon, 4\epsilon]$. We then use KaFFPa to create a k'-partition of G fulfilling the balance constraint $\max c(V_i) \leq (1+\epsilon')c(V)/k'$. In general larger imbalances reduce the cut of a partition which then yields good clusterings for our crossover. To the best of our knowledge there has been no genetic algorithm that performs combine operations combining individuals from different search spaces.

5.1.3. *Natural Cuts.* Delling et al. [**8**] introduced the notion of *natural cuts* as a preprocessing technique for the partitioning of road networks. The preprocessing technique is able to find relatively sparse cuts close to denser areas. We use the computation of natural cuts to provide another combine operator, i.e. combining a k-partition with a clustering generated by the computation of natural cuts. We closely follow their description: The computation of natural cuts works in rounds. Each round picks a center vertex v and grows a breadth-first search (BFS) tree. The BFS is stopped as soon as the weight of the tree, i.e. the sum of the vertex weights of the tree, reaches αU, for some parameters α and U. The set of the neighbors of T in $V\backslash T$ is called the *ring* of v. The *core* of v is the union of all vertices added to T before its size reached $\alpha U/f$ where $f > 1$ is another parameter. The core is then temporarily contracted to a single vertex s and the ring into a single vertex t to compute the minimum s-t-cut between them using the given edge weights as capacities. To assure that every vertex eventually belongs to at least one core, and therefore is inside at least one cut, the vertices v are picked uniformly at random among all vertices that have not yet been part of any core in any round. The process is stopped when there are no such vertices left. In the original work [**8**] each connected component of the graph $G_C = (V, E\backslash C)$, where C is the union of all edges cut by the process above, is contracted to a single vertex. Since we do not use natural cuts as a preprocessing technique at this place we don't contract these components. Instead we build a clustering $\mathcal{C}$ of G such that each connected component of G_C is a block.

This technique yields the third instantiation of the combine framework C_3 which is divided into two stages, i.e. the clustering used for this combine step is dependent on the stage we are currently in. In both stages the partition $\mathcal{P}$ used for the combine step is selected from the population using tournament selection. During the first stage we choose f uniformly at random in $[5, 20]$, α uniformly at random in $[0.75, 1.25]$ and we set $U = |V|/3k$. Using these parameters we obtain a clustering $\mathcal{C}$ of the graph which is then used in the combine framework described above. This kind of clustering is used until we reach an upper bound of ten calls to this combine step. When the upper bound is reached we switch to the second stage. In this stage we use the clusterings computed during the first stage, i.e. we extract elementary natural cuts and use them to quickly compute new clusterings. An *elementary natural cut* (ENC) consists of a set of cut edges and the set of nodes in its core. Moreover, for each node v in the graph, we store the set of ENCs $N(v)$ that contain v in their core. With these data structures it is easy to pick a new clustering $\mathcal{C}$ (see Algorithm 2) which is then used in the combine framework described above.

Algorithm 2 computeNaturalCutClustering (second stage)

```
unmark all nodes in V
for each v ∈ V in random order do
    if v is not marked then
        pick a random ENC C in N(v)
        output C
        mark all nodes in C's core
```

5.2. Mutation Operators. We define two mutation operators, an ordinary and a modified F-cycle. Both mutation operators use a random individual from the current population. The main idea is to iterate coarsening and refinement several times using different seeds for random tie breaking. The first mutation operator M_1 can assure that the quality of the input partition does not decrease. It is basically an ordinary F-cycle which is an algorithm used in KaFFPa. Edges between blocks are not contracted. The given partition is then used as initial partition of the coarsest graph. In contrast to KaFFPa, we now can use the partition as input to the partition in the very beginning. This ensures nondecreasing quality since our refinement algorithms guarantee no worsening. The second mutation operator

M_2 works quite similar with the small difference that the input partition is not used as initial partition of the coarsest graph. That means we obtain very good coarse graphs but we cannot assure that the final individual has a higher quality than the input individual. In both cases the resulting offspring is inserted into the population using the eviction strategy described in Section 5.1.

5.3. Putting Things Together and Parallelization. We now explain the parallelization and describe how everything is put together. Each processing element (PE) basically performs the same operations using different random seeds (see Algorithm 3). First we estimate the population size S: each PE performs a partitioning step and measures the time $\bar{t}$ spent for partitioning. We then choose S such that the time for creating S partitions is approximately t_{total}/f where the fraction f is a tuning parameter and t_{total} is the total running time that the algorithm is given to produce a partition of the graph. Each PE then builds its own population, i.e. KaFFPa is called several times to create S individuals/partitions. Afterwards the algorithm proceeds in rounds as long as time is left. With corresponding probabilities, mutation or combine operations are performed and the new offspring is inserted into the population. We choose a parallelization/communication protocol that is quite similar to *randomized rumor spreading* [**9**]. Let p denote the number of PEs used. A communication step is organized in rounds. In each round, a PE chooses a communication partner and sends her the currently best partition P of the local population. The selection of the communication partner is done uniformly at random among those PEs to which P not already has been sent to. Afterwards, a PE checks if there are incoming individuals and if so inserts them into the local population using the eviction strategy described above. If P is improved, all PEs are again eligible. This is repeated $\log p$ times. Note that the algorithm is implemented *completely asynchronously*, i.e. there is no need for a global synchronisation. The process of creating individuals is parallelized as follows: Each PE makes $s' = |S|/p$ calls to KaFFPa using different seeds to create s' individuals. Afterwards we do the following $S - s'$ times: The root PE computes a random cyclic permutation of all PEs and broadcasts it to all PEs. Each PE then sends a random individual to its successor in the cyclic permutation and receives a individual from its predecessor in the cyclic permutation which is then inserted into the local population. When this particular part of the algorithm (*quick start*) is finished, each PE has $|S|$ partitions.

After some experiments we fixed the ratio of mutation to crossover operations to $1:9$, the ratio of the mutation operators $M_1 : M_2$ to $4:1$ and the ratio of the combine operators $C_1 : C_2 : C_3$ to $3:1:1$. Note that the communication step in the last line of the algorithm could also be performed only every x iterations (where x is a tuning parameter) to save communication time. Since the communication network of our test system is very fast (see Section 6), we perform the communication step in each iteration.

6. Experiments

Implementation. We have implemented the algorithm described above using C++. Overall, our program (including KaFFPa and KaFFPaE) consists of about 22 500 lines of code. We use three configurations of KaFFPa: KaFFPaStrong,

Algorithm 3 All PEs perform the same operations using different random seeds.

```
procedure locallyEvolve
  estimate population size S
  while time left
    if elapsed time < t_total/f then create individual and insert into local population
    else
      flip coin c with corresponding probabilities
      if c shows head then
        perform a mutation operation
      else
        perform a combine operation
      insert offspring into population if possible
    communicate according to communication protocol
```

KaFFPaEco and KaFFPaFast. KaFFPaFast is the fastest configuration, KaFFPaEco is a good tradeoff between quality and speed, and KaFFPaStrong is focused on quality (see [**22**] for more details).

Systems. Experiments have been done on three machines. Machine A is a cluster with 200 nodes where each node is equipped with two Quad-core Intel Xeon processors (X5355) which run at a clock speed of 2.667 GHz. Each node has 2x4 MB of level 2 cache each and runs Suse Linux Enterprise 10 SP 1. All nodes are attached to an InfiniBand 4X DDR interconnect which is characterized by its very low latency of below 2 microseconds and a point to point bandwidth between two nodes of more than 1300 MB/s. Machine B has four Quad-core Opteron 8350 (2.0GHz), 64GB RAM, running Ubuntu 10.04. Machine C has two Intel Xeon X5550, 48GB RAM, running Ubuntu 10.04. Each CPU has 4 cores (8 cores when hyperthreading is active) running at 2.67 GHz. Experiments in Section 6.1 were conducted on machine A. Shortly after these experiments were conducted the machine had a file system crash and was not available for two weeks (and after that the machine was very full). Therefore we switched to the much smaller machines B and C, focused on a small subset of the challenge and restricted further experiments to $k = 8$. Experiments in Section 6.2 have been conducted on machine B, and experiments in Section 6.3 have been conducted on machine C. All programs were compiled using GCC Version 4.4.3 and optimization level 3 using OpenMPI 1.5.3. Henceforth, a PE is one core of a machine.

Instances. We report experiments on a subset of the graphs of the 10th DIMACS Implementation Challenge [**3**]. Experiments in Section 6.1 were done on all graphs of the Walshaw Benchmark. Here we used $k \in \{2, 4, 8, 16, 32, 64\}$ since they are the default values in [**25**]. Experiments in Section 6.2 focus on the graph subset depicted in Table 1 (except the road networks). In Section 6.3 we have a closer look on all road networks of the Challenge. We finish the experimental evaluation with Section 6.4 describing how we obtained the results on the challenge testbed and comparing the performance of Metis and Scotch. Our default value for the allowed imbalance is 3% since this is one of the values used in [**25**] and the default value in Metis. Our default number of PEs is 16.

6.1. Walshaw Benchmark. [1]

We now apply KaFFPaE to Walshaw's benchmark archive [**23**] using the rules used there, i.e., running time is not an issue but

[1] see KaFFPaE [**21**] for more details on this experiment.

TABLE 1. Basic properties of choosen subset (except Walshaw Instances).

graph	n	m
Random Geometric Graphs		
rgg16	2^{16}	≈342 K
rgg17	2^{17}	≈729 K
Delaunay		
delaunay16	2^{16}	≈197 K
delaunay17	2^{17}	≈393 K
Kronecker G500		
kron_simple_16	2^{16}	≈2 M
kron_simple_17	2^{17}	≈5 M
Numerical		
adaptive	≈6 M	≈14 M
channel	≈5 M	≈42 M
venturi	≈4 M	≈8 M
packing	≈2 M	≈17 M
2D Frames		
hugetrace-00000	≈5 M	≈7 M
hugetric-00000	≈6 M	≈9 M
Sparse Matrices		
af_shell9	≈500 K	≈9 M
thermal2	≈1 M	≈4 M
Coauthor Networks		
coAutCiteseer	≈227 K	≈814 K
coAutDBLP	≈299 K	≈978 K
Social Networks		
cnr	≈326 K	≈3 M
caidaRouterLvl	≈192 K	≈609 K
Road Networks		
luxembourg	≈144 K	≈120 K
belgium	≈1 M	≈2 M
netherlands	≈2 M	≈2 M
italy	≈7 M	≈7 M
great-britain	≈8 M	≈8 M
germany	≈12 M	≈12 M
asia	≈12 M	≈13 M
europe	≈51 M	≈54 M

we want to obtain minimal cut values for $k \in \{2,4,8,16,32,64\}$ and balance parameters $\epsilon \in \{0, 0.01, 0.03, 0.05\}$. We focus on $\epsilon \in \{1\%, 3\%, 5\%\}$ since KaFFPaE (more precisely KaFFPa) is not made for the case $\epsilon = 0$. We run KaFFPaE with a time limit of two hours using 16 PEs (two nodes of the cluster) per graph, k and ϵ. On the eight largest graphs of the archive we gave KaFFPaE eight hours per graph, k and ϵ. KaFFPaE computed 300 partitions which are better than previous best partitions reported there: 91 for 1%, 103 for 3% and 106 for 5%. Moreover, it reproduced *equally sized* cuts in 170 of the 312 remaining cases. When only considering the 15 largest graphs and $\epsilon \in \{1.03, 1.05\}$ we are able to reproduce or improve the current result in 224 out of 240 cases. Overall our systems (including

KaPPa, KaSPar, KaFFPa, KaFFPaE) now improved or reproduced the entries in 550 out of 612 cases (for $\epsilon \in \{0.01, 0.03, 0.05\}$).

6.2. Various DIMACS Graphs. In this Section we apply KaFFPaE (and on some graphs KaFFPa) to a meaningful subset of the graphs of the DIMACS Challenge. Here we use all cores of machine B and give KaFFPaE eight hours of time per graph to compute a partition into eight blocks. When using KaFFPa to create a partition we use one core of this machine. The experiments were repeated three times. A summary of the results can be found in Table 2.

TABLE 2. Results achieved for $k = 8$ various graphs of the DIMACS Challenge. Results which were computed by KaFFPa are indicated by *.

graph	best	avg.	graph	best	avg.
rgg16	1 067	1 067	coAutDBLP	94 866	95 866
rgg17	1 777	1 778	channel*	333 396	333 396
delaunay16	1 547	1 547	packing*	108 771	111 255
delaunay17	2 200	2 203	adaptive	8 482	8 482
kron_simple_16*	1 257 512	1 305 207	venturi	5 780	5 788
kron_simple_17*	2 247 116	2 444 925	hugetrace-00000	3 656	3 658
cnr	4 687	4 837	hugetric-00000	4 769	4 785
caidaRouterLevel	42 679	43 659	af_shell9	40 775	40 791
coAutCiteseer	42 875	43 295	thermal2	6 426	6 426

6.3. Road Networks. In this Section we focus on finding partitions of the street networks of the DIMACS Challenge. We implemented a specialized algorithm, *Buffoon*, which is similar to PUNCH [8] in the sense that it also uses natural cuts as a preprocessing technique to obtain a coarser graph on which the graph partitioning problem is solved. For more information on natural cuts, we refer the reader to [8]. Using our (shared memory) parallelized version of natural cut preprocessing we obtain a coarse version of the graph. Note that our preprocessing uses slightly different parameters than PUNCH (using the notation of [8], we use $C = 2, U = (1+\epsilon)\frac{n}{2k}, f = 10, \alpha = 1$). Since partitions of the coarse graph correspond to partitions of the original graph, we use KaFFPaE to partition the coarse version of the graph. After preprocessing, we gave KaFFPaE one hour of time to compute a partition. In both cases we used all 16 cores (hyperthreading active) of machine C for preprocessing and for KaFFPaE. We also used the strong configuration of KaFFPa to partition the road networks. In both cases the experiments were repeated ten times. Table 3 summarizes the results.

6.4. The Challenge Testbed. We now describe how we obtained the results on the challenge testbed and evaluate the performance of kMetis and Scotch on these graphs in the Pareto challenge.

Pareto Challenge. For this particular challenge we run all configurations of KaFFPa (KaFFPaStrong, KaFFPaEco, KaFFPaFast, see [22] for details), KaFFPaE, Metis 5.0 and Scotch 5.1 on machine A. To compute a partition for an instance (graph, k) we repeatedly run the corressponding partitioner (except KaFFPaE) using different random seeds until the resulting partition is feasable. We stopped the

TABLE 3. Results on road networks for $k = 8$: average and best cut results of Buffoon (B) and KaFFPa (K) as well as average runtime [m] (including preprocessing).

grp.	algorithm/runtime t					
	B_{best}	B_{avg}	$t_{avg}[m]$	K_{best}	K_{avg}	$t_{avg}[m]$
lux.	**79**	79	60.1	81	83	0.1
bel.	**307**	307	60.5	320	326	0.9
net.	**191**	193	60.6	207	217	1.2
ita.	**200**	200	64.3	205	210	3.9
gb.	**363**	365	63.0	381	395	6.5
ger.	**473**	475	65.3	482	499	11.3
asia.	**47**	47	67.6	52	55	6.4
eur.	**526**	527	131.5	550	590	76.1

TABLE 4. Pareto challenge results including Metis and Scotch (left hand side) and original Pareto challenge results (right hand side).

Solver	Points
KaFFPaFast	1372
Metis	1265
KaFFPaEco	1174
KaFFPaE	1134
KaFFPaStrong	1085
UMPa [6]	624
Scotch	361
Mondrian [11]	225

Solver	Points
KaFFPaFast	1680
KaFFPaEco	1305
KaFFPaE	1145
KaFFPaStrong	1106
UMPa [6]	782
Mondrian [11]	462

process after one day of computation or after one hundred repetitions yielding unbalanced partitions. The resulting partition was used for both parts of the challenge, i.e. optimizing for edge cut and optimizing for maximum communication volume. The runtime of each iteration was added if more then one iteration was needed to obtain a feasable partition. KaFFPaE was given four nodes of machine A and a time limit of eight hours for each instance. When computing partitions for the objective function maximum communication volume we altered the fitness function to this objective. This ensures that individuals having a better maximum communication volume are more often selected for a combine operation. Using this methodology KaFFPaStrong, KaFFPaEco, KaFFPaFast, KaFFPaE, Metis and Scotch were able to solve 136, 150, 170, 130, 146 and 110 instances respectively. The resulting points achieved in the Pareto challenge can be found in Table 4 (see [1] for a description on how points are computed for the challenges). Note that KaFFPaFast gained more points than KaFFPaEco, KaFFPaStrong and KaFFPaE. Since it is much faster than the other KaFFPa configurations it is almost never dominated by them and therefore scores a lot of points in this particular challenge. For some instances the partitions produced by Metis always exceeded the balance constraint by exactly one vertex. We assume that a small modification of Metis would increase the number of instances solved and most probably also the score achieved.

Quality Challenge. Our quality submission KaPa (Karlsruhe Partitioners) assembles the best solutions of the partitions obtained of our partitioners in the Pareto challenge. Furthermore, on road networks we also run Buffoon to create partitions. The resulting points achieved in the quality challenge can be found in Table 5.

TABLE 5. Original quality challenge results.

Solver	Points
KaPa	1574
UMPa [**6**]	1066
Mondrian [**11**]	616

7. Conclusion and Future Work

We presented two approaches to the graph partitioning problem, KaFFPa and KaFFPaE. KaFFPa uses novel local improvement methods and more sophisticated global search strategies to tackle the problem. KaFFPaE is an distributed evolutionary algorithm which uses KaFFPa as a base case partitioner. Due to new crossover and mutation operators as well as its scalable parallelization it is able to compute the best known partitions for many standard benchmark instances in only a *few minutes* for graphs of moderate size. We therefore believe that KaFFPaE is still helpful in the area of high performance computing. Regarding future work, we want look at more DIMACS Instances, more values of k and more values of ϵ. In particular we want to investigate at the case $\epsilon = 0$.

References

[1] Competition rules and objective functions for the 10th dimacs implementation challenge on graph partitioning and graph clustering, `http://www.cc.gatech.edu/dimacs10/data/dimacs10-rules.pdf`.

[2] Reid Andersen and Kevin J. Lang, *An algorithm for improving graph partitions*, Proceedings of the Nineteenth Annual ACM-SIAM Symposium on Discrete Algorithms, ACM, New York, 2008, pp. 651–660. MR2487634

[3] David Bader, Henning Meyerhenke, Peter Sanders, and Dorothea Wagner. 10th DIMACS Implementation Challenge - Graph Partitioning and Graph Clustering, `http://www.cc.gatech.edu/dimacs10/`.

[4] Una Benlic and Jin-Kao Hao. A multilevel memtetic approach for improving graph k-partitions. In *22nd Intl. Conf. Tools with Artificial Intelligence*, pages 121–128, 2010.

[5] Thang Nguyen Bui and Curt Jones, *Finding good approximate vertex and edge partitions is NP-hard*, Inform. Process. Lett. **42** (1992), no. 3, 153–159, DOI 10.1016/0020-0190(92)90140-Q. MR1168771 (93h:68111)

[6] Ümit V. Çatalyürek, Mehmet Deveci, Kamer Kaya, and Bora Uçar. Umpa: A multi-objective, multi-level partitioner for communication minimization. In *10th DIMACS Impl. Challenge Workshop: Graph Partitioning and Graph Clustering*. Georgia Institute of Technology, Atlanta, GA, February 13-14, 2012.

[7] Kenneth A. De Jong, *Evolutionary computation: a unified approach*, A Bradford Book, MIT Press, Cambridge, MA, 2006. MR2234532 (2007b:68003)

[8] Daniel Delling, Andrew V. Goldberg, Ilya Razenshteyn, and Renato F. Werneck. Graph Partitioning with Natural Cuts. In *25th IPDPS*. IEEE Computer Society, 2011.

[9] Benjamin Doerr and Mahmoud Fouz, *Asymptotically optimal randomized rumor spreading*, Automata, languages and programming. Part II, Lecture Notes in Comput. Sci., vol. 6756, Springer, Heidelberg, 2011, pp. 502–513, DOI 10.1007/978-3-642-22012-8_40. MR2852451

[10] Doratha E. Drake and Stefan Hougardy, *A simple approximation algorithm for the weighted matching problem*, Inform. Process. Lett. **85** (2003), no. 4, 211–213, DOI 10.1016/S0020-0190(02)00393-9. MR1950496 (2003m:68185)

[11] B. O. Fagginger Auer and R. H. Bisseling. Abusing a hypergraph partitioner for unweighted graph partitioning. In *10th DIMACS Impl. Challenge Workshop: Graph Partitioning and Graph Clustering*. Georgia Institute of Technology, Atlanta, GA, February 13-14, 2012.

[12] C. M. Fiduccia and R. M. Mattheyses. A Linear-Time Heuristic for Improving Network Partitions. In *19th Conference on Design Automation*, pages 175–181, 1982.

[13] M. Holtgrewe, P. Sanders, and C. Schulz. Engineering a Scalable High Quality Graph Partitioner. *24th IEEE International Parallal and Distributed Processing Symposium*, 2010.

[14] George Karypis and Vipin Kumar, *Parallel multilevel k-way partitioning scheme for irregular graphs*, SIAM Rev. **41** (1999), no. 2, 278–300 (electronic), DOI 10.1137/S0036144598334138. MR1684545 (2000d:68117)

[15] Jin Kim, Inwook Hwang, Yong-Hyuk Kim, and Byung Ro Moon. Genetic approaches for graph partitioning: a survey. In *GECCO*, pages 473–480. ACM, 2011.

[16] Kevin Lang and Satish Rao, *A flow-based method for improving the expansion or conductance of graph cuts*, Integer programming and combinatorial optimization, Lecture Notes in Comput. Sci., vol. 3064, Springer, Berlin, 2004, pp. 325–337, DOI 10.1007/978-3-540-25960-2_25. MR2144596 (2005m:05181)

[17] J. Maue and P. Sanders. Engineering algorithms for approximate weighted matching. In *6th Workshop on Exp. Alg. (WEA)*, volume 4525 of *LNCS*, pages 242–255. Springer, 2007.

[18] Brad L. Miller and David E. Goldberg, *Genetic algorithms, tournament selection, and the effects of noise*, Complex Systems **9** (1995), no. 3, 193–212. MR1390121 (97c:68136)

[19] Vitaly Osipov and Peter Sanders, *n-level graph partitioning*, Algorithms—ESA 2010. Part I, Lecture Notes in Comput. Sci., vol. 6346, Springer, Berlin, 2010, pp. 278–289, DOI 10.1007/978-3-642-15775-2_24. MR2762861

[20] F. Pellegrini. Scotch home page. `http://www.labri.fr/pelegrin/scotch`.

[21] P. Sanders and C. Schulz. Distributed Evolutionary Graph Partitioning. *12th Workshop on Algorithm Engineering and Experimentation*, 2011.

[22] Peter Sanders and Christian Schulz, *Engineering multilevel graph partitioning algorithms*, Algorithms—ESA 2011, Lecture Notes in Comput. Sci., vol. 6942, Springer, Heidelberg, 2011, pp. 469–480, DOI 10.1007/978-3-642-23719-5_40. MR2893224 (2012k:68259)

[23] A. J. Soper, C. Walshaw, and M. Cross, *A combined evolutionary search and multilevel optimisation approach to graph-partitioning*, J. Global Optim. **29** (2004), no. 2, 225–241, DOI 10.1023/B:JOGO.0000042115.44455.f3. MR2092958 (2005k:05228)

[24] Chris Walshaw, *Multilevel refinement for combinatorial optimisation problems*, Ann. Oper. Res. **131** (2004), 325–372, DOI 10.1023/B:ANOR.0000039525.80601.15. MR2095810

[25] C. Walshaw and M. Cross, *Mesh partitioning: a multilevel balancing and refinement algorithm*, SIAM J. Sci. Comput. **22** (2000), no. 1, 63–80 (electronic), DOI 10.1137/S1064827598337373. MR1769526 (2001b:65153)

[26] C. Walshaw and M. Cross. JOSTLE: Parallel Multilevel Graph-Partitioning Software – An Overview. In F. Magoules, editor, *Mesh Partitioning Techniques and Domain Decomposition Techniques*, pages 27–58. Civil-Comp Ltd., 2007. (Invited chapter).

Karlsruhe Institute of Technology, 76128 Karlsruhe, Germany
E-mail address: `christian.schulz@kit.edu`

Karlsruhe Institute of Technology, 76128 Karlsruhe, Germany
E-mail address: `sanders@kit.edu`

Abusing a hypergraph partitioner for unweighted graph partitioning

B. O. Fagginger Auer and R. H. Bisseling

ABSTRACT. We investigate using the Mondriaan matrix partitioner for unweighted graph partitioning in the communication volume and edge-cut metrics. By converting the unweighted graphs to appropriate matrices, we measure Mondriaan's performance as a graph partitioner for the 10th DIMACS challenge on graph partitioning and clustering. We find that Mondriaan can effectively be used as a graph partitioner: w.r.t. the edge-cut metric, Mondriaan's best results are on average within 13% of the best known results as listed in Chris Walshaw's partitioning archive, but it is an order of magnitude slower than dedicated graph partitioners.

1. Introduction

In this paper, we use the Mondriaan matrix partitioner [**22**] to partition the graphs from the 10th DIMACS challenge on graph partitioning and clustering [**1**]. In this way, we can compare Mondriaan's performance as a graph partitioner with the performance of the state-of-the-art partitioners participating in the challenge.

An *undirected graph* G is a pair (V, E), with vertices V, and edges E that are of the form $\{u, v\}$ for $u, v \in V$ with possibly $u = v$. For vertices $v \in V$, we denote the set of all of v's *neighbours* by

$$V_v := \{u \in V \mid \{u, v\} \in E\}.$$

Note that vertex v is a neighbour of itself precisely when the self-edge $\{v, v\} \in E$.

Hypergraphs are a generalisation of undirected graphs, where edges can contain an arbitrary number of vertices. A *hypergraph* $\mathcal{G}$ is a pair $(\mathcal{V}, \mathcal{N})$, with vertices $\mathcal{V}$, and nets (or hyperedges) $\mathcal{N}$; nets are subsets of $\mathcal{V}$ that can contain any number of vertices.

Let $\epsilon > 0$, $k \in \mathbf{N}$, and $G = (V, E)$ be an undirected graph. Then a valid solution to the *graph partitioning problem* for partitioning G into k parts with imbalance ϵ, is a partitioning $\Pi : V \to \{1, \ldots, k\}$ of the graph's vertices into k parts, each part $\Pi^{-1}(\{i\})$ containing at most

$$|\Pi^{-1}(\{i\})| \leq (1 + \epsilon) \left\lceil \frac{|V|}{k} \right\rceil, \qquad (1 \leq i \leq k) \tag{1.1}$$

vertices.

2010 *Mathematics Subject Classification.* Primary 05C65, 05C70; Secondary 05C85.
Key words and phrases. Hypergraphs, graph partitioning, edge cut, communication volume.

To measure the quality of a valid partitioning we use two different metrics. The *communication volume metric*[1] [1] is defined by

$$\mathrm{CV}(\Pi) := \max_{1\leq i\leq k} \sum_{\substack{v\in V\\ \Pi(v)=i}} |\Pi(V_v) \setminus \{\Pi(v)\}|. \tag{1.2}$$

For each vertex v, we determine the number $\pi(v)$ of different parts in which v has neighbours, except its own part $\Pi(v)$. Then, the communication volume is given by the maximum over i, of the sum of all $\pi(v)$ for vertices v belonging to part i.

The *edge-cut metric* [1], defined as

$$\mathrm{EC}(\Pi) := |\{\{u,v\} \in E \mid \Pi(u) \neq \Pi(v)\}|, \tag{1.3}$$

measures the number of edges between different parts of the partitioning Π.

TABLE 1. Overview of available software for partitioning graphs (left) and hypergraphs (right), from [**3**, Table 12.1].

Name	Ref.	Sequential/ parallel	Name	Ref.	Sequential/ parallel
Chaco	[**13**]	sequential	hMETIS	[**15**]	sequential
METIS	[**14**]	sequential	ML-Part	[**6**]	sequential
Scotch	[**18**]	sequential	Mondriaan	[**22**]	sequential
Jostle	[**23**]	parallel	PaToH	[**8**]	sequential
ParMETIS	[**16**]	parallel	Par*k*way	[**21**]	parallel
PT-Scotch	[**10**]	parallel	Zoltan	[**12**]	parallel

There exist a lot of different (hyper)graph partitioners, which are summarised in Table 1. All partitioners follow a multi-level strategy [**5**], where the (hyper)graph is coarsened by generating a matching of the (hyper)graph's vertices and contracting matched vertices to a single vertex. Doing this recursively creates a hierarchy of increasingly coarser approximations of the original (hyper)graph. After this has been done, an initial partitioning is generated on the coarsest (hyper)graph in the hierarchy, i.e. the one possessing the smallest number of vertices. This partitioning is subsequently propagated to the finer (hyper)graphs in the hierarchy and refined at each level (e.g. using the Kernighan–Lin algorithm [**17**]), until we reach the original (hyper)graph and obtain the final partitioning.

2. Mondriaan

2.1. Mondriaan sparse matrix partitioner. The Mondriaan partitioner has been designed to partition the matrix and the vectors for a parallel sparse matrix–vector multiplication, where a sparse matrix A is multiplied by a dense input vector $\mathbf{v}$ to give a dense output vector $\mathbf{u} = A\mathbf{v}$ as the result. First, the matrix partitioning algorithm is executed to minimise the total communication volume $\mathrm{LV}(\Pi)$ of the partitioning, defined below, and then the vector partitioning algorithm is executed with the aim of balancing the communication among the processors. The matrix partitioning itself does not aim to achieve such balance, but it is not biased in favour of any processor part either.

[1]We forgo custom edge and vertex weights and assume they are all equal to one, because Mondriaan's hypergraph partitioner does not support net weights.

TABLE 2. Available representations of an $m \times n$ matrix $A = (a_{ij})$ by a hypergraph $\mathcal{G} = (\mathcal{V}, \mathcal{N})$ in Mondriaan.

Name	Ref.	$\mathcal{V}$	$\mathcal{N}$
Column-net	[7]	$\{r_1, \ldots, r_m\}$	$\{\{r_i \mid 1 \le i \le m, a_{ij} \ne 0\} \mid 1 \le j \le n\}$
Row-net	[7]	$\{c_1, \ldots, c_n\}$	$\{\{c_j \mid 1 \le j \le n, a_{ij} \ne 0\} \mid 1 \le i \le m\}$
Fine-grain	[9]	$\{v_{ij} \mid a_{ij} \ne 0\}$	$\underbrace{\{\{v_{ij} \mid 1 \le i \le m, a_{ij} \ne 0\} \mid 1 \le j \le n\}}_{\text{column nets}} \cup \underbrace{\{\{v_{ij} \mid 1 \le j \le n, a_{ij} \ne 0\} \mid 1 \le i \le m\}}_{\text{row nets}}$

Mondriaan uses recursive bipartitioning to split the matrix or its submatrices repeatedly into two parts, choosing the best of the row or column direction in the matrix. The current submatrix is translated into a hypergraph by the column-net or row-net model, respectively (see Table 2). Another possibility is to split the submatrix based on the fine-grain model, and if desired the best split of the three methods can be chosen. The outcome of running Mondriaan is a two-dimensional partitioning of the sparse matrix (i.e., a partitioning where both the matrix rows and columns are split). The number of parts is not restricted to a power of two, as Mondriaan can split parts according to a given ratio, such as 2:1. After each split, Mondriaan adjusts the weight balancing goals of the new parts obtained, as the new part that receives the largest fraction of the weight will need to be stricter in allowing an imbalance during further splits than the part with the smaller fraction.

The total communication volume of the parallel sparse matrix–vector multiplication is minimised by Mondriaan in the following manner. Because the total volume is simply the sum of the volumes incurred by every split into two by the recursive bipartitioning [**22**, Theorem 2.2], the minimisation is completely achieved by the bipartitioning. We will explain the procedure for splits in the column direction (the row direction is similar). When using Mondriaan as a hypergraph partitioner, as we do for the DIMACS challenge, see Section 2.2, only the column direction is used.

First, in the bipartitioning, similar columns are merged by matching columns that have a large overlap in their nonzero patterns. A pair of columns j, j' with similar pattern will then be merged and hence will be assigned to the same processor part in the subsequent initial partitioning, thus preventing the communication that would occur if two nonzeros a_{ij} and $a_{ij'}$ from the same row were assigned to different parts. Repeated rounds of merging during this coarsening phase result in a final sparse matrix with far fewer columns, and a whole multilevel hierarchy of intermediate matrices.

Second, the resulting smaller matrix is bipartitioned using the Kernighan–Lin algorithm [**17**]. This local-search algorithm with so-called hill-climbing capabilities starts with a random partitioning of the columns satisfying the load balance constraints, and then tries to improve it by repeated moves of a column from its current processor part to the other part. To enhance the success of the Kernighan–Lin algorithm and to prevent getting stuck in local minima, we limit the number of columns to at most 200 in this stage; the coarsening only stops when this number has been reached. The Kernighan–Lin algorithm is run eight times and the best solution is taken.

Third, the partitioning of the smaller matrix is propagated back to a partitioning of the original matrix, at each level unmerging pairs of columns while trying to refine the partitioning by one run of the Kernighan–Lin algorithm. This further reduces the amount of communication, while still satisfying the load balance constraints.

If the input and output vector can be partitioned independently, the vector partitioning algorithm usually has enough freedom to achieve a reasonable communication balancing. Each component v_i of the input vector can then be assigned to any of the processors that hold nonzeros in the corresponding column, and each component u_i of the output vector to any of the processors that hold nonzeros in the corresponding row. If the matrix is square, and both vectors must be partitioned in the same way, then there is usually little freedom, as the only common element of row i and column i is the diagonal matrix element a_{ii}, which may or may not be zero. If it is zero, it has no owning processor, and the set of processors owning row i and that owning column i may be disjoint. This means that the total communication volume must be increased by one for vector components v_i and u_i. If the matrix diagonal has only nonzero elements, however, the vector partitioning can be achieved without incurring additional communication by assigning vector components v_i and u_i to the same processor as the diagonal matrix element a_{ii}. More details on the matrix and vector partitioning can be found in [**22**]; improved methods for vector partitioning are given in [**4**], see also [**2**].

2.2. Mondriaan hypergraph partitioner. Here, we will use Mondriaan as a hypergraph partitioner, which can be done by choosing the column direction in all splits, so that columns are vertices and rows are nets. This means that we use Mondriaan in one-dimensional mode, as only rows will be split. Figure 1 illustrates this splitting procedure. Mondriaan has the option to use its own, native hypergraph bipartitioner, or link to the external partitioner PaToH [**8**]. In the present work, we use the native partitioner.

For the graph partitioning challenge posed by DIMACS, we try to fit the existing software to the aims of the challenge. One could say that this entails abusing the software, as it was designed for a different purpose, namely matrix and hypergraph partitioning. Using a hypergraph partitioner to partition graphs will be at the cost of some additional, unnecessary overhead. Still, it will be interesting to see how the Mondriaan software performs in this unforeseen mode, and to compare the quality of the generated partitionings to the quality of partitionings generated by other software, in particular by graph partitioning packages.

In the situation of the challenge, we can only use the matrix partitioning of Mondriaan and not the vector partitioning, as the vertex partitioning of the graph is already completely determined by the column partitioning of the matrix. The balance of the communication will then solely depend on the balance achieved by the matrix partitioning.

Internally, Mondriaan's hypergraph partitioner solves the following problem. For a hypergraph $\mathcal{G} = (\mathcal{V}, \mathcal{N})$ with vertex weights $\zeta : \mathcal{V} \to \mathbf{N}$, an imbalance factor $\epsilon > 0$, and a number of parts $k \in \mathbf{N}$, Mondriaan's partitioner produces a partitioning $\Pi : \mathcal{V} \to \{1, \ldots, k\}$ such that

$$\zeta(\Pi^{-1}(\{i\})) \leq (1 + \epsilon) \left\lceil \frac{\zeta(\mathcal{V})}{k} \right\rceil, \qquad (1 \leq i \leq k), \tag{2.1}$$

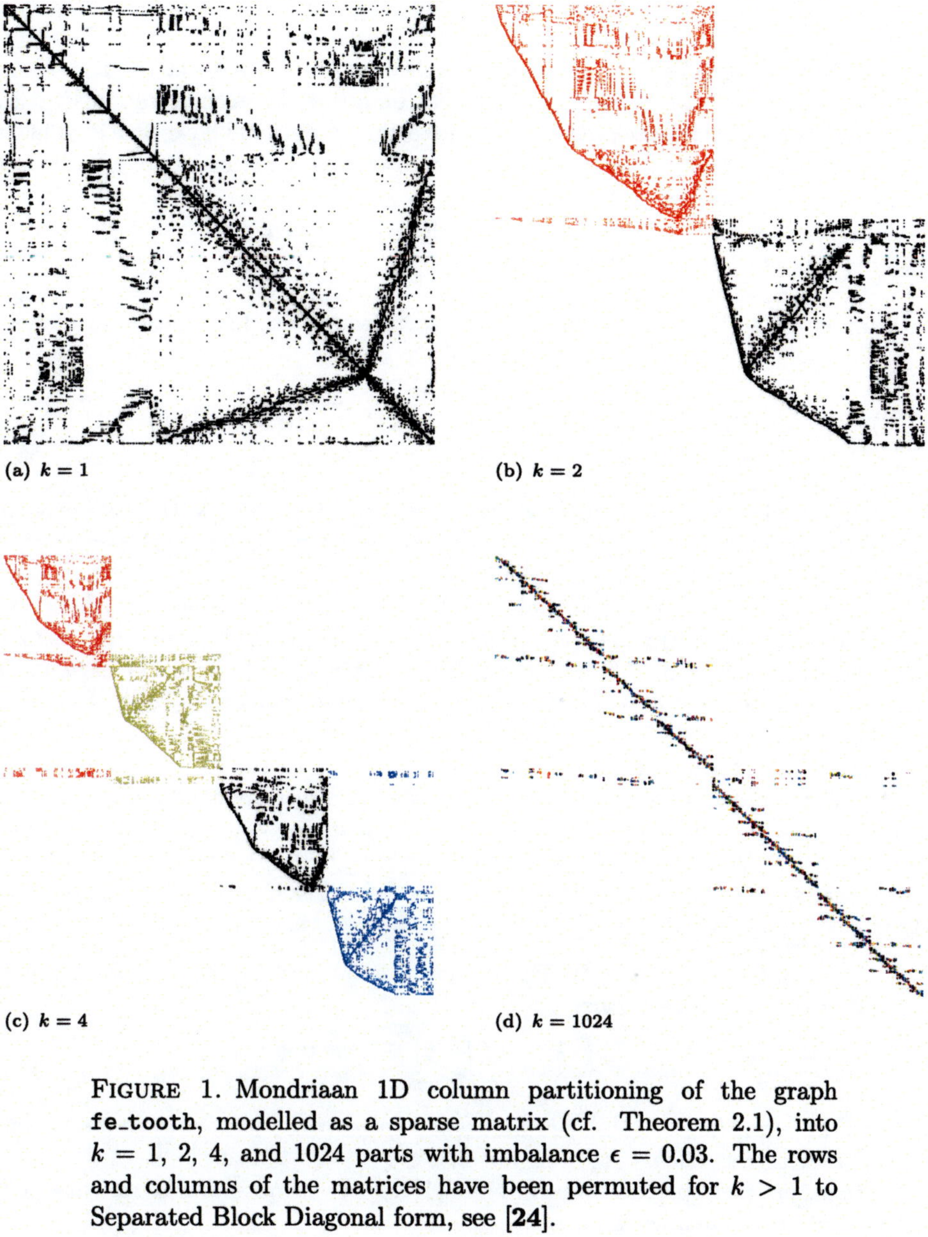

(a) $k = 1$ (b) $k = 2$

(c) $k = 4$ (d) $k = 1024$

FIGURE 1. Mondriaan 1D column partitioning of the graph **fe_tooth**, modelled as a sparse matrix (cf. Theorem 2.1), into $k = 1$, 2, 4, and 1024 parts with imbalance $\epsilon = 0.03$. The rows and columns of the matrices have been permuted for $k > 1$ to Separated Block Diagonal form, see [**24**].

where the partitioner tries to minimise the $(\lambda - 1)$-*volume*

$$\mathrm{LV}(\Pi) := \sum_{n \in \mathcal{N}} (|\Pi(n)| - 1). \tag{2.2}$$

We will now translate the DIMACS partitioning problems from Section 1 to the hypergraph partitioning problem that Mondriaan is designed to solve, by creating a suitable hypergraph $\mathcal{G}$, encoded as a sparse matrix A in the row-net model.

2.3. Minimising communication volume. Let $G = (V, E)$ be a given graph, $k \in \mathbf{N}$, and $\epsilon > 0$. Our aim will be to construct a matrix A from G such that minimising (2.2) subject to (2.1) enforces minimisation of (1.2) subject to (1.1).

To satisfy (1.1), we need to create one column in A for each vertex in V, such that the hypergraph represented by A in the row-net model will have $\mathcal{V} = V$. This is also necessary to have a direct correspondence between partitionings of the vertices V of the graph and the vertices $\mathcal{V}$ of the hypergraph. Setting the weights ζ of all vertices/matrix columns to 1 will then ensure that (1.1) is satisfied if and only if (2.1) is satisfied.

It is a little more tricky to match (1.2) to (2.2). Note that because of the maximum in (1.2), we are not able to create an equivalent formulation. However, as

$$\text{CV}(\Pi) \leq \sum_{i=1}^{k} \sum_{\substack{v \in V \\ \Pi(v)=i}} |\Pi(V_v) \setminus \{\Pi(v)\}| = \sum_{v \in V} |\Pi(V_v) \setminus \{\Pi(v)\}|, \tag{2.3}$$

we can provide an upper bound, which we can use to limit $\text{CV}(\Pi)$. We need to choose the rows of A, corresponding to nets in the row-net hypergraph $\mathcal{G} = (\mathcal{V}, \mathcal{N})$, such that (2.3) and (2.2) are in agreement.

For a net $n \in \mathcal{N}$, we have that $n \subseteq \mathcal{V} = V$ is simply a collection of vertices of G, so $|\Pi(n)|$ in (2.2) equals the number of different parts in which the vertices of n are contained. In (2.3) we count, for a vertex $v \in V$, all parts in which v has a neighbour, except $\Pi(v)$. Note that this number equals $|\Pi(V_v) \setminus \{\Pi(v)\}| = |\Pi(V_v \cup \{v\})| - 1$.

Hence, we should pick $\mathcal{N} := \{V_v \cup \{v\} \mid v \in V\}$ as the set of nets, for (2.3) and (2.2) to agree. In the row-net matrix model, this corresponds to letting A be a matrix with a row for every vertex $v \in V$, filled with nonzeros $a_{v\,v}$ and $a_{u\,v}$ for all $u \in V_v \setminus \{v\}$. Then, for this hypergraph $\mathcal{G}$, we have by (2.3) that $\text{CV}(\Pi) \leq \text{LV}(\Pi)$. Note that since the communication volume is defined as a maximum, we also have that $k\,\text{CV}(\Pi) \geq \text{LV}(\Pi)$.

THEOREM 2.1. *Let $G = (V, E)$ be a given graph, $k \in \mathbf{N}$, and $\epsilon > 0$. Let A be the $|V| \times |V|$ matrix with entries*

$$a_{u\,v} := \begin{cases} 1 & \text{if } \{u, v\} \in E \text{ or } u = v, \\ 0 & \text{otherwise,} \end{cases}$$

for $u, v \in V$, and let $\mathcal{G} = (\mathcal{V}, \mathcal{N})$ be the hypergraph corresponding to A in the row-net model with vertex weights $\zeta(v) = 1$ for all $v \in \mathcal{V}$.

Then, for every partitioning $\Pi : V \to \{1, \ldots, k\}$, we have that Π satisfies (1.1) if and only if Π satisfies (2.1), and

$$\frac{1}{k} LV(\Pi) \leq CV(\Pi) \leq LV(\Pi). \tag{2.4}$$

2.4. Minimising edge cut. We will now follow the same procedure as in Section 2.3 to construct a matrix A such that minimising (2.2) subject to (2.1) is equivalent to minimising (1.3) subject to (1.1).

As in Section 2.3, the columns of A should correspond to the vertices V of G to ensure that (2.1) is equivalent to (1.1).

Equation (1.3) simply counts all of G's edges that contain vertices belonging to two parts of the partitioning Π. Since every edge contains vertices belonging to at least one part, and at most two parts, this yields

$$\mathrm{EC}(\Pi) = \sum_{e \in E} (|\Pi(e)| - 1).$$

Choosing $\mathcal{N} := E$ will therefore give us a direct correspondence between (2.2) and (1.3).

THEOREM 2.2. *Let $G = (V, E)$ be a given graph, $k \in \mathbf{N}$, and $\epsilon > 0$. Let A be the $|E| \times |V|$ matrix with entries*

$$a_{e\,v} := \begin{cases} 1 & \text{if } v \in e, \\ 0 & \text{otherwise,} \end{cases}$$

for $e \in E$, $v \in V$, and let $\mathcal{G} = (\mathcal{V}, \mathcal{N})$ be the hypergraph corresponding to A in the row-net model with vertex weights $\zeta(v) = 1$ for all $v \in \mathcal{V}$.

Then, for every partitioning $\Pi : V \to \{1, \ldots, k\}$, we have that Π satisfies (1.1) *if and only if Π satisfies* (2.1), *and*

$$EC(\Pi) = LV(\Pi). \tag{2.5}$$

With Theorem 2.1 and Theorem 2.2, we know how to translate a given graph G to a hypergraph that Mondriaan can partition to obtain solutions to the DIMACS partitioning challenges.

3. Results

We measure Mondriaan's performance as a graph partitioner by partitioning graphs from the `walshaw/` [20] category, as well as a subset of the specified partitioning instances of the DIMACS challenge test bed [**1**], see Tables 3 and 4. This is done by converting the graphs to matrices, as described by Theorem 2.1 and Theorem 2.2, and partitioning these matrices with Mondriaan 3.11, using the `onedimcol` splitting strategy (since the matrices represent row-net hypergraphs) with the `lambda1` metric (cf. (2.2)). The imbalance is set to $\epsilon = 0.03$, the number of parts k is chosen from $\{2, 4, \ldots, 1024\}$, and we measure the communication volumes and edge cuts over 16 runs of the Mondriaan partitioner (as Mondriaan uses random tie-breaking). All results were recorded on a dual quad-core AMD Opteron 2378 system with 32GiB of main memory and they can be found in Tables 5–8 and Figures 2 and 3. None of the graphs from Table 3 or 4 contain self-edges, edge weights, or vertex weights. Therefore, the values recorded in Tables 5–8 satisfy either (1.2) or (1.3) (which both assume unit weights), and can directly be compared to the results of other DIMACS challenge participants.

Tables 5 and 6 contain the lowest communication volumes and edge cuts obtained by Mondriaan in 16 runs for the graphs from Table 3. The strange dip in the communication volume for `finan512` in Table 5 for $k = 32$ parts can be explained by the fact that the graph `finan512` consists exactly of 32 densely connected parts with few connections between them, see the visualisation of this graph in [**11**], such that there is a natural partitioning with very low communication volume in this case.

To determine how well Mondriaan performs as a graph partitioner, we have also partitioned the graphs from Tables 3 and 4 using METIS 5.0.2 [**14**] and Scotch 5.1.12 [**18**]. For METIS we used the high-quality `PartGraphKway` option, while Scotch was invoked using `graphPart` with the `QUALITY` and `SAFETY` strategies enabled. We furthermore compare the results from Table 6 to the lowest known edge cuts

TABLE 3. Graphs $G = (V, E)$ from the walshaw/ [1, 20] category.

G	$\|V\|$	$\|E\|$	G	$\|V\|$	$\|E\|$
add20	2,395	7,462	bcsstk30	28,924	1,007,284
data	2,851	15,093	bcsstk31	35,588	572,914
3elt	4,720	13,722	fe_pwt	36,519	144,794
uk	4,824	6,837	bcsstk32	44,609	985,046
add32	4,960	9,462	fe_body	45,087	163,734
bcsstk33	8,738	291,583	t60k	60,005	89,440
whitaker3	9,800	28,989	wing	62,032	121,544
crack	10,240	30,380	brack2	62,631	366,559
wing_nodal	10,937	75,488	finan512	74,752	261,120
fe_4elt2	11,143	32,818	fe_tooth	78,136	452,591
vibrobox	12,328	165,250	fe_rotor	99,617	662,431
bcsstk29	13,992	302,748	598a	110,971	741,934
4elt	15,606	45,878	fe_ocean	143,437	409,593
fe_sphere	16,386	49,152	144	144,649	1,074,393
cti	16,840	48,232	wave	156,317	1,059,331
memplus	17,758	54,196	m14b	214,765	1,679,018
cs4	22,499	43,858	auto	448,695	3,314,611

TABLE 4. Graphs $G = (V, E)$ from the 10th DIMACS challenge [1] partitioning instances.

	G	$\|V\|$	$\|E\|$
1	delaunay_n15	32,768	98,274
2	kron_g500-simple-logn17	131,072	5,113,985
3	coAuthorsCiteseer	227,320	814,134
4	rgg_n_2_18_s0	262,144	1,547,283
5	auto	448,695	3,314,611
6	G3_circuit	1,585,478	3,037,674
7	kkt_power	2,063,494	6,482,320
8	M6	3,501,776	10,501,936
9	AS365	3,799,275	11,368,076
10	NLR	4,163,763	12,487,976
11	hugetric-00000	5,824,554	8,733,523
12	great-britain.osm	7,733,822	8,156,517
13	asia.osm	11,950,757	12,711,603
14	hugebubbles-00010	19,458,087	29,179,764

with 3% imbalance for graphs from the walshaw/ category, available from http://staffweb.cms.gre.ac.uk/~wc06/partition/ [20]. These data were retrieved on May 8, 2012 and include results from the KaFFPa partitioner, contributed by Sanders and Schulz [19], who also participated in the DIMACS challenge. Results for graphs from the DIMACS challenge, Tables 7 and 8, are given for the number of parts k specified in the challenge partitioning instances, for a single run of the Mondriaan, METIS, and Scotch partitioners.

TABLE 5. Minimum communication volume, (1.2), over 16 Mondriaan runs, for graphs from the `walshaw/` category, Table 3, divided into $k = 2, 4, \ldots, 64$ parts with imbalance $\epsilon = 0.03$. A '-' indicates that Mondriaan was unable to generate a partitioning satisfying the balancing requirement, (1.1).

G	2	4	8	16	32	64
add20	74	101	118	141	159	-
data	63	84	80	78	65	-
3elt	45	65	59	65	53	49
uk	19	27	36	33	31	24
add32	9	21	29	24	20	22
bcsstk33	454	667	719	630	547	449
whitaker3	64	130	104	98	77	60
crack	95	97	123	100	78	64
wing_nodal	453	593	523	423	362	256
fe_4elt2	66	94	97	85	69	60
vibrobox	996	1,080	966	887	663	482
bcsstk29	180	366	360	336	252	220
4elt	70	90	86	89	88	71
fe_sphere	193	213	178	139	107	83
cti	268	526	496	379	295	200
memplus	2,519	1,689	1,069	720	572	514
cs4	319	492	409	311	228	161
bcsstk30	283	637	611	689	601	559
bcsstk31	358	492	498	490	451	400
fe_pwt	120	122	133	145	148	132
bcsstk32	491	573	733	671	561	442
fe_body	109	143	173	171	145	133
t60k	71	141	154	139	129	96
wing	705	854	759	594	451	324
brack2	231	650	761	635	562	458
finan512	75	76	137	141	84	165
fe_tooth	1,238	1,269	1,282	1,066	844	703
fe_rotor	549	1,437	1,258	1,138	944	749
598a	647	1,400	1,415	1,432	1,064	871
fe_ocean	269	797	1,002	1,000	867	647
144	1,660	2,499	2,047	1,613	1,346	1,184
wave	2,366	2,986	2,755	2,138	1,640	1,222
m14b	921	2,111	2,086	2,016	1,524	1,171
auto	2,526	4,518	4,456	3,982	3,028	2,388

TABLE 6. Minimum edge cut, (1.3), over 16 Mondriaan runs, for graphs from the **walshaw/** category, Table 3, divided into $k = 2, 4, \ldots, 64$ parts with imbalance $\epsilon = 0.03$. A '-' indicates that Mondriaan was unable to generate a partitioning satisfying the balancing requirement, (1.1).

G	2	4	8	16	32	64
add20	680	1,197	1,776	2,247	2,561	-
data	195	408	676	1,233	2,006	-
3elt	87	206	368	639	1,078	1,966
uk	20	43	98	177	299	529
add32	21	86	167	247	441	700
bcsstk33	10,068	21,993	37,054	58,188	82,102	114,483
whitaker3	126	385	692	1,172	1,825	2,769
crack	186	372	716	1,169	1,851	2,788
wing_nodal	1,703	3,694	5,845	8,963	12,870	17,458
fe_4elt2	130	350	616	1,091	1,770	2,760
vibrobox	10,310	19,401	28,690	37,038	45,877	53,560
bcsstk29	2,846	8,508	16,714	25,954	39,508	59,873
4elt	137	335	543	1,040	1,724	2,896
fe_sphere	404	822	1,258	1,972	2,857	4,082
cti	318	934	1,786	2,887	4,302	6,027
memplus	5,507	9,666	12,147	14,077	15,737	17,698
cs4	389	1,042	1,654	2,411	3,407	4,639
bcsstk30	6,324	16,698	35,046	77,589	123,766	186,084
bcsstk31	2,677	7,731	14,299	25,212	40,641	65,893
fe_pwt	347	720	1,435	2,855	5,888	9,146
bcsstk32	4,779	9,146	23,040	41,214	66,606	102,977
fe_body	271	668	1,153	2,011	3,450	5,614
t60k	77	227	506	952	1,592	2,483
wing	845	1,832	2,843	4,451	6,558	8,929
brack2	690	2,905	7,314	12,181	19,100	28,509
finan512	162	324	891	1,539	2,592	10,593
fe_tooth	3,991	7,434	12,736	19,709	27,670	38,477
fe_rotor	1,970	7,716	13,643	22,304	34,515	50,540
598a	2,434	8,170	16,736	27,895	43,192	63,056
fe_ocean	317	1,772	4,316	8,457	13,936	21,522
144	6,628	16,822	27,629	41,947	62,157	86,647
wave	8,883	18,949	32,025	47,835	69,236	94,099
m14b	3,862	13,464	26,962	46,430	73,177	107,293
auto	9,973	27,297	49,087	83,505	132,998	191,429

TABLE 7. Communication volume, (1.2), for graphs from Table 4, divided into k parts with imbalance $\epsilon = 0.03$ for one run of Mondriaan, METIS, and Scotch. The numbering of the graphs is given by Table 4.

G	k	Mon.	MET.	Sco.	G	k	Mon.	MET.	Sco.
1	8	228	238	250	8	2	1,392	1,420	1,416
	16	180	169	202		8	2,999	2,242	2,434
	32	154	134	137		32	1,852	1,497	1,611
	64	110	112	94		128	1,029	783	814
	128	94	72	88		256	737	553	606
2	2	38,565	46,225	49,273	9	64	1,375	1,099	1,266
	4	38,188	61,833	56,503		128	1,037	814	837
	8	73,739	62,418	60,600		256	761	555	639
	16	82,356	47,988	61,469		512	552	419	481
	32	88,273	43,990	74,956		1024	374	299	330
3	4	11,063	10,790	20,018	10	8	2,508	2,707	3,104
	8	9,652	9,951	14,004		32	1,659	1,620	1,763
	16	7,216	6,507	9,928		128	1,056	820	895
	32	4,732	4,480	6,684		256	728	624	713
	64	3,298	3,111	4,273		512	596	464	478
4	8	749	710	837	11	2	1,222	1,328	1,408
	16	522	640	665		4	2,536	2,668	2,693
	32	524	437	455		32	1,175	1,224	1,168
	64	342	359	348		64	1,022	985	893
	128	285	238	326		256	594	467	510
5	64	2,423	2,407	2,569	12	32	235	214	191
	128	1,774	1,634	1,766		64	228	133	149
	256	1,111	1,120	1,248		128	194	130	138
	512	786	717	824		256	135	95	115
	1024	552	519	540		1024	102	78	83
6	2	1,219	1,267	1,308	13	64	139	53	84
	4	1,887	1,630	2,144		128	139	58	73
	32	1,304	1,285	1,291		256	145	65	104
	64	1,190	1,111	1,228		512	157	110	90
	256	668	566	702		1024	127	124	109
7	16	6,752	9,303	36,875	14	4	3,359	3,283	3,620
	32	7,057	9,123	20,232		32	2,452	2,139	2,462
	64	7,255	9,244	10,669		64	1,864	1,592	1,797
	256	4,379	4,198	4,842		256	1,143	847	1,040
	512	3,280	2,589	3,265		512	737	621	704

TABLE 8. Edge cut, (1.3), for graphs from Table 4, divided into k parts with imbalance $\epsilon = 0.03$ for one run of Mondriaan, METIS, and Scotch. The numbering of the graphs is given by Table 4.

G	k	Mon.	MET.	Sco.	G	k	Mon.	MET.	Sco.
1	8	1,367	1,358	1,386	8	2	2,949	2,869	2,827
	16	2,164	2,170	2,121		8	15,052	14,206	14,622
	32	3,217	3,267	3,283		32	39,756	35,906	36,795
	64	4,840	4,943	4,726		128	81,934	78,824	80,157
	128	7,134	6,979	7,000		256	117,197	114,413	114,800
2	2	208,227	1,972,153	773,367	9	64	56,009	53,557	54,835
	4	835,098	2,402,130	2,614,571		128	81,768	78,055	79,193
	8	1,789,048	2,988,293	3,417,254		256	119,394	113,171	114,758
	16	2,791,475	3,393,061	3,886,568		512	167,820	163,673	165,078
	32	3,587,053	3,936,154	4,319,148		1024	239,947	234,301	234,439
3	4	37,975	37,151	67,513	10	8	16,881	16,992	17,172
	8	54,573	53,502	81,556		32	42,523	40,130	40,967
	16	67,308	66,040	92,992		128	90,105	86,332	86,760
	32	77,443	75,448	104,050		256	129,635	124,737	126,233
	64	85,610	84,111	111,090		512	186,016	178,324	179,779
4	8	4,327	4,381	4,682	11	2	1,345	1,328	1,408
	16	7,718	7,107	7,879		4	4,197	3,143	3,693
	32	13,207	10,386	11,304		32	16,659	13,981	14,434
	64	20,546	16,160	16,630		64	24,031	20,525	21,597
	128	32,039	24,644	25,749		256	50,605	44,082	44,634
5	64	192,783	188,424	196,385	12	32	2,213	1,622	1,770
	128	266,541	257,800	265,941		64	3,274	2,461	2,891
	256	359,123	346,655	366,258		128	5,309	3,948	4,439
	512	475,284	455,321	479,379		256	8,719	6,001	6,710
	1024	621,339	591,928	629,085		1024	19,922	14,692	15,577
6	2	1,370	1,371	1,339	13	64	1,875	623	1,028
	4	3,174	3,163	3,398		128	3,246	1,106	1,637
	32	14,326	14,054	14,040		256	5,381	2,175	2,938
	64	24,095	22,913	25,434		512	9,439	4,157	5,133
	256	58,164	57,255	60,411		1024	15,842	7,987	9,196
7	16	136,555	132,431	279,808	14	4	6,290	5,631	6,340
	32	204,688	219,370	370,494		32	29,137	25,049	27,693
	64	339,620	351,913	462,030		64	43,795	38,596	41,442
	256	653,613	662,569	694,692		256	90,849	82,566	86,554
	512	774,477	755,994	814,142		512	131,481	118,974	124,694

TABLE 9. Comparison of the minimum communication volume, (1.2), and edge cut, (1.3), for graphs from Table 3 (`walshaw/` collection) and Table 4 (DIMACS challenge collection). We compare the Mondriaan, METIS, and Scotch partitioners using (3.1) with $\mathcal{X}$ consisting of the graphs from either Table 3 or 4 and using either the communication volume or the edge cut metric.

	Communication volume				Edge cut			
		Mon.	MET.	Sco.		Mon.	MET.	Sco.
Walshaw	Mon.	-	0.98	0.95	Mon.	-	1.02	1.01
	MET.	1.02	-	0.98	MET.	0.98	-	1.00
	Sco.	1.05	1.02	-	Sco.	0.99	1.00	-
		Mon.	MET.	Sco.		Mon.	MET.	Sco.
DIMACS	Mon.	-	1.15	0.99	Mon.	-	1.08	0.98
	MET.	0.87	-	0.86	MET.	0.93	-	0.91
	Sco.	1.01	1.16	-	Sco.	1.02	1.10	-

Table 9 gives a summary of each partitioner's relative performance with respect to the others. To illustrate how we compare the quality of the partitionings generated by Mondriaan, METIS, and Scotch, consider the following example. Let $\mathcal{X}$ be a collection of graphs (e.g. the graphs from Table 3) on which we would like to compare the quality of the Mondriaan and METIS partitioners in the communication volume metric. Let Π_G^{Mon} and Π_G^{MET} denote the partitionings found for the graph $G \in \mathcal{X}$ by Mondriaan and METIS, respectively. Then, we determine how much better Mondriaan performs than METIS by looking at the average logarithm of the ratios of the communication volumes for all partitionings of graphs in $\mathcal{X}$,

$$\kappa_{\mathrm{Mon,MET}}(\mathcal{X}) := \exp\left(\frac{1}{|\mathcal{X}|}\sum_{G\in\mathcal{X}} \log\frac{\mathrm{CV}(\Pi_G^{\mathrm{Mon}})}{\mathrm{CV}(\Pi_G^{\mathrm{MET}})}\right), \tag{3.1}$$

which is equal to 0.98 in Table 9 for $\mathcal{X} = \{\text{graphs from Table 3}\}$. If the value from (3.1) is smaller than 1, Mondriaan outperforms METIS, while METIS outperforms Mondriaan if it is larger than 1. We use this quality measure instead of simply calculating the average of all $\mathrm{CV}(\Pi_G^{\mathrm{Mon}})/\mathrm{CV}(\Pi_G^{\mathrm{MET}})$ ratios, because it gives us a symmetric comparison of all partitioners, in the following sense:

$$\kappa_{\mathrm{Mon,MET}}(\mathcal{X}) = 1/\kappa_{\mathrm{MET,Mon}}(\mathcal{X}).$$

Scotch is unable to optimise for the communication volume metric directly and therefore it is not surprising that Scotch is outperformed by both Mondriaan and METIS in this metric. Surprisingly, Mondriaan outperforms Scotch in terms of edge cut for the graphs from Table 4. The more extreme results for the graphs from Table 4 could be caused by the fact that they have been recorded for a single run of the partitioners, while the results for graphs from Table 3 are the best in 16 runs. METIS yields lower average communication volumes and edge cuts than both Mondriaan and Scotch in almost all DIMACS cases.

If we compare the edge cuts for graphs from Table 3 to the best-known results from **[20]**, we find that Mondriaan's, METIS', and Scotch's best edge cuts obtained in 16 runs are on average 13%, 10%, and 10% larger, respectively, than those from **[20]**.

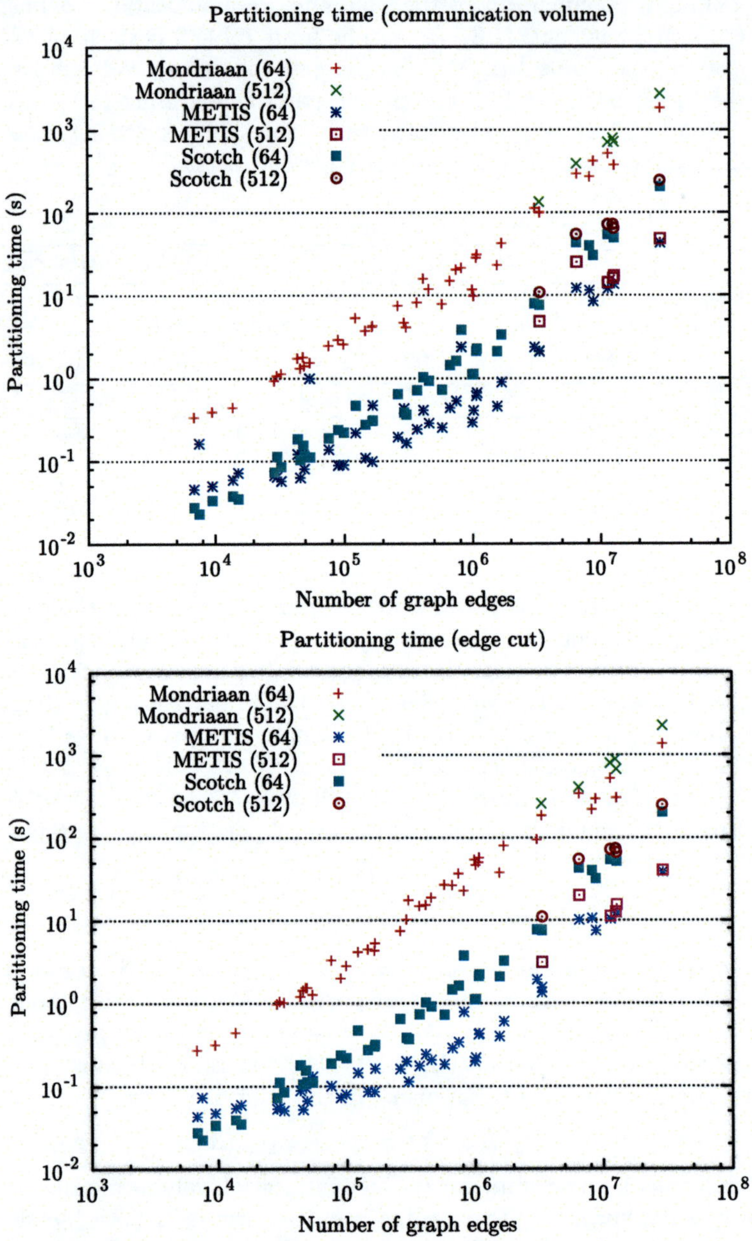

FIGURE 2. The average partitioning time required by the Mondriaan, METIS, and Scotch partitioners to generate the partitionings from Table 5–8 (for 64 and 512 parts).

In Figure 2, we plot the time required by Mondriaan, METIS, and Scotch to create a partitioning for both communication volume and edge cut. Note that the partitioning times are almost the same for both communication volume and edge cut minimisation. METIS is on average 29× faster than Mondriaan for 64 parts

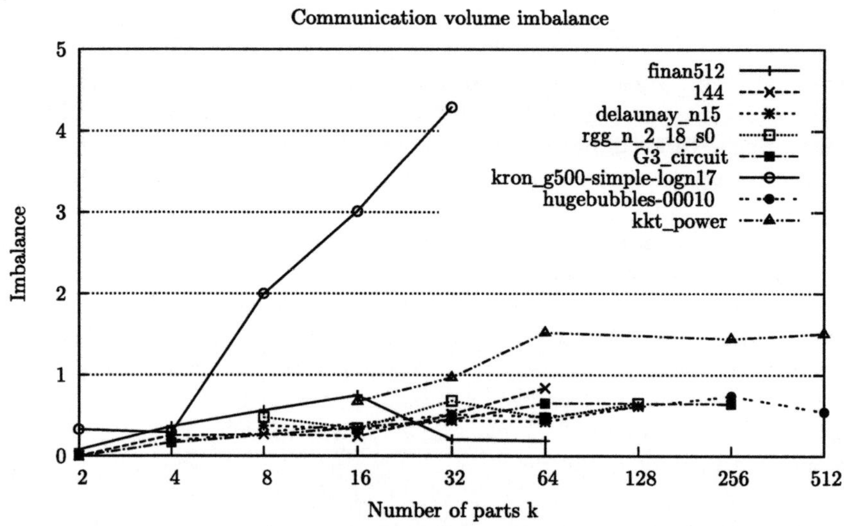

FIGURE 3. The communication volume imbalance given by (3.2), plotted for several graphs.

and Scotch is 12× faster. Note that only six (large) matrices are partitioned into 512 parts.

In the absence of self-edges, the number of nonzeros in the matrices from Theorem 2.1 and Theorem 2.2 equals $2|E| + |V|$ and $2|E|$, respectively. However, the matrix sizes are equal to $|V| \times |V|$ and $|E| \times |V|$, respectively. Therefore, the number of nonzeros in matrices from Theorem 2.2 is smaller, but the larger number of nets (typically $|E| > |V|$, e.g. `rgg_n_2_18_s0`) will lead to increased memory requirements for the edge-cut matrices.

We have also investigated Mondriaan's communication volume imbalance, defined for a partitioning Π of G into k parts as

$$\frac{\mathrm{CV}(\Pi)}{\mathrm{LV}(\Pi)/k} - 1. \tag{3.2}$$

This equation measures the imbalance in communication volume and can be compared to the factor ϵ for vertex imbalance in (1.1). We plot (3.2) for a selection of graphs in Figure 3, where we see that the deviation of the communication volume $\mathrm{CV}(\Pi)$ from perfect balance, i.e. from $\mathrm{LV}(\Pi)/k$, is very small compared to the theoretical upper bound of $k - 1$ (via (2.4)), for all graphs except `kron_g500-simple-logn17`. This means that for most graphs, at most a factor of 2–3 in communication volume per processor can still be gained by improving the communication balance. Therefore, as the number of parts increases, the different parts of the partitionings generated by Mondriaan are not only balanced in terms of vertices, cf. (1.1), but also in terms of communication volume.

4. Conclusion

We have shown that it is possible to use the Mondriaan matrix partitioner as a graph partitioner by constructing appropriate matrices of a given graph for either the communication volume or edge-cut metric. Mondriaan's performance was measured by partitioning graphs from the 10th DIMACS challenge on graph partitioning and clustering with Mondriaan, METIS, and Scotch, as well as comparing obtained edge cuts with the best known results from [**20**]: here Mondriaan's best edge cut in 16 runs was, on average, 13% higher than the best known. Mondriaan is competitive in terms of partitioning quality (METIS' and Scotch's best edge cuts are, on average, 10% higher than the best known), but it is an order of magnitude slower (Figure 2). METIS is the overall winner, both in quality and performance. In conclusion, it is possible to perform graph partitioning with a hypergraph partitioner, but graph partitioners are much faster.

To our surprise, the partitionings generated by Mondriaan are reasonably balanced in terms of communication volume, as shown in Figure 3, even though Mondriaan does not perform explicit communication volume balancing during matrix partitioning. We attribute the observed balancing to the fact that the Mondriaan algorithm performs random tie-breaking, without any preference for a specific part of the partitioning.

Fortunately, for the given test set of the DIMACS challenge, we did not need to consider edge weights. However, for Mondriaan to be useful as graph partitioner also for weighted graphs, we have to extend Mondriaan to take hypergraph net weights into account for the $(\lambda - 1)$-metric, (2.2). We intend to add this feature in a next version of Mondriaan.

References

[1] D. A. Bader, P. Sanders, D. Wagner, H. Meyerhenke, B. Hendrickson, D. S. Johnson, C. Walshaw, and T. G. Mattson, *10th DIMACS implementation challenge - graph partitioning and graph clustering*, 2012; `http://www.cc.gatech.edu/dimacs10`.

[2] Rob H. Bisseling, *Parallel scientific computation: A structured approach using BSP and MPI*, Oxford University Press, Oxford, 2004. MR2059580

[3] Rob H. Bisseling, Bas O. Fagginger Auer, A. N. Yzelman, Tristan van Leeuwen, and Ümit V. Çatalyürek, *Two-dimensional approaches to sparse matrix partitioning*, Combinatorial scientific computing, Chapman & Hall/CRC Comput. Sci. Ser., CRC Press, Boca Raton, FL, 2012, pp. 321–349, DOI 10.1201/b11644-13. MR2952757

[4] Rob H. Bisseling and Wouter Meesen, *Communication balancing in parallel sparse matrix-vector multiplication*, Electron. Trans. Numer. Anal. **21** (2005), 47–65 (electronic). MR2195104 (2007c:65040)

[5] T. Bui and C. Jones, *A heuristic for reducing fill-in in sparse matrix factorization*, Proceedings Sixth SIAM Conference on Parallel Processing for Scientific Computing, SIAM, Philadelphia, PA, 1993, pp. 445–452.

[6] A. E. Caldwell, A. B. Kahng, and I. L. Markov, *Improved algorithms for hypergraph bipartitioning*, Proceedings Asia and South Pacific Design Automation Conference, ACM Press, New York, 2000, pp. 661–666. DOI 10.1145/368434.368864.

[7] Ü. V. Çatalyürek and C. Aykanat, *Hypergraph-partitioning-based decomposition for parallel sparse-matrix vector multiplication*, IEEE Transactions on Parallel and Distributed Systems **10** (1999), no. 7, 673–693. DOI 10.1109/71.780863.

[8] ———, *PaToH: A multilevel hypergraph partitioning tool, version 3.0*, Bilkent University, Department of Computer Engineering, Ankara, 06533 Turkey. PaToH is available at `http://bmi.osu.edu/~umit/software.htm`, 1999.

[9] ———, *A fine-grain hypergraph model for 2D decomposition of sparse matrices*, Proceedings Eighth International Workshop on Solving Irregularly Structured Problems in Parallel (Irregular 2001), IEEE Press, Los Alamitos, CA, 2001, p. 118.

[10] C. Chevalier and F. Pellegrini, *PT-Scotch: a tool for efficient parallel graph ordering*, Parallel Comput. **34** (2008), no. 6-8, 318–331, DOI 10.1016/j.parco.2007.12.001. MR2428880

[11] Timothy A. Davis and Yifan Hu, *The University of Florida sparse matrix collection*, ACM Trans. Math. Software **38** (2011), no. 1, Art. 1, 25pp, DOI 10.1145/2049662.2049663. MR2865011 (2012k:65051)

[12] K. D. Devine, E. G. Boman, R. T. Heaphy, R. H. Bisseling, and U. V. Catalyurek, *Parallel hypergraph partitioning for scientific computing*, Proceedings IEEE International Parallel and Distributed Processing Symposium 2006, IEEE Press, p. 102, 2006. DOI 10.1109/IPDPS.2006.1639359.

[13] Bruce Hendrickson and Robert Leland, *An improved spectral graph partitioning algorithm for mapping parallel computations*, SIAM J. Sci. Comput. **16** (1995), no. 2, 452–469, DOI 10.1137/0916028. MR1317066 (96b:68140)

[14] George Karypis and Vipin Kumar, *A fast and high quality multilevel scheme for partitioning irregular graphs*, SIAM J. Sci. Comput. **20** (1998), no. 1, 359–392 (electronic), DOI 10.1137/S1064827595287997. MR1639073 (99f:68158)

[15] ______ *Multilevel k-way hypergraph partitioning*, Proceedings 36th ACM/IEEE Conference on Design Automation, ACM Press, New York, 1999, pp. 343–348.

[16] ______, *Parallel multilevel k-way partitioning scheme for irregular graphs*, SIAM Review **41** (1999), no. 2, 278–300. DOI 10.1145/309847.309954.

[17] B. W. Kernighan and S. Lin, *An efficient heuristic procedure for partitioning graphs*, Bell System Technical Journal **49** (2) (1970), 291–307.

[18] F. Pellegrini and J. Roman, *Scotch: A software package for static mapping by dual recursive bipartitioning of process and architecture graphs*, Proceedings High Performance Computing and Networking Europe, Lecture Notes in Computer Science, vol. 1067, Springer, 1996, pp. 493–498. DOI 10.1007/3-540-61142-8_588.

[19] Peter Sanders and Christian Schulz, *Engineering multilevel graph partitioning algorithms*, Algorithms—ESA 2011, Lecture Notes in Comput. Sci., vol. 6942, Springer, Heidelberg, 2011, pp. 469–480, DOI 10.1007/978-3-642-23719-5_40. MR2893224 (2012k:68259)

[20] A. J. Soper, C. Walshaw, and M. Cross, *A combined evolutionary search and multilevel optimisation approach to graph-partitioning*, J. Global Optim. **29** (2004), no. 2, 225–241, DOI 10.1023/B:JOGO.0000042115.44455.f3. MR2092958 (2005k:05228)

[21] A. Trifunović and W. J. Knottenbelt, *Parallel multilevel algorithms for hypergraph partitioning*, Journal of Parallel and Distributed Computing **68** (2008), no. 5, 563–581. DOI 10.1016/j.jpdc.2007.11.002.

[22] Brendan Vastenhouw and Rob H. Bisseling, *A two-dimensional data distribution method for parallel sparse matrix-vector multiplication*, SIAM Rev. **47** (2005), no. 1, 67–95 (electronic), DOI 10.1137/S0036144502409019. MR2149102 (2006a:65070)

[23] C. Walshaw and M. Cross, *JOSTLE: Parallel Multilevel Graph-Partitioning Software – An Overview*, Mesh Partitioning Techniques and Domain Decomposition Techniques (F. Magoules, ed.), Civil-Comp Ltd., 2007, pp. 27–58.

[24] A. N. Yzelman and Rob H. Bisseling, *Cache-oblivious sparse matrix-vector multiplication by using sparse matrix partitioning methods*, SIAM J. Sci. Comput. **31** (2009), no. 4, 3128–3154, DOI 10.1137/080733243. MR2529783 (2011a:65111)

Mathematics Institute, Utrecht University, Budapestlaan 6, 3584 CD, Utrecht, the Netherlands

E-mail address: `B.O.FaggingerAuer@uu.nl`

Mathematics Institute, Utrecht University, Budapestlaan 6, 3584 CD, Utrecht, the Netherlands

E-mail address: `R.H.Bisseling@uu.nl`

Parallel partitioning with Zoltan: Is hypergraph partitioning worth it?

Sivasankaran Rajamanickam and Erik G. Boman

ABSTRACT. Graph partitioning is an important and well studied problem in combinatorial scientific computing, and is commonly used to reduce communication in parallel computing. Different models (graph, hypergraph) and objectives (edge cut, boundary vertices) have been proposed. Hypergraph partitioning has become increasingly popular over the last decade. Its main strength is that it accurately captures communication volume, but it is slower to compute than graph partitioning. We present an empirical study of the Zoltan parallel hypergraph and graph (PHG) partitioner on graphs from the 10th DIMACS implementation challenge and some directed (nonsymmetric) graphs. We show that hypergraph partitioning is superior to graph partitioning on directed graphs (nonsymmetric matrices), where the communication volume is reduced in several cases by over an order of magnitude, but has no significant benefit on undirected graphs (symmetric matrices) using current parallel software tools.

1. Introduction

Graph partitioning is a well studied problem in combinatorial scientific computing. An important application is the mapping of data and/or tasks on a parallel computer, where the goals are to balance the load and to minimize communication [**12**]. There are several variations of graph partitioning, but they are all NP-hard problems. Fortunately, good heuristic algorithms exist. Naturally, there is a trade-off between run-time and solution quality. In parallel computing, partitioning may be performed either once (static partitioning) or many times (dynamic load balancing). In the latter case, it is crucial that the partitioning itself is fast. Furthermore, the rapid growth of problem sizes in scientific computing dictates that partitioning algorithms must be scalable. The multilevel approach developed in the 1990s [**3, 11, 17**] provides a good compromise between run-time (complexity)

1991 *Mathematics Subject Classification.* Primary 68R10; Secondary 05C65, 68W10, 68Q85.

Key words and phrases. Graph partitioning, hypergraph partitioning, parallel computing.

We thank the U.S. Department of Energy's Office of Science, the Advanced Scientific Computing Research (ASCR) office, and the National Nuclear Security Administration's ASC program for financial support. Sandia is a multiprogram laboratory managed and operated by Sandia Corporation, a wholly owned subsidiary of Lockheed Martin, for the U.S. Department of Energy's National Nuclear Security Administration under contract DE-AC04-94AL85000. This research used resources of the National Energy Research Scientific Computing Center (NERSC), which is supported by the Office of Science of the DOE under Contract No. DE-AC02-05CH11231.

and quality. Software packages based on this approach (Chaco [13], Metis [14], and Scotch [19]) have been extremely successful. Even today, all the major parallel software packages for partitioning in scientific computing (ParMetis [15], PT-Scotch [20], and Zoltan [8, 9]) use variations of the multilevel graph partitioning algorithm.

The 10th DIMACS implementation challenge offers an opportunity to evaluate the current (2012) state-of-the-art in partitioning software. This is a daunting task, as there are several variations of the partitioning problem (e.g., objectives), several software codes, and a large number of data sets. In this paper we limit the scope in the following ways: We only consider parallel software since our focus is high-performance computing. We focus on the Zoltan toolkit since its partitioner can be used to minimize either the edge cut (graph partitioning) or the communication volume (hypergraph partitioning). We include some baseline comparisons with ParMetis, since that is the most widely used parallel partitioning software. We limit the experiments to a subset of the DIMACS graphs. One may view this paper as a follow-up to the 2006 paper that introduced the Zoltan PHG partitioner [9].

Contributions: We compare graph and hypergraph partitioners for both symmetric and unsymmetric inputs and obtain results that are quite different than in [4]. For nonsymmetric matrices we see a big difference in communication volume (orders of magnitude), while there is virtually no difference among the partitioners for symmetric matrices. We exercise Zoltan PHG on larger number of processors than before (up to 1024). We present results for impact of partitioning on an iterative solver. We also include results for the maximum communication volume, which is important in practice but not an objective directly modeled by any current partitioner.

2. Models and Metrics

The term "graph partitioning" can refer to several different problems. Most often, it refers to the edge cut metric, though in practice the communication volume metric is often more important. For the latter objective, it is useful to extend graphs to hypergraphs. Here, we review the different models and metrics and explain how they relate.

2.1. Graph Models. Given an undirected graph $G = (V, E)$, the classic version of graph partitioning is to partition V into k disjoint subsets (parts) such that all the parts are approximately the same size and the total number of edges between parts are minimized. More formally, let $\Pi = \{\pi_0, \ldots, \pi_{k-1}\}$ be a balanced partition such that

$$|V(\pi_i)| \leq (1+\epsilon)\,\frac{|V|}{k} \quad \forall i, \tag{1}$$

for a given $\epsilon > 0$. The edge cut problem (EC) is then to minimize the cut set

$$C(G, \Pi) = \{\{(u, v) \in E\} | \Pi(u) \neq \Pi(v)\}\,. \tag{2}$$

There are straight-forward generalizations for edge weights (minimize weighted cuts) and vertex weights (balance is weighted).

Most algorithms and software attempt to minimize the edge cut. However, several authors have shown that the edge cut does not represent communication in parallel computing [4, 12]. A key insight was that the communication is proportional to the *vertices* along the part boundaries, not the cut edges. A more relevant metric is therefore the communication volume, which roughly corresponds to the boundary vertices. Formally, let the communication volume for part p be

$$comm(\pi_p) = \sum_{v \in \pi(p)} (\lambda(v, \Pi) - 1), \tag{3}$$

where $\lambda(v, \Pi)$ denotes the number of parts that v or any of its neighbors belong to, with respect to the partition Π.

We then obtain the following two metrics:

$$CV_{max}(G, \Pi) = \max_p comm(\pi_p) \tag{4}$$

$$CV_{sum}(G, \Pi) = \sum_p comm(\pi_p) \tag{5}$$

In parallel computing, this corresponds to the maximum communication volume for any process and the total sum of communication volumes, respectively.

2.2. Hypergraph Models. A *hypergraph* $H = (V, E)$ extends a graph since now E denotes a set of hyperedges. An hyperedge is any non-empty subset of the vertices V. A graph is just a special case of a hypergraph where each hyperedge has cardinality two (since a graph edge always connects two vertices). Hyperedges are sometimes called *nets*, a term commonly used in the (VLSI) circuit design community.

Analogous to graph partitioning, one can define several hypergraph partitioning problems. As before, the balance constraint is on the vertices. Several different cut metrics have been proposed. The most straight-forward generalization of edge cut to hypergraphs is:

$$C(H, \Pi) = \{\{e \in E\} | \Pi(u) \neq \Pi(v) \text{ where } u \in e, v \in e\}. \tag{6}$$

However, a more popular metric is the so-called $(\lambda - 1)$ metric:

$$CV(H, \Pi) = \sum_{e \in E} (\lambda(e, \Pi) - 1), \tag{7}$$

where $\lambda(e, \Pi)$ is the number of distinct parts that contain any vertex in e.

While graphs are restricted to structurally symmetric problems (undirected graphs), hypergraphs make no such assumption. Furthermore, the number of vertices and hyperedges may differ, making the model suitable for rectangular matrices. The key advantage of the hypergraph model is that the hyperedge $(\lambda - 1)$ cut (CV) accurately models the total communication volume. This was first observed in **[4]** in the context of sparse matrix-vector multiplication. The limitations of the graph model were described in detail in **[12]**. This realization led to a shift from the graph model to the hypergraph model. Today, many partitioning packages use the hypergraph model: PaToH **[4]**, hMetis **[16]**, Mondriaan **[21]**, and Zoltan-PHG **[9]**.

Hypergraphs are often used to represent sparse matrices. For example, using row-based storage (CSR), each row becomes a vertex and each column becomes a hyperedge. Other hypergraph models exist: in the "fine-grain" model, each nonzero is a vertex **[5]**. For the DIMACS challenge, all input is symmetric and given as undirected graphs. Given a graph $G(V, E)$, we will use the following derived hypergraph $H(V, E')$: for each vertex $v \in V$, create an hyperedge $e \in E'$ that contains v and all its neighbors. In this case, it is easy to see that $CV(H, \Pi) =$

$CV_{sum}(G, \Pi)$. Thus, we do not need to distinguish between communication volume in the graph and hypergraph models.

2.3. Relevance of the Metrics. Most partitioners minimize either the total edge cut (EC) or the total communication volume (CV-sum). A main reason for this choice is that algorithms for these metrics are well developed. Less work has been done to minimize the maximum communication volume (CV-max), though in a parallel computing setting this may be more relevant as it corresponds to the maximum communication for any one process.

In order to compare the three metrics and how they correspond to the actual performance we use conjugate gradient (CG) iteration (from the Belos package [**1**]) as a test case. We used the matrices from the UF sparse matrix collection group of the DIMACS challenge. As the goal is to compare the matrix-vector multiply time in the CG iteration, we used no preconditioner as the performance characteristics will be different depending on the preconditioners. As there is no preconditioner and some of these problems are ill-conditioned the CG iteration might not converge at all, so we report the solve time for 1000 iterations. We compare four different row-based partitionings (on 12 processors): natural (block) partitioning, random partitioning, graph partitioning with ParMetis, and hypergraph partitioning with Zoltan hypergraph partitioner. We only change the data distribution, and do not reorder the matrix, so the convergence of CG is not affected. The results are shown in Table 1. As expected, random partitioning is worst since it just balances the load but has high communication. In all but one case, we see that both graph and hypergraph partitioning beat the simple natural (block) partitioning (which is the default in Trilinos). For the audikw1 test matrix, the time is cut to less than half. For these symmetric problems, the difference between graph and hypergraph partitioning is very small in terms of real performance gains. We will show in Section 4.2 that the partitioners actually differ in terms of the measured performance metrics for three of the problems shown in Table 1. However, the difference in the metrics do not translate to measurable real performance gain in the time for the matrix-vector multiply.

TABLE 1. Solve time (seconds) for 1000 iterations of CG for different row partitioning options.

Matrix Name	Natural	Random	ParMetis	Zoltan PHG
audikw1	62.90	98.58	27.71	27.52
ldoor	22.18	72.09	18.24	18.08
G3_circuit	11.26	25.78	8.13	8.62
af_shell10	20.09	84.51	21.29	21.17
bone010	24.33	84.07	24.92	25.39
geo_1438	25.35	106.36	25.53	25.78
inline_1	22.47	44.57	13.54	13.90
pwtk	4.30	11.88	4.34	4.37

3. Overview of the Zoltan Hypergraph Partitioner

Zoltan was originally designed as a toolkit for dynamic load-balancing [**8**]. It included several geometric partitioning algorithms, plus interfaces to external (third-party) graph partitioners, such as ParMetis. Later, a native parallel hypergraph

$$\begin{array}{c} \text{vertices} \\ \begin{bmatrix} P_0 & P_1 & P_2 \\ P_3 & P_4 & P_5 \end{bmatrix} \end{array} \text{hyperedges}$$

FIGURE 1. Example of the 2D layout for 2×3 processes.

partitioner (PHG) was developed **[9]** and added to Zoltan. While PHG was designed for hypergraph partitioning, it can also be used for graph partitioning but it is not optimized for this use case. (Note: "PHG" now stands for Parallel Hypergraph and Graph partitioner.) Zoltan also supports other combinatorial problems such as graph ordering and graph coloring **[2]**.

Zoltan PHG is a parallel multilevel partitioner, consisting of the usual *coarsening*, *initial partitioning*, and *refinement* phases. The algorithm is similar to the serial partitioners PaToH **[4]**, hMetis **[16]** and Mondriaan **[21]**, but Zoltan PHG is parallel (based on MPI) so can run on both shared-memory and distributed-memory systems. Note that Zoltan can partition data into k parts using p processes, where $k \neq p$. Neither k nor p need be powers of two. We briefly describe the algorithm in Zoltan PHG, with emphasis on the parallel computing aspects. For further details on PHG, we refer to **[9]**. The basic algorithm remains the same, though several improvements have been made over the years.

3.1. 2D Data Distribution. A novel feature of Zoltan PHG is that internally, the hypergraph is mapped to processes in a 2D block (checkerboard) fashion. That is, the processes are logically mapped to a p_x by p_y grid, where $p = p_x p_y$. The hypergraph is partitioned accordingly, when viewed as a sparse matrix (Fig. 1). We do not attempt an optimal 2D Cartesian (checkerboard) distribution: The best known algorithm requires multiconstraint hypergraph partitioning **[6]**, which is even harder than the partitioning problem we wish to solve.

The goal of this design is to reduce communication within the partitioner itself. Instead of expensive all-to-all or any-to-any communication, all communication is limited to process rows or columns. Thus, the collective communication is limited to communicators of size p_x or p_y, which is $O(\sqrt{p})$ for squarish configurations. The drawback of this design is that there are more synchronization points than if an 1D distribution had been used. Furthermore, neither vertices nor hyperedges have unique owners, but are spread over multiple processes. This made the 2D parallel implementation quite complex and challenging. 2D data distributions have recently been used in several applications, such as sparse matrix-vector multiplication in eigensolvers **[22]**. SpMV is a fairly simple kernel to parallelize. 2D distributions are still rarely used in graph algorithms, probably due to the complexity of implementation and the lack of payoff for small numbers of processors.

3.2. Coarsening. The coarsening phase approximates the original hypergraph via a succession of smaller hypergraphs. When the smallest hypergraph has fewer vertices than some threshold (e.g., 100), the coarsening stops. Several methods have been proposed for constructing coarser representations of graphs and hypergraphs. The most popular methods merge pairs of vertices, but one can also aggregate more than two vertices at a time. Intuitively, we wish to merge vertices that are similar and therefore more likely to be in the same partition in a good partitioning. Catalyurek and Aykanat [**4**] suggested a heavy-connectivity matching, which measures a similarity metric between pairs of vertices. Their preferred similarity metric, which was also adopted by hMETIS [**16**] and Mondriaan [**21**], is known as the *inner product* or simply, *heavy connectivity.* The inner product between two vertices is defined as the Euclidean inner product between their binary hyperedge incidence vectors, that is, the number of hyperedges they have in common. (Edge weights can be incorporated in a straight-forward way.) Zoltan PHG also uses the heavy-connectivity (inner-product) metric in the coarsening. Originally only pairs of vertices were merged (matched) but later vertex aggregation (clustering) that allows more than two vertices to be merged was made the default as it produces slightly better results.

Previous work have shown that greedy strategies work well in practice so optimal matching based on similarity scores (inner products) is not necessary. The sequential greedy algorithm works as follows. Pick a (random) unmatched vertex v. For each unmatched neighbor vertex u, compute the inner product $< v, u >$. Select the vertex with the highest non-zero inner product value and match it with v. Repeat until all vertices have been considered. If we consider the hypergraph as a sparse matrix A, we essentially need to compute the matrix product $A^T A$. We can use the sparsity of A to compute only entries of $A^T A$ that may be nonzero. Since we use a greedy strategy, we save work and compute only a subset of the nonzero entries in $A^T A$. This strategy has been used (successfully) in several serial partitioners.

With Zoltan's 2D data layout, this fairly simple algorithm becomes much more complicated. Each processor knows about only a subset of the vertices and the hyperedges. Computing the inner products requires communication. Even if A is typically very sparse, $A^T A$ may be fairly dense. Therefore we cannot compute all of $A^T A$ at once, but instead compute parts of it in separate *rounds.* In each round, each processor selects a (random) subset of its vertices that we call *candidates.* These candidates are broadcast to all other processors in the processor row. This requires horizontal communication in the 2D layout. Each processor then computes the inner products between its local vertices and the external candidates received. Note that these inner products are only partial inner products; vertical communication along processor columns is required to obtain the full (global) inner products. One could let a single processor within a column accumulate these full inner products, but this processor may run out of memory. So to improve load balance, we accumulate inner products in a distributed way, where each processor is responsible for a subset of the vertices.

At this point, the potential matches in a processor column are sent to the *master row* of processors (row 0). The master row first greedily decides the best local vertex for each candidate. These local vertices are then *locked*, meaning they can match only to the desired candidate (in this round). This locking prevents conflicts between candidates, which could otherwise occur when the same local vertex is the best match for several candidates. Horizontal communication along the master row is used to find the best global match for each candidate. Due to our locking scheme, the desired vertex for each match is guaranteed to be available so no conflicts arise between vertices. The full algorithm is given in [**9**].

Observe that the full heavy connectivity matching is computationally intensive and requires several communication phases along both processor rows and columns. Empirically, we observed that the matching usually takes more time than the other parts of the algorithm. Potentially, one could save substantial time in the coarsening phase by using a cheaper heuristic that gives preference to local data. We have experimented with several such strategies, but the faster run time comes at the expense of the partitioning quality. Therefore, the default in Zoltan PHG is heavy connectivity aggregation, which was also used in our experiments.

After the matching or aggregation has been computed, we build the coarser hypergraph by merging matched vertices. Note that hyperedges are not contracted, leading to unsymmetry. The matrix corresponding to the hypergraph becomes more rectangular at every level of coarsening. The number of hyperedges is only reduced in two ways: (a) hyperedges that become internal to a coarse vertex are simply deleted, and (b) identical hyperedges are collapsed into a single hyperedge with adjusted weight.

3.3. Initial Partitioning. The coarsening stops when the hypergraph is smaller than a certain threshold. Since the coarsest hypergraph is small, we replicate it on every process. Each processor runs a randomized greedy algorithm to compute a different partitioning. We then evaluate the desired cut metric on each processor and pick the globally best partitioning, which is broadcast to all processes.

3.4. Refinement. The refinement phase takes a partition assignment projected from a coarser hypergraph and improves it using a local optimization method. The most successful refinement methods are variations of Kernighan–Lin (KL) [**18**] and Fiduccia–Mattheyses (FM) [**10**]. These are iterative methods that move (or swap) vertices from one partition to another based on *gain* values, that is, how much the cut weight decreases by the move. While greedy algorithms are often preferred in parallel because they are simpler and faster, they generally do not produce partition quality as good as KL/FM. Thus, Zoltan PHG uses an FM-like approach but made some changes to accomodate the 2D data layout.

Since Zoltan PHG uses recursive bisection, only two-way refinement is needed. The main challenge with the 2D layout is that each vertex is shared among several processes, making it difficult to compute gain values and also to decide which moves to actually perform (as processes may have conflicting local information). The strategy used in PHG is a compromise between staying faithful to the FM algorithm and accomodating more concurrency in the 2D parallel setting. See [**9**] for further details. Although the refinement in PHG works well on moderate number of processors, the quality will degrade for very large number of processes.

3.5. Recursive Bisection. Zoltan PHG uses recursive bisection to partition into k parts. Note that k can be any integer greater than one, and does not need to be a power of two. Also, Zoltan can run on p processes, where $k \neq p$. However, the typical use case is $k = p$.

An important design choice in the recursive bisection is whether the data is left in-place or moved onto separate subsets of processors. The first approach avoids some data movement but the latter reduces communication in the partitioner and allows more parallelism. Initial experiments indicated that moving the data and splitting into independent subproblems gave better performance, so this is the default in Zoltan PHG.

3.6. PHG as a Graph Partitioner. PHG was designed as a hypergraph partitioner but can also do graph partitioning since a graph is just a special case of a hypergraph. When PHG is used as a graph partitioner, each hyperedge is of size two. When we coarsen the hypergraph, only vertices are coarsened, not hyperedges. This means that the symmetry of graphs is destroyed already after the first level of coarsening. We conjecture that PHG is not particularly efficient as a graph partitioner because it does not take advantage of the special structure of graphs (in particular, symmetry and constant size hyperedges). Still, we believe it is interesting (and fair) to compare PHG as a graph partitioner because it uses exactly the same code as the hypergraph partitioner, so any performance difference is due to the model not the implementation.

4. Experiments

4.1. Software, Platform, and Data. Our primary goal is to study the behavior of Zoltan PHG as a graph and a hypergraph partitioner, using different objectives and a range of data sets. We use Zoltan 3.5 (Trilinos 10.8) and ParMetis 4.0 as a reference for all the tests. The compute platform was mainly Hopper, a Cray XE6 at NERSC. Hopper has 6,384 compute nodes, each with 24 cores (two 12-core AMD MagnyCours) and 32 GB of memory. The graphs for the tests are from five test families of the DIMACS collection that are relevant to the computational problems we have encountered at Sandia. Within each family, we selected some of the largest graphs that were not too similar. In addition we picked four other graphs, two each from the street networks and clustering instances (which also happened to be road networks), to compile our diverse 22 test problems.

The graphs are partitioned into 16, 64, 256, and 1024 parts. In the parallel computing context, this covers everything from a multicore workstation to a medium-sized parallel computer. Except where stated otherwise, the partitioner had the same number of MPI ranks as the target number of parts.

Zoltan uses randomization, so results may vary slightly from run to run. However, for large graphs, the random variation is relatively small. Due to limited compute time on Hopper, each partitioning test was run only once. Even with the randomization, it is fair to draw conlusions based on several data sets, though one should be cautious about overinterpreting any single data point.

4.2. Zoltan vs. ParMetis. In this section, we compare Zoltan's graph and hypergraph partitioning with ParMetis's graph partitioning. We partition the graphs into 256 parts with 256 MPI processes. The performance profile of the three metrics – total edge cut (EC), the maximum communication volume (CV-max) and the total communication volume (CV-sum) – for the 22 matrices is shown in Figure 2.

The main advantage of the hypergraph partitioners is the ability to handle unsymmetric problems and to reduce the communication volume for such problems directly (without symmetrizing the problems). However, all the 22 problems used for the comparisons in Figure 2 are symmetric problems from the DIMACS challenge set. We take this opportunity to compare graph and hypergraph partitioners even for symmetric problems.

In terms of the edge cut metric ParMetis does better than Zoltan for 20 of the matrices and Zoltan's graph model does better for just two matrices. However,

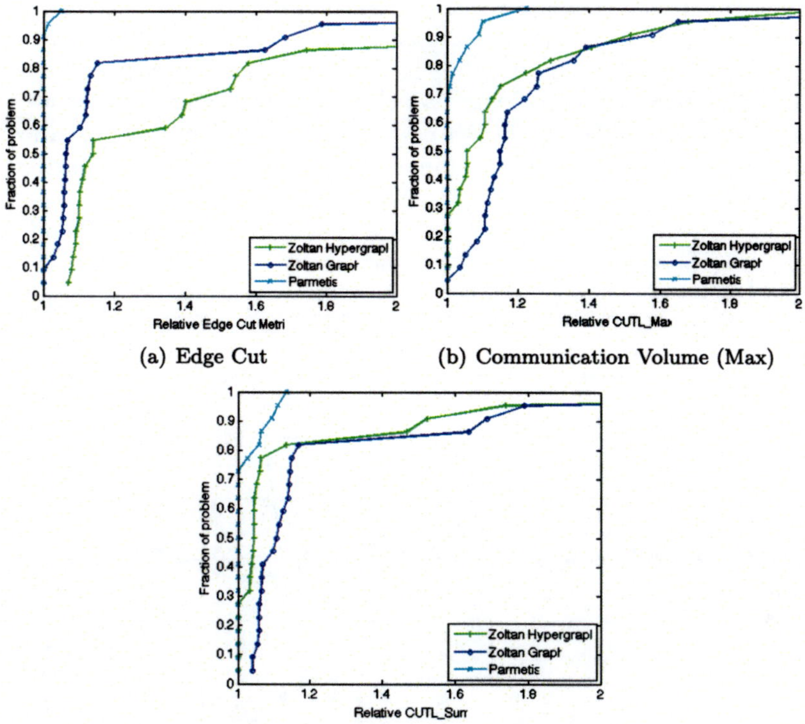

(a) Edge Cut

(b) Communication Volume (Max)

(c) Communication Volume (Sum)

FIGURE 2. **Zoltan Vs Parmetis**: Comparing Zoltan's partitioning with graph and hypergraph model with Parmetis for symmetric problems for 256 parts and 256 MPI processes.

Zoltan's graph model is within 15% of ParMetis's edge cuts for 82% of the problems (see Figure 2(a)). The four problems that cause trouble to Zoltan's graph model are the problems from the street networks and clustering instances.

In terms of the CV-sum metric Zoltan's partitioning with the hypergraph model, is able to do better than Zoltan's graph model in all the instances, and is better than ParMetis for 33% of the problems, and is within 6% or better of CV-sum of the ParMetis for another 44% of the problems (see Figure 2(c)). Again the street networks and the clustering instances are the ones that cause problems for the hypergraph partitioning. In terms of the CV-max metric Zoltan's hypergraph partitioning is better than the other two methods for 27% of the problems, and within 15% of the CV-max for another 42% of the problems (see Figure 2(b)).

From our results, we can see that even for symmetric problems hypergraph partitioners can perform nearly as well as (or even better than) the graph partitioners depending on the problems and the metrics one cares about. We also note that three of these 22 instances (af_shell10, audikw1 and G3_circuit) come from the same problems we used in Section 2.3 and Zoltan does better in one problem and ParMetis does better on other two problems in terms of the CV-max metric. In terms of EC metric ParMetis does better for all these four problems. However, as we can see from Table 1 the actual solution time is slightly better when we use the hypergraph partitioning for the three problems irrespective of which method is better in terms of the metrics we compute. To be precise, we should again note that the differences in actual solve time between graph and hypergraph partitioning are minor for those three problems. We would like to emphasize that we are not able to observe any difference in the performance of the actual application when the difference in the metrics is a small percentage. We study the characteristics of Zoltan's graph and hypergraph partitioning in the rest of this paper.

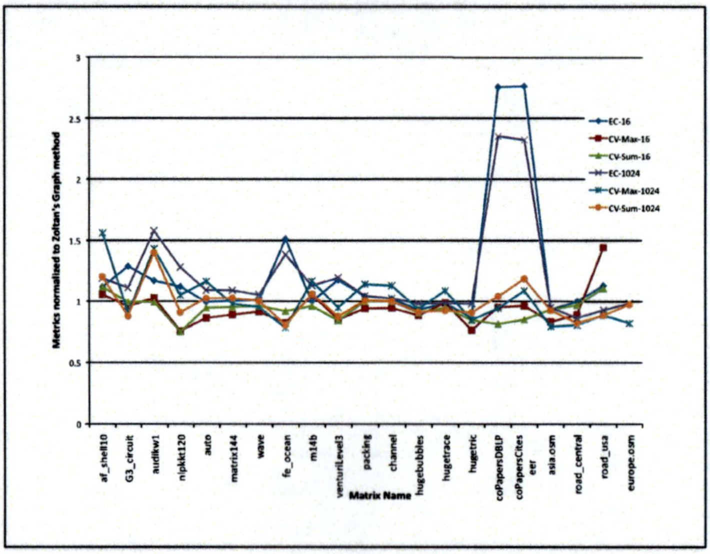

FIGURE 3. Comparing Zoltan's quality metrics with graph and hypergraph models for 16 and 1024 parts.

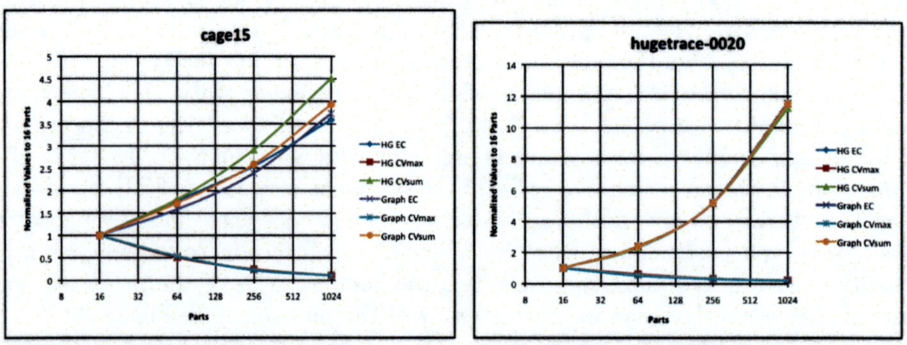

(a) (b)

FIGURE 4. Comparing Zoltan's partitioning with graph and hypergraph (HG) model quality metrics for different part sizes.

4.3. Zoltan Graph vs. Hypergraph model. We did more extensive experiments on the symmetric problems with the graph and hypergraph partitioning of

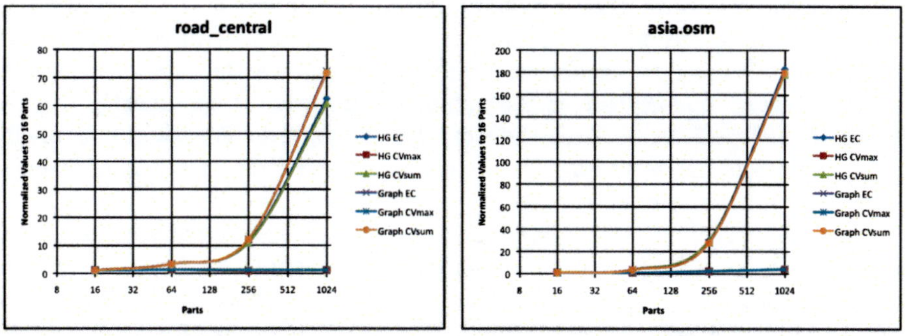

(a) (b)

FIGURE 5. Comparing Zoltan's partitioning with graph and hypergraph (HG) model quality metrics for different part sizes.

Zoltan. For each of the test problems from we compute the three metrics (EC, CV-max, CV-sum) for part sizes 16, 1024. All the experiments use the same number of MPI processes as the part sizes. Figure 3 shows the three metrics for hypergraph partitioning normalized to graph partitioner results for both 16 and 1024 parts. The results show that based on the EC metric, Zoltan's graph partitioning is the best for most problems. In terms of the CV-sum metric the hypergraph partitioning fares better. Neither of the algorithms optimize, CV-max metric and as expected the results are mixed for this metric. The results for 64 and 256 parts were not different from the results presented in Figure 3 and are not presented here.

Figure 4 shows the change in the partitioning quality with respect to the three metrics for both graph and hypergraph partitionings for two problems – cage15 and hugetrace-0020. The metrics are normalized with respect to the values for the 16 parts case in these figures. These results are for the "good" problems and from the results we can see why we call these problems the "good" problems - EC and CV-sum go up by a factor of 3.5 to 4.5 when going from 16 parts to 1024 parts. In contrast, we also show the change in the metrics from one problem from the street networks and clustering set each (road_central and asia.osm) in Figure 5. Note that for the some of these problems the metrics scale with similar values that the lines overlap in the graph. These second set of problems are challenging for both our graph and hypergraph partitioners as EC and CV-max go up by a factor 60-70 going from 16-1024 processes (for road_central). The changes in these values are mainly because of the structure of the graphs.

4.4. Zoltan scalability. Many of Zoltan's users use Zoltan within their parallel applications dynamically, where the number of parts equals the number of MPI processes. As a result it is important for Zoltan to have a scalable parallel hypergraph partitioner. We have made several improvements within Zoltan over the past few years and we evaluate our parallel scalabilty for the DIMACS problems instances in this section. Note that having a parallel hypegraph partitioner also enables us to solve large problems that does not fit into the memory of a compute node. However, we were able to partition all the DIMACS instances except the matrix europe.osm with 16 cores. We omit the europe.osm matrix and three small matrices from the Walshaw group that get partitioned within two seconds even with

16 cores, from these tests. The scalability results for the rest of the 18 matrices are shown in Figure 6. We normalize the time for all the runs with time to compute 16 parts. Note that even though the matrix size remains the same, this is not a traditional strong scaling test as the number of parts increases linearly with the number of MPI processes. Since the work for the partitioner grows, it is unclear what "perfect scaling" would be, but we believe this is a reasonable experiment as it reflects a typical use case.

Even with the increase in the amount of work for large matrices like cage15 and hugebubbles-0020 we see performance improvements as we go to 1024 MPI processes. However, for smaller problems like the auto or m14b the performance remains flat (or degrades) as we go from 256 MPI processes to 1024 MPI processes.

The scalability of Zoltan's graph partitioners is shown in Figure 7. We see that the graph partitioner tends to scale well for most problems. Surprisingly, the PHG hypergraph partitioner is faster than our graph partitioner in terms of actual execution time for several of the problems. This may in part be due to the fact that there are only n hyperedges in the hypergraph model compared to m edges in the graph model. Recall that PHG treats graphs as hypergraphs, without exploiting the special structure.

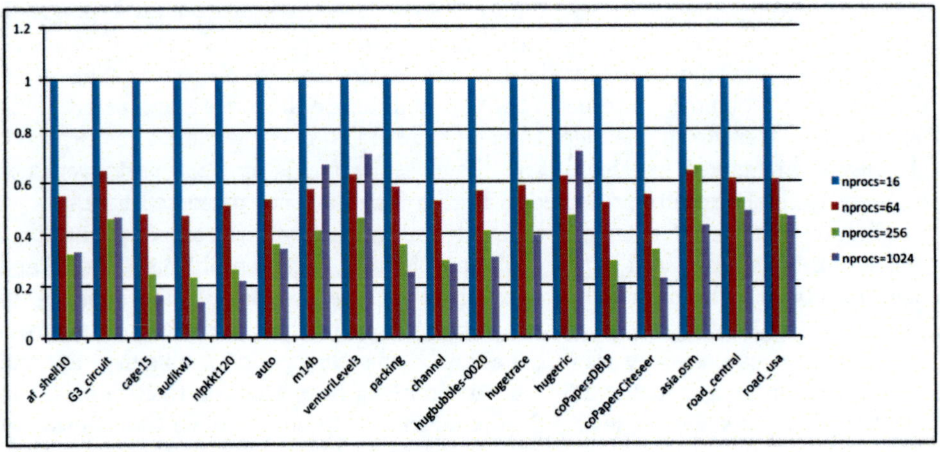

FIGURE 6. Scalability of Zoltan Hypergraph Partitioning time for DIMACS challenge matrices normalized to the time for 16 MPI processes and 16 parts.

4.5. Partitioning on a single node to improve quality. As discussed before, Zoltan can compute a partitioning statically with different number of MPI processes than number of parts. One strategy to obtain better partitioning quality is therefore to partition for p parts on k cores, where $k < p$. This often results in better quality than the dynamic approach where $k = p$. However, the users have to retain the partition in this case for future use. We evaluate this case for the symmetric matrices from the DIMACS collection for just the hypergraph partitioning. We compute 1024 parts with 24 MPI processes. The assumption is that the user will be willing to devote one compute node to compute the partition he needs. The results of these experiments for the 22 DIMACS graphs in Figure 8. On an average the edge cuts gets reduced by 10% and the CV-sum gets reduced

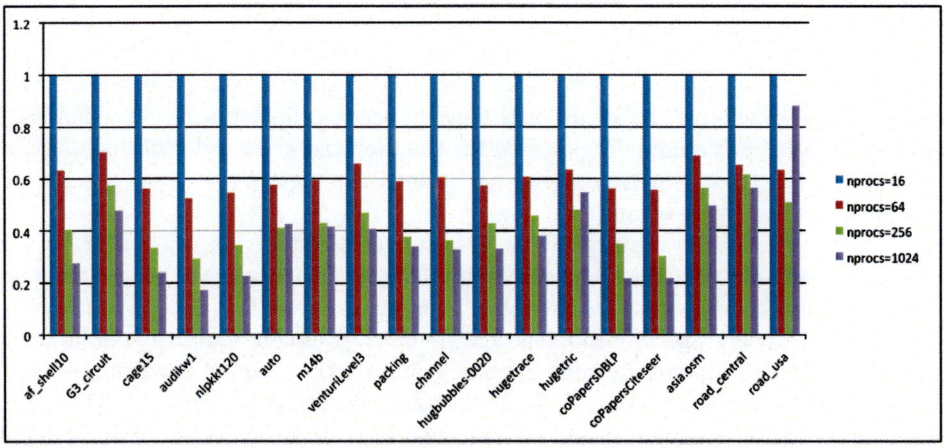

FIGURE 7. Scalability of Zoltan Graph Partitioning time for DIMACS challenge matrices normalized to the time for 16 MPI processes and 16 parts.

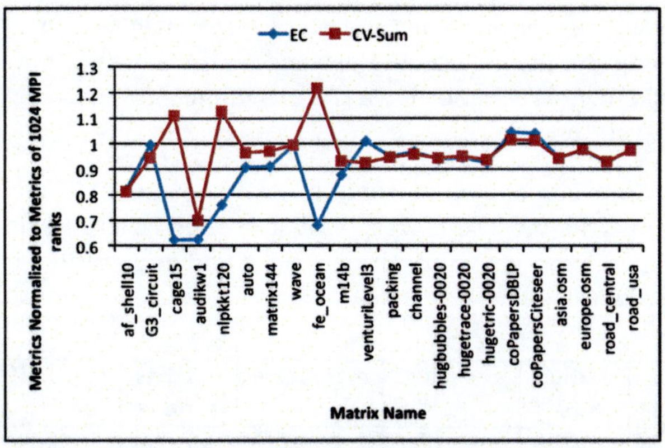

FIGURE 8. Improvement in the partitioning quality when computing 1024 parts with 24 MPI ranks instead of 1024 MPI ranks.

by 4% when partitioners 1024 parts with just 24 MPI processes instead of using the 1024 MPI processes. This confirms our conjecture that using fewer cores (MPI processes) gives higher quality results, and raises the possibility of using shared-memory techniques to improve the quality in the future.

4.6. Nonsymmetric Data (Directed Graphs). The 10th DIMACS implementation challenge includes only undirected graphs, corresponding to structurally symmetric matrices. This clearly favors graph partitioners. Many real-world problems are nonsymmetric, e.g., web link graphs, term-by-document matrices, and circuit simulation. For such applications, it is well known in the partitioning community that it is better to work directly on the original (nonsymmetric) problem [**4, 12**]. Remarkably, applications people who are not experts still overwhelmingly use a graph partitioner with symmetrization ($A + A^T$ or $A^T A$) and apply the result to the original unsymmetric problem. The difference in terms of the communication volume is presumed to be small. We compare hypergraph partitioning on A against graph partitioning on the symmetrized graph/matrix. We measure the communication volume on the original nonsymmetric problem, since this typically corresponds to the communication cost for the user and show order of magnitudes difference. For these experiments we partitioned the matrix rows, but results for column partitioning were similar.

These experiments were run on a 12-core workstation. We ran Zoltan on 12 MPI processes and partitioned into 12 parts. The test matrices were taken from the UF collection [7], and vary in their degree of symmetry from 0 to 95%. We see from Table 2 that hypergraph partitioning directly on A gives communication volume at least one order of magnitude smaller than graph partitioning on the symmetrized version in half the test cases. This is substantially different from the 30 – 38% average reduction observed in [4]. We arranged the matrices in decreasing degree of symmetry. Observe that hypergraph partitioning performs relatively better on the highly nonsymmetric matrices. Also note that there is essentially no difference in quality between Zoltan PHG as a graph partitioner and ParMetis for these cases. We conjecture the difference is neglible because the error made in the model by symmetrizing the matrix is far greater than differences in the implementation.

Note that some of the problems in the 22 symmetric test problems were originally unsymmetric problems (like citeseer and DBLP data) but were symmetrized for graph partitioning. We do not have the unsymmetric versions of these problems so we could not use those here.

TABLE 2. Comparison of communication volume (CV-sum) for nonsymmetric and the corresponding symmetrized matrices.PHG was used as hypergraph partitioner on A and as a graph partitioner on $A_{sym} \equiv (A + A^T)$

Matrix	dim. ($\times 10^3$)	avg. deg.	symmetry	PHG on A	PHG on A_{sym}	ParMetis on A_{sym}
torso3	259	17.1	95%	27,083	48,034	51,193
stomach	213	14.2	85%	15,128	20,742	21,619
rajat21	411	4.6	76%	112,273	174,717	158,296
amazon0312	400	8.0	53%	81,957	846,011	851,793
web-stanford	281	8.2	28%	2,307	543,446	543,547
twotone	120	9.9	24%	6,364	19,771	20,145
wiki-Talk	2,394	2.1	14%	0	53,009	–
hamrle3	1,447	3.8	0%	18,748	1,446,541	1,447,388

5. Conclusions

We have evaluated the parallel performance of Zoltan PHG, both as a graph and hypergraph partitioner on test graphs from the DIMACS challenge data set. We also made comparisons to ParMetis, a popular graph partitioner. We observed that ParMetis consistently obtained best edge cut (EC), as we expected. Surprisingly, ParMetis also obtained lower communication volume (CV) in lot of the symmetric problems. This raises the question: Is hypergraph partitioning worth it? A key advantage of hypergraph partitioning is that it accurately minimizes communication volume [**4**, **12**]. It appears that the superiority of the hypergraph model is not reflected in current software. We believe that one reason Zoltan PHG does relatively poorly on undirected graphs, is that symmetry is not preserved during coarsening, unlike graph partitioners. Future research should consider hypergraph partitioners with symmetric coarsening, to try combine the best of both methods.

We further showed that hypergraph partitioners are superior to graph partitioners on nonsymmetric data. The reduction in communication volume can be one or two orders of magnitude. This is a much larger difference than previously observed. This may in part be due to the selection of data sets, which included some new areas such as weblink matrices. A common approach today is to symmetrize a nonsymmetric matrix and partition $A + A^T$. We demonstrated this is often a poor approach, and with the availability of the PHG parallel hypergraph partitioner in Zoltan, we believe many applications could benefit from using hypergraph partitioners without any symmetrization.

Our results confirm that it is important to use a hypergraph partitioner on directed graphs (nonsymmetric matrices). However, for naturally undirected graphs (symmetric matrices) graph partitioners perform better. If a single partitioner for all cases is desired, then Zoltan-PHG is a reasonable universal partitioner.

Acknowledgements

We thank Karen Devine for helpful discussions.

References

[1] E. Bavier, M. Hoemmen, S. Rajamanickam, and H. Thornquist, *Amesos2 and Belos: Direct and iterative solvers for large sparse linear systems*, Scientific Programming **20** (2012), no. 3, 241–255.

[2] E. G. Boman, U. V. Catalyurek, C. Chevalier, and K. D. Devine, *The Zoltan and Isorropia parallel toolkits for combinatorial scientific computing: Partitioning, ordering and coloring*, Scientific Programming **20** (2012), no. 2.

[3] T. Bui and C. Jones, *A heuristic for reducing fill in sparse matrix factorization*, Proc. 6th SIAM Conf. Parallel Processing for Scientific Computing, SIAM, 1993, pp. 445–452.

[4] Ü. Çatalyürek and C. Aykanat, *Hypergraph-partitioning-based decomposition for parallel sparse-matrix vector multiplication*, IEEE Trans. Parallel Dist. Systems **10** (1999), no. 7, 673–693.

[5] ———, *A fine-grain hypergraph model for 2d decomposition of sparse matrices*, Proc. IPDPS 8th Int'l Workshop on Solving Irregularly Structured Problems in Parallel (Irregular 2001), April 2001.

[6] ———, *A hypergraph-partitioning approach for coarse-grain decomposition*, Proc. Supercomputing 2001, ACM, 2001.

[7] Timothy A. Davis and Yifan Hu, *The University of Florida sparse matrix collection*, ACM Trans. Math. Software **38** (2011), no. 1, Art. 1, 25, DOI 10.1145/2049662.2049663. MR2865011 (2012k:65051)

[8] Karen Devine, Erik Boman, Robert Heaphy, Bruce Hendrickson, and Courtenay Vaughan, *Zoltan data management services for parallel dynamic applications*, Computing in Science and Engineering **4** (2002), no. 2, 90–97.

[9] K.D. Devine, E.G. Boman, R.T. Heaphy, R.H. Bisseling, and U.V. Catalyurek, *Parallel hypergraph partitioning for scientific computing*, Proc. of 20th International Parallel and Distributed Processing Symposium (IPDPS'06), IEEE, 2006.

[10] C. M. Fiduccia and R. M. Mattheyses, *A linear-time heuristic for improving network partitions*, Proc. 19th IEEE Design Automation Conf., 1982, pp. 175–181.

[11] B. Hendrickson and R. Leland, *A multilevel algorithm for partitioning graphs*, Proc. Supercomputing '95, ACM, December 1995.

[12] Bruce Hendrickson and Tamara G. Kolda, *Graph partitioning models for parallel computing*, Parallel Comput. **26** (2000), no. 12, 1519–1534, DOI 10.1016/S0167-8191(00)00048-X. MR1786938

[13] Bruce Hendrickson and Robert Leland, *The Chaco user's guide, version 1.0*, Tech. Report SAND93-2339, Sandia National Laboratories, 1993.

[14] G. Karypis and V. Kumar, *METIS: Unstructured graph partitioning and sparse matrix ordering system*, Tech. report, Dept. Computer Science, University of Minnesota, 1995, *http://www.cs.umn.edu/~karypis/metis.*

[15] ———, *Parmetis: Parallel graph partitioning and sparse matrix ordering library*, Tech. Report 97-060, Dept. Computer Science, University of Minnesota, 1997, `http://www.cs.umn.edu/~metis`.

[16] George Karypis, Rajat Aggarwal, Vipin Kumar, and Shashi Shekhar, *Multilevel hypergraph partitioning: Applications in VLSI domain*, Proc. 34th Design Automation Conf., ACM, 1997, pp. 526 – 529.

[17] George Karypis and Vipin Kumar, *A fast and high quality multilevel scheme for partitioning irregular graphs*, SIAM J. Sci. Comput. **20** (1998), no. 1, 359–392 (electronic), DOI 10.1137/S1064827595287997. MR1639073 (99f:68158)

[18] B. W. Kernighan and S. Lin, *An efficient heuristic procedure for partitioning graphs*, Bell System Technical Journal **49** (1970), 291–307.

[19] F. Pelligrini, *SCOTCH 3.4 user's guide*, Research Rep. RR-1264-01, LaBRI, Nov. 2001.

[20] ———, *PT-SCOTCH 5.1 user's guide*, Research rep., LaBRI, 2008.

[21] Brendan Vastenhouw and Rob H. Bisseling, *A two-dimensional data distribution method for parallel sparse matrix-vector multiplication*, SIAM Rev. **47** (2005), no. 1, 67–95 (electronic), DOI 10.1137/S0036144502409019. MR2149102 (2006a:65070)

[22] Andy Yoo, Allison H. Baker, Roger Pearce, and Van Emden Henson, *A scalable eigensolver for large scale-free graphs using 2d graph partitioning*, Proceedings of 2011 International Conference for High Performance Computing, Networking, Storage and Analysis (New York, NY, USA), SC '11, ACM, 2011, pp. 63:1–63:11.

SANDIA NATIONAL LABORATORIES, ALBUQUERQUE, NEW MEXICO
E-mail address: `srajama@sandia.gov`

SANDIA NATIONAL LABORATORIES, ALBUQUERQUE, NEW MEXICO
E-mail address: `egboman@sandia.gov`

UMPa: A multi-objective, multi-level partitioner for communication minimization

Ümit V. Çatalyürek, Mehmet Deveci, Kamer Kaya, and Bora Uçar

Abstract. We propose a directed hypergraph model and a refinement heuristic to distribute communicating tasks among the processing units in a distributed memory setting. The aim is to achieve load balance and minimize the maximum data sent by a processing unit. We also take two other communication metrics into account with a tie-breaking scheme. With this approach, task distributions causing an excessive use of network or a bottleneck processor which participates in almost all of the communication are avoided. We show on a large number of problem instances that our model improves the maximum data sent by a processor up to 34% for parallel environments with $4, 16, 64$ and 256 processing units compared to the state of the art which only minimizes the total communication volume.

1. Introduction

In parallel computing, the problem of distributing communicating tasks among the available processing units is important. To solve this problem, several graph and hypergraph models are proposed [**6, 7, 9, 12, 20**]. These models transform the problem at hand to a balanced partitioning problem. The balance restriction on part weights in conventional partitioning corresponds to the load balance in the parallel environment, and the minimized objective function corresponds to the total communication volume between processing units. Both criteria are crucial in practice for obtaining short execution times, using less power, and utilizing the computation and communication resources better.

In addition to the total data transfer, there are other communication metrics investigated before, e.g., the total number of messages sent [**19**], or maximum volume of messages sent and/or received by a processor [**4, 19**]. Even with perfect load balancing and minimized total data transfer, there can be a bottleneck processing unit which participates in most of the data transfers. This can create a problem especially for data intensive applications where reducing the amount of data transferred by the bottleneck processing unit can improve the total execution time significantly.

2010 *Mathematics Subject Classification.* Primary 05C65, 05C70; Secondary 90C35.

Key words and phrases. Hypergraph partitioning, multi-level partitioning, communication minimization.

This work was supported in parts by the DOE grant DE-FC02-06ER2775 and by the NSF grants CNS-0643969, OCI-0904809, and OCI-0904802.

In this work, given a task graph, our main objective is distributing its tasks evenly and minimizing the maximum amount of data sent by a processing unit. Previous studies addressing different communication cost metrics (such as [**4**, **19**]) work in two phases. In the first phase, the total volume of communication is reduced, and in the second phase the other metrics are addressed. We propose a directed hypergraph model and partition the related hypergraph with a multi-level approach and a novel K-way refinement heuristic. While minimizing the primary objective function, our refinement heuristic also takes the maximum data sent and received by a processing unit and the total amount of data transfer into account by employing a tie-breaking scheme. Therefore, our approach is different from the existing studies in that the objective functions are minimized all at the same time.

The organization of the paper is as follows. In Section 2, the background material on graph and hypergraph partitioning is given. Section 2.3 shows the differences of the graph and hypergraph models and describes the proposed directed hypergraph model. In Section 3, we present our multi-level, multi-objective partitioning tool UMPa (pronounced as "Oompa"). Section 4 presents the experimental results, and Section 5 concludes the paper.

2. Background

2.1. Hypergraph partitioning. A hypergraph $\mathcal{H}=(\mathcal{V},\mathcal{N})$ is defined as a set of vertices $\mathcal{V}$ and a set of nets (hyperedges) $\mathcal{N}$ among those vertices. A net $n \in \mathcal{N}$ is a subset of vertices and the vertices in n are called its *pins*. The number of pins of a net is called the *size* of it, and the *degree* of a vertex is equal to the number of nets it is connected to. In this paper, we will use $\mathsf{pins}[n]$ and $\mathsf{nets}[v]$ to represent the pin set of a net n and the set of nets vertex v is connected to, respectively. The vertices can be associated with weights, denoted with $\mathsf{w}[\cdot]$, and the nets can be associated with costs, denoted with $\mathsf{c}[\cdot]$.

A K-*way partition* of a hypergraph $\mathcal{H}$ is denoted as $\Pi=\{\mathcal{V}_1,\mathcal{V}_2,\ldots,\mathcal{V}_K\}$ where

- parts are pairwise disjoint, i.e., $\mathcal{V}_k \cap \mathcal{V}_\ell = \emptyset$ for all $1 \le k < \ell \le K$,
- each part $\mathcal{V}_k$ is a nonempty subset of $\mathcal{V}$, i.e., $\mathcal{V}_k \subseteq \mathcal{V}$ and $\mathcal{V}_k \neq \emptyset$ for $1 \le k \le K$,
- union of K parts is equal to $\mathcal{V}$, i.e., $\bigcup_{k=1}^{K} \mathcal{V}_k = \mathcal{V}$.

Let W_k denote the total vertex weight in $\mathcal{V}_k$ (i.e., $W_k=\sum_{v\in\mathcal{V}_k}\mathsf{w}[v]$) and W_{avg} denote the weight of each part when the total vertex weight is equally distributed (i.e., $W_{avg}=(\sum_{v\in\mathcal{V}}\mathsf{w}[v])/K$). If each part $\mathcal{V}_k \in \Pi$ satisfies the *balance criterion*

$$W_k \le W_{avg}(1+\varepsilon), \quad \text{for } k = 1,2,\ldots,K \tag{2.1}$$

we say that Π is ϵ-*balanced* where ε represents the maximum allowed imbalance ratio.

For a K-way partition Π, a net that has at least one pin (vertex) in a part is said to *connect* that part. The number of parts connected by a net n, i.e., *connectivity*, is denoted as λ_n. A net n is said to be *uncut* (*internal*) if it connects exactly one part (i.e., $\lambda_n = 1$), and *cut* (*external*), otherwise (i.e., $\lambda_n > 1$).

The set of external nets of a partition Π is denoted as $\mathcal{N}_E$. There are various cutsize definitions [**16**] for hypergraph partitioning. The one that will be used in this work, which is shown to accurately model the total communication volume [**7**], is called the *connectivity* metric and defined as:

$$\chi(\Pi) = \sum_{n \in \mathcal{N}} \mathsf{c}[n](\lambda_n - 1)\,. \tag{2.2}$$

In this metric, each cut net n contributes $\mathsf{c}[n](\lambda_n - 1)$ to the cutsize. The hypergraph partitioning problem can be defined as the task of finding a balanced partition Π with K parts such that $\chi(\Pi)$ is minimized. This problem is also NP-hard [**16**].

2.2. K-way partitioning and multi-level framework. Arguably, the multi-level approach [**3**] is the most successful heuristic for the hypergraph partitioning problem. Although, it has been first proposed for recursive-bisection based graph partitioning, it also works well for hypergraphs [**2, 5, 7, 13, 17**]. In the multi-level approach, a given hypergraph is coarsened to a much smaller one, a partition is obtained on the the smallest hypergraph, and that partition is projected to the original hypergraph. These three phases will be called the coarsening, initial partitioning, and uncoarsening phases, respectively. The coarsening and uncoarsening phases have multiple levels. In a coarsening level, similar vertices are merged to make the hypergraph smaller. In the corresponding uncoarsening level, the merged vertices are split, and the partition of the coarser hypergraph is refined for the finer one.

Most of the multi-level partitioning tools used in practice are based on recursive bisection. In recursive bisection, the multi-level approach is used to partition a given hypergraph into two. Each of these parts is further partitioned into two recursively until K parts are obtained in total. Hence, to partition a hypergraph into $K = 2^k$, the recursive bisection approach uses $K - 1$ coarsening, initial partitioning, and uncoarsening phases.

Several successful clustering heuristics are proposed to coarsen a hypergraph. Although their similarity metrics aim to reduce the cutsize, they cannot find an optimal solution, since the problem is NP-hard. Hence, an optimal partition of the coarser hypergraph may not be optimal for the finer one. To obtain better partitions, iterative-improvement-based heuristics are used to refine the coarser's partition after projecting it to finer. In practice, Kernighan-Lin (KL) [**15**] and Fiduccia-Mattheyses (FM) [**11**] based refinement heuristics that depend on vertex swaps and moves between two parts are used.

2.3. Task graph and communication volume metrics. Let $\mathcal{A} = (\mathcal{T}, \mathcal{C})$ be a task graph where $\mathcal{T}$ is the set of tasks to be executed, and $\mathcal{C}$ is the set of communications between pairs of tasks. We assume that the execution time of each task may differ, hence each task $t \in \mathcal{T}$ is associated with an execution time $exec(t)$. Each task $t_i \in \mathcal{T}$ sends a different amount of data $data(t_i)$ to each t_j such that $t_it_j \in \mathcal{C}$. The communications between tasks may be uni-directional, That is $t_it_j \in \mathcal{C}$ does not imply $t_jt_i \in \mathcal{C}$. In our parallel setting, we assume owner computes rule and hence, each task of $\mathcal{A}$ is executed by the processing unit to which it is assigned. Let $\mathcal{T}_i$ be the set of tasks assigned to processing unit P_i. Since it is desirable to distribute the tasks evenly, the computational load $\sum_{t \in \mathcal{T}_i} exec(t)$ should be almost the same for each P_i. In addition to that, two heavily communicating tasks should be assigned to the same processing unit since less data transfer over the network is needed in this case. The total amount of data transfer throughout the execution of the tasks is called the *total communication volume* (*totV*). Note that when a task $t \in \mathcal{T}_i$ needs to send data to a set of tasks in $\mathcal{T}_j$, the contribution to $totV$ is $data(t)$, since it is enough to send t's data to P_j only once.

Although minimizing the total communication volume is important, it is sometimes preferable to reduce other communication metrics [**12**]. For example, in the context of one-dimensional partitioning of structurally unsymmetric sparse matrices for parallel matrix-vector multiplies, Uçar and Aykanat used a communication hypergraph model to reduce the maximum of number of messages and the maximum amount of data sent and received by a processor [**19**] (see also [**4**] and [**18**] for other communication metrics).

Let $SV[i]$ and $RV[i]$ be the volumes of communication sent and received by P_i, respectively. Hence, the total communication volume equals to $totV = \sum_i \mathsf{SV}[i] = \sum_i \mathsf{RV}[i]$. In addition to $totV$, we are interested in two other communication metrics: *maximum send volume* ($maxSV$), which equals to $\max_i(\mathsf{SV}[i])$; and *maximum send-receive volume* ($maxSRV$), which is $\max_i(\mathsf{SV}[i] + \mathsf{RV}[i])$.

3. UMPa: A multi-objective partitioning tool for communication minimization

3.1. Directed hypergraph model. We propose modeling the task graphs with directed hypergraphs. Given a task graph $\mathcal{A}$, we construct the directed hypergraph model $\mathcal{H} = (\mathcal{V}, \mathcal{N})$ as follows. For each task $t_i \in \mathcal{T}$, we have a corresponding vertex $v_i \in \mathcal{V}$ and a net $n_i \in \mathcal{N}$ where $\mathsf{pins}[n_i] = \{v_i\} \cup \{v_j \mid t_i t_j \in \mathcal{C}\}$, $\mathsf{w}[v_i] = exec(t_i)$, and $\mathsf{c}[n_i] = data(t_i)$. In this directed hypergraph model, the communication represented by a net n is flowing from its source vertex, which will be denoted as $s(n)$, to the target vertices $\mathsf{pins}[n] \setminus \{s(n)\}$. Given a partition Π, let $\delta(n, \mathcal{V}_i) = 1$ if $n \cap \mathcal{V}_i \neq \emptyset$, and 0, otherwise. Then the data sent and received by P_i are equal to $\mathsf{SV}[i] = \sum_{n, s(n) \in \mathcal{V}_i} \mathsf{c}[n](\lambda_n - 1)$ and $\mathsf{RV}[i] = \sum_{n, s(n) \notin \mathcal{V}_i} \mathsf{c}[n]\delta(n, \mathcal{V}_i)$, respectively. Our primary objective is to minimize $maxSV$, the maximum send volume. While doing this, we also take the maximum send-receive volume and the total communication volume into account. The total volume of communication corresponds to the cutsize definition (2.2) as in the standard hypergraph model. In other words, the sense of direction is not important for the total communication volume $totV$. On the other hand, the directions of the flow is crucial while minimizing $maxSV$ and $maxSRV$.

To optimize its metrics, UMPa follows the multi-level approach. Instead of a recursive bisection, it adopts a direct K-way partitioning. Given the hypergraph, UMPa gradually coarsens it, obtains an initial K-way partition for the coarsest hypergraph, and projects it into the original one by uncoarsening and refinement steps at each level.

3.2. Multi-level coarsening phase. In this phase, the original hypergraph is gradually coarsened in multiple levels by clustering subsets of vertices at each level. There are two types of clustering algorithms: matching-based ones and agglomerative ones. The matching-based ones put at most two similar vertices in a cluster, whereas the agglomerative ones allow any number of similar vertices. There are various similarity metrics—see for example [**1, 7, 14**]. All these metrics are defined only on two adjacent vertices (one of them can be a vertex cluster). Two vertices are adjacent if they share a net and they can be in the same cluster if the are adjacent.

In this work, we use an agglomerative algorithm and the absorption clustering metric using pins [**1**, **8**]. For this metric, the similarity between two adjacent vertices u and v is

$$\sum_{n \in \mathsf{nets}[u] \cap \mathsf{nets}[v]} \frac{\mathsf{c}[n]}{|\mathsf{pins}[n]| - 1}$$

This is also the default metric in PaToH [**8**], a well-known hypergraph partitioner. In each level ℓ, we start with a finer hypergraph $\mathcal{H}^\ell$ and obtain a coarser one $\mathcal{H}^{\ell+1}$. If $\mathcal{V}_C \subset \mathcal{V}^\ell$ is a subset of vertices deemed to be clustered, we create the cluster vertex $u \in \mathcal{V}^{\ell+1}$ where $\mathsf{nets}[u] = \cup_{v \in \mathcal{V}_C} \mathsf{nets}[v]$. We also update the pin sets of the nets in $\mathsf{nets}[u]$ accordingly.

Since we need the direction, i.e., source vertex information for each net to minimize $maxSV$ and $maxSRV$, we always store the source vertex of a net $n \in \mathcal{N}$ as the first pin in $\mathsf{pins}[n]$. To maintain this information, when a cluster vertex u is formed in the coarsening phase, we put u to the head of $\mathsf{pins}[n]$ for each net n whose source vertex is in the cluster.

3.3. Initial partitioning phase. To obtain an initial partition for the coarsest hypergraph, we use PaToH [**8**], which is proven to produce high quality partitions with respect to the total communication volume metric [**7**]. We execute PaToH ten times and get the best partition according to the $maxSV$ metric. We have several reasons to use PaToH. First, although our main objective is minimizing $maxSV$, since we also take $totV$ into account, it is better to start with an initial partition having a good total communication volume. Second, since $totV$ is the sum of the send volumes of all parts, as we observed in our preliminary experiments, minimizing it may also be good for both $maxSV$ and $maxSRV$. Also, as stated in [**2**], using recursive bisection and FM-based improvement heuristics for partitioning the coarsest hypergraph is favorable due to small net sizes and high vertex degrees.

3.4. K-way refinement of communication volume metrics. In an uncoarsening level, which corresponds to the ℓth coarsening level, we project the partition $\Pi^{\ell+1}$ obtained for $\mathcal{H}^{\ell+1}$ to $\mathcal{H}^\ell$. Then, we refine it by using a novel K-way refinement heuristic which is described below.

Given a partition Π, let a vertex be a *boundary vertex* if it is in the pin set of at least one cutnet. Let $\Lambda(n,p) = |\mathsf{pins}[n] \cap \mathcal{V}_p|$ be the number of pins of net n in part p, and $\mathsf{part}[u]$ be the current part of u. The proposed heuristic runs in multiple passes where in a pass it visits each boundary vertex u and either leaves it in $\mathsf{part}[u]$, or moves it to another part according to some move selection policy. Algorithm 1 shows a pass of the proposed refinement heuristic. For each visited boundary vertex u and for each available part p other than $\mathsf{part}[u]$, the heuristic computes how the communication metrics are affected when u is moved to p. This is accomplished in three steps. First, u is removed from $\mathsf{part}[u]$, and the leave gains on the send/receive volumes of the parts are computed (after line 1). Second, u is put into a candidate part p and the arrival losses on the send/receive volumes are computed (after line 2). Last, the maximum send, maximum send-receive, and total volumes are computed for this move (after line 4).

3.4.1. *Move selection policy and tie-breaking scheme.* Our move selection policy given in Algorithm 2 favors the moves with the maximum gain on $maxSV$ and never allows a move with negative gain on the same metric. To take other metrics into

Algorithm 1: A pass for K-way refinement

```
  Data: H = (V, N), boundary[], part[], SV[], RV[], λ, Λ
  for each unlocked u ∈ boundary do
      receiveGain ← 0
      uToPartU ← 0
      sendGain[] ← 0
1     for each n ∈ nets[u] do
          if s(n) = u then
              sendGain[part[u]] ← sendGain[part[u]] + (λ_n − 1)c[n]
              if Λ(n, part[u]) > 1 then
                  receiveGain ← receiveGain − c[n] uToPartU ← uToPartU + c[n]
          else if Λ(n, part[u]) = 1 then
              sendGain[part[s(n)]] ← sendGain[part[s(n)]] + c[n]
              receiveGain ← receiveGain + c[n]
      (bestMaxSV, bestMaxSRV, bestTotV) ← (maxSV, maxSRV, totV)
      bestPart ← part[u]
      for each part p other than part[u] do
          if p has enough space for vertex u then
              receiveLoss ← 0
              sendLoss[] ← 0
2             sendLoss[p] ← sendGain[part[u]] + uToPartU
3             for each n ∈ nets[u] do
                  if s(n) = u then
                      if Λ(n, p) > 0 then
                          sendLoss[p] ← sendLoss[p] − c[n]
                          receiveLoss ← receiveLoss − c[n]
                  else if Λ(n, p) = 0 then
                      sendLoss[part[s(n)]] ← sendLoss[part[s(n)]] + c[n]
                      receiveLoss ← receiveLoss + c[n]
4             (moveSV, moveSRV) ← (−∞, −∞)
5             for each part q do
                  Δ_S ← sendLoss[q] − sendGain[q]
                  Δ_R ← 0
                  if q = part[u] then
                      Δ_R ← receiveGain
                  else if q = p then
                      Δ_R ← receiveLoss
                  moveSV ← max(moveSV, SV[q] + Δ_S)
                  moveSRV ← max(moveSRV, SV[q] + Δ_S + RV[q] + Δ_R)
              moveV ← totV + receiveLoss − receiveGain
6             MoveSelect(moveSV, moveSRV, moveV, p,
                          bestMaxSV, bestMaxSRV, bestTotV, bestPart)
      if bestPart ≠ part[u] then
          move u to bestPart and update data structures accordingly
```

account, we use a tie-breaking scheme which is enabled when two different moves of a vertex u have the same $maxSV$ gain. In this case, the move with $maxSRV$ gain is selected as the best move. If the gains on $maxSRV$ are also equal then the move with maximum gain on $totV$ is selected. We do not allow a vertex move without a positive gain on any of the communication metrics. As the experimental results show, this move selection policy and tie-breaking scheme have positive impact on all the metrics.

Algorithm 2: MOVESELECT

```
Data: moveSV, moveSRV, moveV, p,
      bestMaxSV, bestMaxSRV, bestTotV, bestPart
select ← 0
if moveSV < bestMaxSV then
    select ← 1                                  ▷Main objective
else if moveSV = bestMaxSV then
    if moveSRV < bestMaxSRV then
        select ← 1                              ▷First tie break
else if moveSV = bestMaxSV then
    if moveSRV = bestMaxSRV then
        if moveV < bestTotV then
            select ← 1                          ▷Second tie break
if select = 1 then
    bestMaxSV ← moveSV
    bestMaxSRV ← moveSRV
    bestTotV ← moveV
    bestPart ← p
```

Figure 1 shows a sample graph with 8 vertices and 13 edges partitioned into 3 parts. Assume that this is a partial illustration of boundary vertices, and any move will not violate the balance criteria. Each row in the table contains a possible vertex move and the changes on the communication volume metrics. In the initial configuration, $maxSV = 6$, $maxSRV = 9$, and $totV = 12$. If we move v_3 from the partition $\mathcal{V}_2$ to the partition $\mathcal{V}_3$, we reduce all metrics by 1. On the other hand, if we move v_3 to $\mathcal{V}_1$, we decrease $maxSV$ and $maxSRV$, but $totV$ does not change. In this case, since its gain on $totV$ is better, the tie-breaking scheme favors the move v_3 to $\mathcal{V}_3$. Moreover, the moves v_4 to $\mathcal{V}_1$, v_6 to $\mathcal{V}_3$ and v_7 to $\mathcal{V}_3$ are other move examples where tie-breaking scheme is used. Note that we allow all the moves in the first 13 rows of the table including these two. However, we do not allow the ones in the last three rows.

3.4.2. *Implementation details.* During the gain computations, the heuristic uses the connectivity information between nets and parts stored in data structures λ and Λ. These structures are constructed after the initial partitioning phase, and then maintained by the uncoarsening phase. Since the connectivity information changes after each vertex move, when a vertex u is moved, we visit the nets of u and update the data structures accordingly. Also, when new vertices become boundary vertices, they are inserted into boundary array and visited in the same pass.

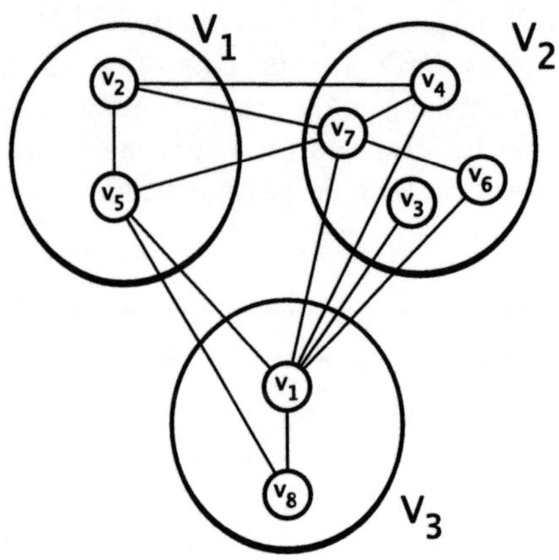

Vertex	Part	*maxSV*	*maxSRV*	*totV*
v_1	$\mathcal{V}_1$	−1	+1	−2
v_1	$\mathcal{V}_2$	−2	−2	−3
v_2	$\mathcal{V}_2$	0	−1	−1
v_2	$\mathcal{V}_3$	−1	+1	0
v_3	$\mathcal{V}_1$	−1	−1	0
v_3	$\mathcal{V}_3$	−1	−1	−1
v_4	$\mathcal{V}_1$	−1	−1	0
v_4	$\mathcal{V}_3$	−1	+1	+1
v_5	$\mathcal{V}_3$	0	0	−1
v_6	$\mathcal{V}_1$	−1	0	+1
v_6	$\mathcal{V}_3$	−1	0	0
v_7	$\mathcal{V}_1$	−1	+1	0
v_7	$\mathcal{V}_3$	−1	−1	0
v_5	$\mathcal{V}_2$	+2	+2	−1
v_8	$\mathcal{V}_1$	0	0	0
v_8	$\mathcal{V}_2$	+2	+2	+1

FIGURE 1. A sample partitioning and some potential moves with their effects on the communication volume metrics. The initial values are $maxSV = 6$, $maxSRV = 9$ and $totV = 12$. A negative value in a column indicates a reduction on the corresponding metric.

If at least one move with a positive gain on $maxSV$ is realized in a refinement pass, the heuristic continues with the next pass. Otherwise, it stops. For efficiency purposes, throughout the execution of a pass, we restrict the number of moves for each vertex u. If this number is reached, we lock the vertex and remove it from the boundary. In our experiments, the maximum number of moves per vertex is 4.

Let $\rho = \sum_{n \in \mathcal{N}} |\mathsf{pins}[n]|$ be the number of pins in a hypergraph. The time complexity of a pass of the proposed refinement heuristic is $\mathcal{O}(\rho K + |\mathcal{V}|K^2)$ due to the gain computation loops at lines 3 and 5. To store the numbers of pins per part for each net, Λ, we use a 2-dimensional array. Hence, the space complexity is $\mathcal{O}(K|\mathcal{N}|)$. This can be improved as shown in [**2**].

4. Experimental results

UMPa is tested on a computer with 2.27GHz dual quad-core Intel Xeon CPUs and 48GB main memory. It is implemented in C++ and compiled with g++ version 4.5.2.

To obtain our data set, we used several graphs from the testbed of 10th DIMACS implementation challenge [**10**]. We remove relatively small graphs containing less than 10^4 vertices, and also extremely large ones. There are 123 graphs in our data set from 10 graph classes. For each graph, we execute UMPa and other algorithms 10 times. The results in the tables are the averages of these 10 executions.

To see the effect of UMPa's K-way partitioning structure and its tie-breaking scheme, we compare it with two different refinement approaches and PaToH. The first approach is partitioning the hypergraph into K with PaToH's recursive bisection scheme and refining it by using the proposed K-way refinement algorithm

TABLE 1. The relative performance of UMPa and PaToH+refinement without tie breaking. The performance are computed with respect to that of PaToH.

	PaToH + Refinement No tie breaking			UMPa No tie breaking			UMPa With tie breaking		
K	*maxSV*	*maxSRV*	*totV*	*maxSV*	*maxSRV*	*totV*	*maxSV*	*maxSRV*	*totV*
4	0.93	1.05	1.06	0.73	0.83	0.93	0.66	0.77	0.84
16	0.93	1.06	1.04	0.84	0.94	1.11	0.73	0.83	0.98
64	0.91	1.04	1.02	0.86	0.98	1.12	0.76	0.87	1.00
256	0.91	1.03	1.01	0.89	1.00	1.10	0.81	0.91	1.02
Avg.	0.92	1.05	1.03	0.83	0.93	1.06	0.74	0.84	0.96

without employing the tie-breaking scheme. The second approach is using UMPa but again without tie breaking. To remove tie breaking, we remove the **else** statements at lines labeled with 1 and 2 of Algorithm 2.

Table 1 gives the average performance of all these approaches normalized with respect to PaToH's performance. Without tie breaking, refining PaToH's output reduces the maximum send volume by 8%. However, it increases the maximum send-receive and total volumes by 5% and 3%, respectively. Hence, we do not suggest using the refinement heuristic alone and without tie breaking. On the other hand, if it is used in the multi-level structure of UMPa, we obtain better results even without a tie-breaking scheme.

Table 1 shows that UMPa's multi-level structure helps to obtain 17% and 7% less volumes than PaToH's partitions in terms of *maxSV* and *maxSRV*, respectively. But since PaToH minimizes the total communication volume, there is a 6% overhead on the *totV*. Considering 17% reduction on *maxSV*, this overhead is acceptable. However, we can still reduce all the communication metrics 9%-to-10% more by employing the proposed tie-breaking scheme. For $K = 4$, this leads us to a 34% better maximum send volume, which is impressive since even the total communication volume is 16% less compared with PaToH. Actually, for all K values, UMPa manages to reduce *maxSV* and *maxSRV* on the average. The percent of improvement reduces with the increasing K. This may be expected since when K is large, the total volume will be distributed into more parts, and the maximum send or send-receive volume will be less. Still, on the average, the reductions on *maxSV*, *maxSRV*, and *totV* are 26%, 16%, and 4%, respectively.

Tables 2 and 3 show performance of PaToH and UMPa in terms of the communication metrics and time. There are 20 graphs in each table selected from 10 graph class in DIMACS testbed. For each graph class, we select the two (displayed consecutively in the tables) for which UMPa obtains the best and worst improvements on *maxSV*. The numbers given in the tables are averages of 10 different executions. For all experiments with $K = 16$ parts, as Table 2 shows, UMPa obtains a better *maxSV* value than PaToH on the average. When $K = 4, 64$, and 256, PaToH obtains a better average *maxSV* only for 16, 4, and 1 graphs, out of 123, respectively.

There are some instances in the tables for which UMPa improves *maxSV* significantly. For example, for graph ut2010 in Table 2, the *maxSV* value is reduced from 1506 to 330 with approximately 78% improvement. Furthermore, for the same graph, the improvements on *maxSRV* and *totV* are 75% and 67%, respectively.

TABLE 2. The maximum send and send-receive volumes, and the total volume for PaToH and UMPa when $K = 16$. The times are given in seconds. There are 20 graphs in the table where two graphs with the best and the worst improvements on *maxSV* are selected from each class. Each number is the average of 10 different executions.

	PaToH				UMPa			
Graph	*maxSV*	*maxSRV*	*totV*	Time	*maxSV*	*maxSRV*	*totV*	Time
coPapersDBLP	62,174	139,600	673,302	91.45	53,619	117,907	842,954	145.47
as-22july06	1,506	5,063	12,956	0.63	1,144	3,986	13,162	2.70
road_central	500	999	3,926	112.64	279	576	2,810	27.85
smallworld	12,043	24,020	188,269	3.09	10,920	21,844	174,645	19.27
delaunay_n14	119	235	1,500	0.19	115	236	1,529	0.88
delaunay_n17	351	706	4,100	1.09	322	655	4,237	2.54
hugetrace-00010	2,113	4,225	25,809	93.99	2,070	4,144	28,572	43.39
hugetric-00020	1,660	3,320	20,479	60.96	1,601	3,202	22,019	29.51
venturiLevel3	1,774	3,548	19,020	27.41	1,640	3,282	20,394	16.01
adaptive	2,483	4,967	27,715	54.00	2,345	4,692	29,444	29.33
rgg_n_2_15_s0	146	293	1,519	0.34	119	254	1,492	1.03
rgg_n_2_21_s0	1,697	3,387	19,627	37.86	1,560	3,215	20,220	16.66
tn2010	2,010	3,666	13,473	1.26	1,684	3,895	56,780	1.54
ut2010	1,506	2,673	3,977	0.43	330	677	1,303	0.82
af_shell9	1,643	3,287	17,306	14.83	1,621	3,242	18,430	8.64
audikw1	15,119	29,280	145,976	161.23	11,900	24,182	159,640	77.16
asia.osm	63	125	409	40.43	30	62	323	7.67
belgium.osm	141	281	1,420	4.80	120.6	243	1,406	1.96
memplus	986	7,138	7,958	0.23	686	3,726	10,082	0.72
t60k	155	310	1,792	0.29	148.5	297	1,890	0.99

When $K = 256$ (Table 3) for the graph memplus, UMPa obtains approximately 50% improvement on *maxSV* and *maxSRV*. Although *totV* increases 26% at the same time, this is acceptable considering the improvements on the first two metrics.

Table 4 shows the relative performance of UMPa in terms of execution time with respect to PaToH. As expected, due to the complexity of K-way refinement heuristic, UMPa is slower than PaToH especially when the number of parts is large.

5. Conclusions and future work

We proposed a directed hypergraph model and a multi-level partitioner UMPa for obtaining good partitions in terms of multiple communication metrics where the maximum amount of data sent by a processing unit is the main objective function to be minimized. UMPa uses a novel K-way refinement heuristic employing a tie-breaking scheme to handle multiple communication metrics. We obtain significant improvements on a large number of graphs for all K values.

We are planning to speed up UMPa and the proposed refinement approach by implementing them on modern parallel architectures. We are also planning to investigate partitioning for hierarchical memory systems, such as cluster of multi-socket, multi-core machines with accelerators.

TABLE 3. The maximum send and send-receive volumes, and the total volume for PaToH and UMPa when $K = 256$. The times are given in seconds. There are 20 graphs in the table where two graphs with the best and the worst improvements on $maxSV$ are selected from each class. Each number is the average of 10 different executions.

	PaToH				UMPa			
Graph	$maxSV$	$maxSRV$	$totV$	Time	$maxSV$	$maxSRV$	$totV$	Time
coPapersCiteseer	7,854	16,765	577,278	224.09	5,448	11,615	579,979	658.21
coPapersDBLP	14,568	34,381	1,410,966	143.97	10,629	23,740	1,371,425	1038.86
as-22july06	1,555	7,128	28,246	1.01	617	4,543	33,347	12.62
smallworld	1,045	2,078	232,255	4.55	877	1,751	208,860	36.24
delaunay_n20	301	600	57,089	17.98	279	566	58,454	68.85
delaunay_n21	420	844	80,603	35.01	398	813	83,234	107.35
hugetrace-00000	407	814	74,563	55.51	415	831	80,176	123.66
hugetric-00010	502	1,004	91,318	92.45	477	955	97,263	167.69
adaptive	753	1,505	143,856	96.60	735	1,472	152,859	224.30
venturiLevel3	568	1,137	107,920	49.97	564	1,132	114,119	132.02
rgg_n_2_22_s0	799	1,589	145,902	151.30	724	1,495	147,331	249.23
rgg_n_2_23_s0	1,232	2,432	219,404	347.32	1,062	2,168	221,454	446.78
ri2010	3206	5,989	281,638	0.72	2,777	5,782	279,941	8.66
tx2010	5,139	9,230	124,033	8.47	3,011	7,534	117,960	15.55
af_shell10	898	1,792	174,624	89.90	885	1,769	184,330	158.04
audikw1	4,318	8,299	680,590	322.57	3,865	7,607	692,714	822.73
asia.osm	72	146	4,535	72.37	66	135	4,484	18.79
great-britain.osm	104	209	11,829	50.52	82	168	11,797	25.51
finan512	199	420	36,023	2.75	192	437	36,827	27.70
memplus	1,860	7,982	15,785	0.49	946	4,318	19,945	8.25

TABLE 4. The relative performance of UMPa with respect to PaToH in terms of execution time. The numbers are computed by using the results of 10 executions for each of the 123 graphs in our data set.

K	4	16	64	256	Avg.
Relative time	1.02	1.29	2.01	5.76	1.98

References

[1] C. J. Alpert and A. B. Kahng, *Recent directions in netlist partitioning: A survey*, VLSI Journal **19** (1995), no. 1–2, 1–81.

[2] Cevdet Aykanat, B. Barla Cambazoglu, and Bora Uçar, *Multi-level direct k-way hypergraph partitioning with multiple constraints and fixed vertices*, Journal of Parallel and Distributed Computing **68** (2008), no. 5, 609 625.

[3] S. T. Barnhard and H. D. Simon, *Fast multilevel implementation of recursive spectral bisection for partitioning unstructured problems*, Concurrency: Practice and Experience **6** (1994), no. 2, 67–95.

[4] Rob H. Bisseling and Wouter Meesen, *Communication balancing in parallel sparse matrix-vector multiplication*, Electron. Trans. Numer. Anal. **21** (2005), 47–65 (electronic). MR2195104 (2007c:65040)

[5] T. N. Bui and C. Jones, *A heuristic for reducing fill-in sparse matrix factorization*, Proc. 6th SIAM Conf. Parallel Processing for Scientific Computing, SIAM, 1993, pp. 445–452.

[6] Ü. V. Çatalyürek and C. Aykanat, *A hypergraph model for mapping repeated sparse matrix-vector product computations onto multicomputers*, Proc. International Conference on High Performance Computing, December 1995.

[7] ———, *Hypergraph-partitioning based decomposition for parallel sparse-matrix vector multiplication*, IEEE Transactions on Parallel and Distributed Systems **10** (1999), no. 7, 673–693.

[8] Ü. V. Çatalyürek and C. Aykanat, *Patoh: A multilevel hypergraph partitioning tool, version 3.0*, Bilkent University, Department of Computer Engineering, Ankara, 06533 Turkey. PaToH is available at http://bmi.osu.edu/~umit/software.htm, 1999.

[9] Ümit V. Çatalyürek, Cevdet Aykanat, and Bora Uçar, *On two-dimensional sparse matrix partitioning: models, methods, and a recipe*, SIAM J. Sci. Comput. **32** (2010), no. 2, 656–683, DOI 10.1137/080737770. MR2609335 (2011g:05176)

[10] *10th DIMACS implementation challenge: Graph partitioning and graph clustering*, 2011, http://www.cc.gatech.edu/dimacs10/.

[11] C. M. Fiduccia and R. M. Mattheyses, *A linear-time heuristic for improving network partitions*, Proc. 19th Design Automation Conference, 1982, pp. 175–181.

[12] Bruce Hendrickson and Tamara G. Kolda, *Graph partitioning models for parallel computing*, Parallel Comput. **26** (2000), no. 12, 1519–1534, DOI 10.1016/S0167-8191(00)00048-X. MR1786938

[13] Bruce Hendrickson and Robert Leland, *A multilevel algorithm for partitioning graphs*, Proc. Supercomputing (New York, NY, USA), ACM, 1995.

[14] George Karypis, *Multilevel hypergraph partitioning*, Multilevel optimization in VLSICAD, Comb. Optim., vol. 14, Kluwer Acad. Publ., Dordrecht, 2003, pp. 125–154. MR2021997

[15] B. W. Kernighan and S. Lin, *An efficient heuristic procedure for partitioning graphs*, The Bell System Technical Journal **49** (1970), no. 2, 291–307.

[16] Thomas Lengauer, *Combinatorial algorithms for integrated circuit layout*, Applicable Theory in Computer Science, John Wiley & Sons Ltd., Chichester, 1990. With a foreword by Bryan Preas. MR1071382 (91h:68089)

[17] Aleksandar Trifunovic and William Knottenbelt, *Parkway 2.0: A parallel multilevel hypergraph partitioning tool*, Proc. ISCIS, LNCS, vol. 3280, Springer Berlin / Heidelberg, 2004, pp. 789–800.

[18] Bora Uçar and Cevdet Aykanat, *Minimizing communication cost in fine-grain partitioning of sparse matrices*, Computer and Information Sciences - ISCIS 2003 (A. Yazici and C. Şener, eds.), Lecture Notes in Computer Science, vol. 2869, Springer Berlin / Heidelberg, 2003, pp. 926–933.

[19] Bora Uçar and Cevdet Aykanat, *Encapsulating multiple communication-cost metrics in partitioning sparse rectangular matrices for parallel matrix-vector multiplies*, SIAM J. Sci. Comput. **25** (2004), no. 6, 1837–1859 (electronic), DOI 10.1137/S1064827502410463. MR2086821 (2005g:05092)

[20] C. Walshaw, M. G. Everett, and M. Cross, *Parallel dynamic graph partitioning for adaptive unstructured meshes*, Journal of Parallel Distributed Computing **47** (1997), 102–108.

Appendix A. DIMACS Challenge Results

TABLE 5. The best maximum send volumes for UMPa for the challenge instances. X means UMPa failed to obtain a partition with the desired imbalance value.

	Parts									
Graph	2	4	8	16	32	64	128	256	512	1,024
as365						1,080	790	590	421	316
asia.osm						41	46	60	92	93
auto						2,044	1,497	1,070	733	501
coauthorsciteseer		10,066	7,773	5,313	3,216	2,006				
delaunay_n15			189	154	121	90	70			
er-fact1.5-scale23				5,707,503	3,933,216	2,091,986	1,154,276	622,913		
g3_circuit	1,266	1,630			1,151	938		536		
great-britain.osm					134	114	92	78		58
hugebubbles-00010		3,012			1,948	1,522		822	609	
hugetric-00000	1,274	2,206			1,117	804		458		
kkt_power				6,162	6,069	4,508		3,078	2,088	
kron_g500-logn17	36,656	53,381	55,314	49,657	42,272					
kron_g500-logn21						459,454	351,785	245,355	168,870	X
m6	1,487		2,034		1,427		762	568		
nlpkkt160		71,708	55,235	49,700	36,483	25,107				
nlr			2,380		1,563		847	623	447	
rgg_n_2_18_s0			516	431	330	248	195			

DEPARTMENT OF BIOMEDICAL INFORMATICS, THE OHIO STATE UNIVERSITY
E-mail address: umit@bmi.osu.edu

DEPARTMENT OF BIOMEDICAL INFORMATICS, THE OHIO STATE UNIVERSITY
E-mail address: mdeveci@bmi.osu.edu

DEPARTMENT OF BIOMEDICAL INFORMATICS, THE OHIO STATE UNIVERSITY
E-mail address: kamer@bmi.osu.edu

LIP, ENS LYON, LYON 69364, FRANCE
E-mail address: bora.ucar@ens-lyon.fr

Shape optimizing load balancing for MPI-parallel adaptive numerical simulations

Henning Meyerhenke

ABSTRACT. Load balancing is important for the efficient execution of numerical simulations on parallel computers. In particular when the simulation domain changes over time, the mapping of computational tasks to processors needs to be modified accordingly. Most state-of-the-art libraries addressing this problem are based on graph repartitioning with a parallel variant of the Kernighan-Lin (KL) heuristic. The KL approach has a number of drawbacks, including the optimized metric and solutions with undesirable properties.

Here we further explore the promising diffusion-based multilevel graph partitioning algorithm DIBAP. We describe the evolution of the algorithm and report on its MPI implementation PDIBAP for parallelism with distributed memory. PDIBAP is targeted at small to medium scale parallelism with dozens of processors. The presented experiments use graph sequences that imitate adaptive numerical simulations. They demonstrate the applicability and quality of PDIBAP for load balancing by repartitioning on this scale. Compared to the faster PARMETIS, PDIBAP's solutions often have partitions with fewer external edges and a smaller communication volume in an underlying numerical simulation.

1. Introduction

Numerical simulations are very important tools in science and engineering for the analysis of physical processes modeled by partial differential equations (PDEs). To make the PDEs solvable on a computer, they are discretized within the simulation domain, e. g., by the finite element method (FEM). Such a discretization yields a mesh, which can be regarded as a graph with geometric (and possibly other) information. Application areas of such simulations are fluid dynamics, structural mechanics, nuclear physics, and many others [10].

The solutions of discretized PDEs are usually computed by iterative numerical solvers, which have become classical applications for parallel computers. For efficiency reasons the computational tasks, represented by the mesh elements, must be distributed onto the processors evenly. Moreover, neighboring elements of the mesh need to exchange their values in every iteration to update their own value.

2000 *Mathematics Subject Classification.* Primary 68W10, 90C35.

Key words and phrases. Dynamic load balancing, graph partitioning and repartitioning, parallel adaptive numerical simulations, disturbed diffusion.

A preliminary version of this article appeared in the proceedings of the 15th International Conference on Parallel and Distributed Systems 2009 [22].

Due to the high cost of inter-processor communication, neighboring mesh elements should reside on the same processor. A good initial assignment of subdomains to processors can be found by solving the graph partitioning problem (GPP) [**31**]. The most common GPP formulation for an undirected graph $G = (V, E)$ asks for a division of V into k pairwise disjoint subsets (*parts*) such that all parts are no larger than $(1+\epsilon) \cdot \lceil \frac{|V|}{k} \rceil$ (for small $\epsilon \geq 0$) and the *edge-cut*, i. e., the total number of edges having their incident vertices in different subdomains, is minimized.

In many numerical simulations some areas of the mesh are of higher interest than others. For instance, during the simulation of the interaction of a gas bubble with a surrounding liquid, one is interested in the conditions close to the boundary of the fluids. Another application among many others is the simulation of the dynamic behavior of biomolecular systems [**3**]. To obtain an accurate solution, a high resolution of the mesh is required in the areas of interest. To use the available memory efficiently, one has to work with different resolutions in different areas. Moreover, the areas of interest may change during the simulation, which requires *adaptations* in the mesh and may result in undesirable load imbalances. Hence, after the mesh has been adapted, its elements need to be redistributed such that every processor has a similar computational effort again. While this can be done by solving the GPP for the new mesh, the *repartitioning* process not only needs to find new partitions of high quality. Also as few vertices as possible should be moved to other processors since this *migration* causes high communication costs and changes in the local mesh data structure.

Motivation. The most popular graph partitioning and repartitioning libraries (for details see Section 2) use local vertex-exchanging heuristics like Kernighan-Lin (KL) [**18**] within a multilevel improvement process to compute solutions with low edge cuts very quickly. Yet, their deployment can have certain drawbacks. First of all, minimizing the edge-cut with these tools does not necessarily mean to minimize the total running time of parallel numerical simulations [**13, 37**]. While the total communication volume can be minimized by hypergraph partitioning [**4**], synchronous parallel applications need to wait for the processor computing longest. Hence, the *maximum norm* (i. e., the worst part in a partition) of the simulation's communication costs is of higher importance. Moreover, for some applications, the *shape* of the subdomains plays a significant role. It can be assessed by various measures such as aspect ratio [**8**], maximum diameter [**26**], connectedness, or smooth boundaries. Optimizing partition shapes, however, requires additional techniques (e. g., [**8, 23, 26**]), which are far from being mature. Finally, due to their sequential nature, the most popular repartitioning heuristics are difficult to parallelize—although significant progress has been made (see Section 2).

Our previously developed partitioning algorithm DIBAP aims at computing well-shaped partitions and uses disturbed diffusive schemes to decide not only *how many* vertices move to other parts, but also *which* ones. It contains inherent parallelism and overcomes many of the above mentioned difficulties, as could be shown experimentally for static graph partitioning [**23**]. While it is much slower than state-of-the-art partitioners, it often obtains better results.

Contribution. In this work we further explore the disturbed diffusive approach and focus on repartitioning for load balancing. First we present how the implementation of PDIBAP has been improved and adapted for MPI-parallel repartitioning. With this implementation we perform various repartitioning experiments with benchmark graph sequences. These experiments are the first using PDIBAP for repartitioning and show the suitability of the disturbed diffusive approach. The average quality of the partitions computed by PDIBAP is clearly better than that of the state-of-the-art repartitioners PARMETIS and parallel JOSTLE, while PDIBAP's migration volume is usually comparable. It is important to note that PDIBAP's improvement concerning the partition quality for the graph sequences is even higher than in the case of static partitioning.

2. Related Work

We give a short introduction to the state-of-the-art of practical graph repartitioning algorithms and libraries which only require the adjacency information about the graph and no additional problem-related information. For a broader overview the reader is referred to Schloegel et al. [**31**]. The most recent advances in graph partitioning are probably best covered in their entirety by the proceedings volume [**2**] the present article is part of.

2.1. Graph Partitioning. To employ local improvement heuristics effectively, they need to start with a reasonably good initial solution. If such a solution is not provided as input, the multilevel approach [**12**] is a very powerful technique. It consists of three phases: First, one computes a hierarchy of graphs $G_0, \dots, G_l$ by recursive coarsening in the first phase. G_l ought to be very small in size, but similar in structure to the input graph G_0. A very good initial solution for G_l is computed in the second phase. After that, the solution is interpolated to the next-finer graph recursively. In this final phase each interpolated solution is refined using the desired local improvement algorithm. A very common local improvement algorithm for the third phase of the multilevel process is based on the method by Fiduccia and Mattheyses (FM) [**9**], a variant of the well-known local search heuristic by Kernighan and Lin (KL) [**18**] with improved running time. The main idea of both is to exchange vertices between parts in the order of the cost reductions possible, while maintaining balanced partition sizes. After every vertex has been moved once, the solution with the best gain is chosen. This is repeated several times until no further improvements are found.

State-of-the-art graph partitioning libraries such as METIS [**16, 17**] and JOSTLE [**38**] use KL/FM for local improvement and edge-contractions based on matchings for coarsening. Recently, Holtgrewe et al. [**14**] presented a parallel library for static partitioning called KAPPA. It attains very good edge cut results, mainly by controlling the multilevel process using so-called edge ratings for approximate matchings. Recently Sanders and Osipov [**25**] and Sanders and Schulz [**27, 28**] have presented new sequential approaches for cut-based graph partitioning. They mainly employ a radical multilevel strategy, flow-based local improvement, and evolutionary algorithms, respectively.

2.2. Load Balancing by Repartitioning. To consider both a small edge-cut *and* small migration costs when repartitioning dynamic graphs, different strategies have been explored in the literature. To overcome the limitations of simple scratch-remap and rebalance approaches, Schloegel et al. [**32, 33**] combine both methods. They propose a multilevel algorithm with three main features. In the local improvement phase, two algorithms are used. On the coarse hierarchy levels, a diffusive scheme takes care of balancing the subdomain sizes. Since this might affect the partition quality negatively, a refinement algorithm is employed on the finer levels. It aims at edge-cut minimization by profitable swaps of boundary vertices.

To address the load balancing problem in parallel applications, distributed versions of the partitioners METIS, JOSTLE, and SCOTCH [**6, 34, 39**] have been developed. Also, the tools PARKWAY [**36**], a parallel hypergraph partitioner, and ZOLTAN [**5**], a suite of load balancing algorithms with focus on hypergraph partitioning, need to be mentioned although they concentrate (mostly) on hypergraphs. An efficient parallelization of the KL/FM heuristic that these parallel (hyper)graph partitioners use is complex due to inherently sequential parts in this heuristic. For example, one needs to ensure that during the KL/FM improvement no two neighboring vertices change their partition simultaneously and destroy data consistency. A coloring of the graph's vertices is used by the parallel libraries PARMETIS [**32**] and KAPPA [**14**] for this purpose.

2.3. Diffusive Methods for Shape Optimization. Some applications profit from good partition shapes. As an example, the convergence rate of certain iterative linear solvers can depend on the geometric shape of a partition [**8**]. That is why in previous work [**21, 24**] we have developed shape-optimizing algorithms based on diffusion. Before that, repartitioning methods employed diffusion mostly for computing *how much* load needs to be migrated between subdomains [**30**], not *which* elements should be migrated. Generally speaking, a diffusion problem consists of distributing load from some given seed vertex (or several seed vertices) into the whole graph by iterative load exchanges between neighbor vertices. Typical diffusion schemes have the property to result in the balanced load distribution, in which every vertex has the same amount of load. This is one reason why diffusion has been studied extensively for load balancing [**40**]. Our algorithms BUBBLE-FOS/C [**24**] and the much faster DIBAP [**23**] (also see Section 3) as well as a combination of KL/FM and diffusion by Pellegrini [**26**] exploit that diffusion sends load entities faster into densely connected subgraphs. This fact is used to distinguish dense from sparse graph regions. In the field of graph-based image segmentation, similar arguments are used to find well-shaped segments [**11**].

3. Diffusion-based Repartitioning with DibaP

The algorithm DIBAP, which we have developed and implemented with shared memory parallelism previously [**23**], is a hybrid multilevel combination of the two (re)partitioning methods BUBBLE-FOS/C and TRUNCCONS, which are both based on disturbed diffusion. We call a diffusion scheme *disturbed* if it is modified such that its steady state does not result in the balanced distribution. Disturbed diffusion schemes can be helpful to determine if two graph vertices or regions are densely connected to each other, i. e., if they are connected by many paths of small length. This property is due to the similarity of diffusion to random walks and the notion that a random walk is more likely to stay in a dense region for a long time before leaving it via one of the few external edges. Before we explain the whole algorithm DIBAP, we describe its two main components for (re-)partitioning in more detail.

3.1. Bubble-FOS/C. In contrast to Lloyd's related k-means algorithm [**19**], BUBBLE-FOS/C partitions or clusters graphs instead of geometric inputs. Given a graph $G = (V, E)$ and $k \geq 2$, initial partition representatives (centers) are chosen in the first step of the algorithm, one center for each of the k parts. All remaining vertices are assigned to their closest center vertex. While for k-means one usually uses Euclidean distance, BUBBLE-FOS/C employs the disturbed diffusion scheme FOS/C **[24]** as distance measure (or, more precisely, as similarity measure). The similarity of a vertex v to a non-empty vertex subset S is computed by solving the linear system $\mathbf{L}w = d$ for w, where $\mathbf{L}$ is the Laplacian matrix of the graph and d a suitably chosen vector that disturbs the underlying diffusion system.[1]

After the assignment step, each part computes its new center for the next iteration – again using FOS/C, but with a different right-hand side vector d. The two operations *assigning vertices to parts* and *computing new centers* are repeated alternately a fixed number of times or until a stable state is reached. Each operation requires the solution of k linear systems with the matrix $\mathbf{L}$, one for each partition.

It turns out that this iteration of two alternating operations yields very good partitions. Apart from the distinction of dense and sparse regions, the final partitions are very compact and have short boundaries. However, the repeated solution of linear systems makes BUBBLE-FOS/C slow.

3.2. TruncCons. The algorithm TRUNCCONS **[23]** (for *truncated consolidations*) is also an iterative method for the diffusion-based local improvement of partitions, but it is much faster than BUBBLE-FOS/C. Within each TRUNCCONS iteration, the following is performed independently for each partition π_c: First, the initial load vector $w^{(0)}$ is set. Vertices of π_c receive an equal amount of initial load $|V|/|\pi_c|$, while the other vertices' initial load is set to 0. Then, this load is distributed within the graph by performing a small number ψ of FOS (first order diffusion scheme) [7] iterations. The final load vector w is computed as $w = \mathbf{M}^{\psi} w^{(0)}$, where $\mathbf{M} = \mathbf{I} - \alpha \mathbf{L}$ denotes the diffusion matrix [7] of G. A common choice for α is $\alpha := \frac{1}{(1+\deg(G))}$. The computation $w = \mathbf{M}^{\psi} w^{(0)}$ could be realized by ψ matrix-vector products. A more localized view of its realization is given by iterative load exchanges on each vertex v with its neighbors. Then we get for $1 \leq t \leq \psi$:

$$w_v^{(t)} = w_v^{(t-1)} - \alpha \sum_{\{u,v\} \in E} (w_v^{(t-1)} - w_u^{(t-1)}).$$

After the load vectors have been computed this way independently for all k parts, each vertex v is assigned to the partition it has obtained the highest load from. This completes one TRUNCCONS iteration, which can be repeated several times (the total number is denoted by Λ subsequently) to facilitate sufficiently large movements of the parts.

A vertex with the same amount of load as all its neighbors does not change its load in the next FOS iteration. Due to the choice of initial loads, there are many such *inactive* vertices in the beginning. In fact, only vertices incident to the cut edges of the part under consideration are active initially. In principle each new FOS iteration adds a new layer of active vertices similar to BFS frontiers. We keep

[1]In general $\mathbf{L}$ represents the whole graph. Yet, sparsifying the matrix in certain areas (also called *partial graph coarsening*) is possible and leads to a significant acceleration without sacrificing partitioning quality considerably **[24]**. While the influence of partial graph coarsening on the partitioning quality is low, the solutions of the linear systems become distorted and more difficult to analyze. Moreover, the programming overhead is immense. As the next section introduces a simpler and faster way of diffusive partitioning, we do not consider partial graph coarsening further here.

track which vertices are active and which are not. Thereby, it is possible to forego the inactive vertices when performing the local FOS calculations.

In our implementation the size of the matrix **M** for which we compute a matrix-vector product locally in each iteration is not changed. Instead, inner products involving inactive rows are not computed as we know their respective result does not change in the current iteration. That way the computational effort is restricted to areas close to the partition boundaries.

3.3. The Hybrid Algorithm PDibaP.

The main components of PDIBAP, the MPI-parallel version of the original implementation of DIBAP, are depicted in Figure 1. To build a multilevel hierarchy, the fine levels are coarsened (1) by approximate maximum weight matchings. Once the graphs are sufficiently small, the construction mechanism can be changed. In our sequential DIBAP implementation, we switch the construction mechanism (2) to the more expensive coarsening based on algebraic multigrid (AMG)—for an overview on AMG cf. [**35**]. This is advantageous regarding running time because, after computing an initial partition (3), BUBBLE-FOS/C is used as local improvement algorithm on the coarse levels (4). Since AMG is well-suited as a linear solver within BUBBLE-FOS/C, such a hierarchy would be required for AMG anyway. In our parallel implementation PDIBAP (cf. Section 4), however, due to a significant reduction of the parallel programming effort, we decided to coarsen by matchings, use a conjugate gradient solver, and leave AMG to future work.

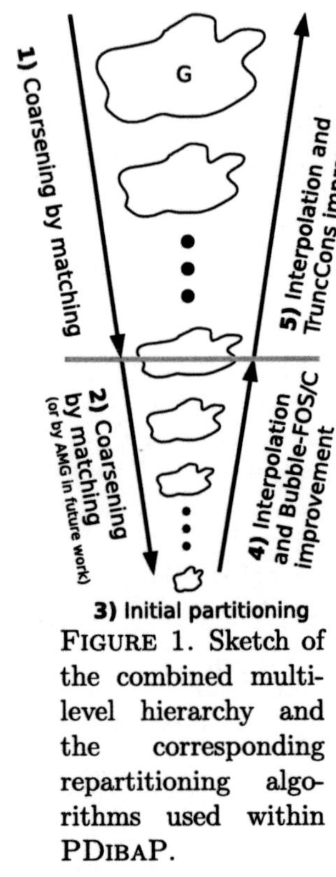

FIGURE 1. Sketch of the combined multilevel hierarchy and the corresponding repartitioning algorithms used within PDIBAP.

Eventually, the partitions on the fine levels are improved by the local improvement scheme TRUNCCONS (5). PDIBAP includes additional components, e. g., for balancing partition sizes and smoothing partition boundaries, see Section 4.3.

The rationale behind PDIBAP can be explained as follows. While BUBBLE-FOS/C computes high-quality graph partitions with good shapes, its similarity measure FOS/C is very expensive to compute compared to established partitioning heuristics. To overcome this problem, we use the simpler process TRUNCCONS, a truly local algorithm to improve partitions generated in a multilevel process. It exploits the observation that, once a reasonably good solution has been found, alterations during a local improvement step take place mostly at the partition boundaries. The disturbing truncation within TRUNCCONS allows for a concentration of the computations around the partition boundaries, where the changes in subdomain affiliation occur. Moreover, since TRUNCCONS is also based on disturbed diffusion, the good properties of the partitions generated by BUBBLE-FOS/C are mostly preserved.

4. PDibaP: Parallel DibaP for Repartitioning

In this section we describe our parallel implementation of DIBAP using MPI. In particular we highlight some differences to the sequential (and thread-parallel) version used for static partitioning [**23**].

4.1. Distributed Memory Parallelism. The foundation of our PDIBAP implementation (data structure, linear algebra routines, auxiliary functions) is to a large extent based on the code described in more detail in our previous work [**24**] and in Schamberger's thesis [**29**]. PDIBAP employs as graph data structure the standard distributed compressed sparse row (CSR) format with ghost (or halo) vertices. The linear systems within BUBBLE-FOS/C are solved with a conjugate gradient (CG) solver using the traditional domain decomposition approach for distributed parallelism. That means that each system is distributed over all processors and solved by all of them in parallel at the same time, which requires three communication operations per iteration within CG. The TRUNCCONS process is executed in a similar manner. To account for the inactive vertices, however, we do not perform complete matrix-vector multiplications, but perform local load exchanges only if an active vertex is involved. Both CG and TRUNCCONS require a halo update after each iteration. This communication routine is rather expensive, so that the number of iterations should be kept small. The linear algebra routines within PDIBAP do not make use of external libraries. This is due to the fact that the solution process in BUBBLE-FOS/C is very specialized [**24, 29**].

4.2. Repartitioning. So far, PDIBAP is targeted at repartitioning dynamic graphs. The option for parallel static partitioning is still in its infancy due to a limitation in the multilevel process, which we explain later on in this section.

When PDIBAP is used for repartitioning instead of partitioning, one part of its input is an initial partition. Based on this partition, the graph is distributed onto the processors. We can assume that this partition is probably more unbalanced than advisable. It might also contain some undesirable artifacts. Nevertheless, its quality is not likely to be extremely bad. It is therefore reasonable to improve the initial partition instead of starting from scratch. Moreover, a refinement limits the number of migrated vertices as well, an important feature of dynamic repartitioning methods.

In particular if the imbalance is higher than allowed, it is advisable to employ the multilevel paradigm. Local improvements on the input graph would not result in sufficiently large movements to a high quality solution. Therefore, a matching hierarchy is constructed until only a few thousand vertices remain in the coarsest graph. So far, only edges whose endpoints lie in the same part are considered to be part of the matching. This simplifies the parallel implementation and is a viable approach when repartitioning.

After constructing the hierarchy, the initial partition is projected downwards the hierarchy onto the coarsest level. On the coarsest level the graph is repartitioned with BUBBLE-FOS/C, starting with the projected initial solution. Going up the multilevel hierarchy recursively, the result is then improved with either BUBBLE-FOS/C or TRUNCCONS, depending on the size of the level. After the refinement, the current solution is interpolated to the next level until the process stops at the input level. Sometimes the matching algorithm has hardly coarsened a level. This happens for example to avoid star-like subgraphs with strongly varying vertex degrees. Limited coarsening results in two very similar adjacent levels. Local improvement with TRUNCCONS on both of these levels would result in similar solutions with an unnecessary running time investment. That is why in such a case TRUNCCONS is skipped on the finer level of the two.

For static partitioning, which is still an ongoing effort, edges in the cut between parts on different processors should be considered as matching edges as well. Otherwise, the multilevel hierarchy contains only a few levels after which no more edges are found for the matching. The development and/or integration of such a more general matching is part of future work.

4.3. Balancing Procedures. In general the diffusive processes employed by PDIBAP do not guarentee the nearly perfect balance required by numerical simulations (say, for example, no part should be larger than the average part size plus 3%). That is why we employ two balancing procedures within PDIBAP. The first one called `ScaleBalance` is an iterative procedure that tries to determine for every part $1 \leq p \leq k$ a scalar β_p with which the diffusion load values are scaled. Suitable scalars are searched such that the assignment of vertices to parts based on the load vector entries $\beta_p w_p$ results in a balanced partition. More details can be found in Meyerhenke et al. [**24**, p. 554]. While `ScaleBalance` works surprisingly well in many cases within PDIBAP, it also happens that it is not fully effective even after a fairly large number of iterations. Then we employ a second approach, called `FlowBalance`, whose basic idea is described in previous work as well [**24**, p. 554]. Here we highlight recent changes necessary to adapt the approach to the distributed parallelism in PDIBAP.

First, we solve a load balancing problem on the quotient graph of the partition Π. The quotient graph Q contains a vertex for each part in Π and two vertices are connected by an edge in Q if and only if their corresponding parts share a common boundary in Π. The load balancing problem can be solved with diffusion [**15**]. The solution yields the migrating flow that balances the partition. Hence, we know *how many* vertices have to be moved from π_i to π_j, let us call this number n_{ij}. It remains to be determined *which* vertices take this move. For quality reasons, this decision should be based on the diffusion values in the respective load vectors computed by BUBBLE-FOS/C or TRUNCCONS. That is why we want to migrate the n_{ij} vertices with the highest values in the load vector w_j.

In our sequential and thread-parallel version of DIBAP, we use a binary heap as priority queue to perform the necessary selection, migration, and resulting updates to the partition. Since parallel priority queues require a considerable effort to obtain good scalability, we opt for a different approach in PDIBAP. For ease of implementation (and because the amount of computation and communication is relatively small), each processor preselects its local vertices with the highest n_{ij} load values in w_j. These preselected load values are sent to processor p_j, which performs a sequential selection. The threshold value found this way is broadcast back to all processors. Finally, all processors assign their vertices whose diffusion loads in w_j is higher than the threshold to part π_j.

This approach might experience problems when the selected threshold value occurs multiple times among the preselected candidate values. In such a case, the next larger candidate value is chosen as threshold. Another problem could be the scheduled order in which migration takes place. It could happen that a processor needs to move a number of vertices that it is about to obtain by a later move. To address this, we employ a conservative approach and move rather fewer vertices than too many. As a compensation, the whole procedure is repeated iteratively until a balanced partition is found.

5. Experiments

Here we present some of our experimental results comparing our PDIBAP implementation to the KL/FM-based load balancers PARMETIS and parallel JOSTLE.

5.1. Benchmark Data. Our benchmark set comprises two types of graph sequences. The first one consists of three smaller graph sequences with 51 frames each, having between approximately $1M$ and $3M$ vertices, respectively. The second group contains two larger sequences of 36 frames each. Each frame in this group has approximately $4.5M$ to $16M$ vertices. These sequences result in 50 and 35 repartitioning steps, respectively. We choose to (re)partition the smaller sequences into $k = 36$ and $k = 60$ parts, while the larger ones are divided into $k = 60$ and $k = 84$ parts. These values have been chosen as multiples of 12 because one of our main test machines has 12 cores per node.

All graphs of these five sequences have a two-dimensional geometry and have been generated to resemble adaptive numerical simulations such as those occurring in computational fluid dynamics. A visual impression of some of the data (in smaller versions) is available in previous work [**24**, p. 562f.]. The graph of frame $i + 1$ in a given sequence is obtained from the graph of frame i by changes restricted to local areas. As an example, some areas are coarsened, whereas others are refined. These changes are in most cases due to the movement of an object in the simulation domain and often result in unbalanced subdomain sizes. For more details the reader is referred to Marquardt and Schamberger [**20**], who have provided the generator for the sequence data.[2] Some of these frames are also part of the archive of the 10th DIMACS Implementation Challenge [**1**].

5.2. Hardware and Software Settings. We have conducted our experiments on a cluster with 60 Fujitsu RX200S6 nodes each having 2 Intel Xeon X5650 processors at 2.66 GHz (results in 12 compute cores per node). Moreover, each node has 36 GB of main memory. The interconnect is InfiniBand HCA 4x SDR HCA PCI-e, the operating system Cent OS 5.4. PDIBAP is implemented in C/C++. PDIBAP as well as PARMETIS and parallel JOSTLE have been compiled with Intel C/C++ compiler 11.1 and MVAPICH2 1.5.1 as MPI library. The number of MPI processes always equals the number of parts k in the partition to be computed.

The main parameters controlling the running time and quality of the DIBAP algorithm are the number of iterations in the (re)partitioning algorithms BUBBLE-FOS/C and TRUNCCONS. For our experiments we perform 3 iterations within BUBBLE-FOS/C, with one `AssignPartition` and one `ComputeCenters` operation, respectively. The faster local approach TRUNCCONS is used on all multilevel hierarchy levels with graph sizes above 12,000 vertices. For TRUNCCONS, the parameter settings $\Lambda = 9$ and $\psi = 14$ for the outer and inner iteration, respectively. These

[2]Some of the input data can be downloaded from the website `http://www.upb.de/cs/henningm/graph.html`.

settings provide a good trade-off between running time and quality. The allowed imbalance is set to the default value 3% for all tools.

5.3. Results. In addition to the graph partitioning metrics edge-cut and communication volume (of the underlying application based on the computed partition), we are also interested in migration costs. These costs result from data changing their processor after repartitioning. We count the number of vertices that change their subdomain from one frame to the next as a measure of these costs. One could also assign cost weights to the partitioning objectives and the migration volume to evaluate the linear combination of both. Since these weights depend both on the underlying application and the parallel architecture, we have not pursued this here. We compare PDIBAP to the state-of-the-art repartitioning tools PARMETIS and parallel JOSTLE. Both competitors are mainly based on the vertex-exchanging KL heuristic for local improvement. The load balancing toolkit ZOLTAN [**5**], whose integrated KL/FM partitioner is based on the hypergraph concept, is not included in the detailed presentation. Our experiments with it indicate that it is not as suitable for our benchmark set of FEM graphs, in particular because it yields disconnected parts which propagate and worsen in the course of the sequence. We conclude that currently the dedicated graph (as opposed to hypergraph) partitioners seem more suitable for this problem type.

The partitioning quality is measured in our experiments by the edge cut (EC, a summation norm) and the maximum communication volume (CV_{max}). CV_{max} is the sum of the maximum incoming communication volume and the maximum outgoing communication volume, taken over all parts, respectively. The values are displayed in Table 1, averaged over the whole sequence and aggregated by the different k. Very similar results are obtained for the geometric mean in nearly all cases, which is why we do not show these data as well. The migration costs are recorded in both norms and shown for each sequence (again aggregated) in Table 2. Missing values for parallel JOSTLE (—) indicate program crashes on the corresponding instance(s).

TABLE 1. Average edge cut and communication volume (max norm) for repartitionings computed by PARMETIS, JOSTLE, and PDIBAP. Lower values are better, best values per instance are written in bold.

Sequence	PARMETIS		Par. JOSTLE		PDIBAP	
	EC	CV_{max}	EC	CV_{max}	EC	CV_{max}
biggerslowtric	11873.5	1486.7	9875.1	1131.9	**8985.5**	**981.8**
biggerbubbles	16956.8	2205.3	14113.2	1638.7	**12768.3**	**1443.5**
biggertrace	17795.6	2391.1	14121.3	1687.0	**12229.2**	**1367.5**
hugetric	34168.5	2903.0	28208.3	2117.6	**24974.4**	**1766.2**
hugetrace	54045.8	5239.7	–	–	**34147.4**	**2459.4**

The aggregated graph partitioning metrics show that PDIBAP is able to compute the best partitions consistently. PDIBAP's advance is highest for the communication volume. With about 12–19% on parallel JOSTLE and about 34–53% on PARMETIS these improvements are clearly higher than the approximately 7%

TABLE 2. Average migration volume in the ℓ_1- and ℓ_∞-norm for repartitionings computed by PARMETIS, JOSTLE, and PDIBAP. Lower values are better, best values per instance are written in bold.

Sequence	PARMETIS		Par. JOSTLE		PDIBAP	
	ℓ_∞	ℓ_1	ℓ_∞	ℓ_1	ℓ_∞	ℓ_1
biggerslowtric	**60314.3**	606419.1	64252.2	557608.7	65376.1	**550427.0**
biggerbubbles	77420.0	1249424.3	**68865.1**	**791723.6**	93767.5	1328116.1
biggertrace	54131.2	733750.4	49997.8	**533809.2**	**46620.4**	613071.2
hugetric	**231072.8**	2877441.8	244082.5	2932607.6	232382.6	**2875302.5**
hugetrace	**175795.8**	**3235984.1**	–	–	189085.3	3308461.4

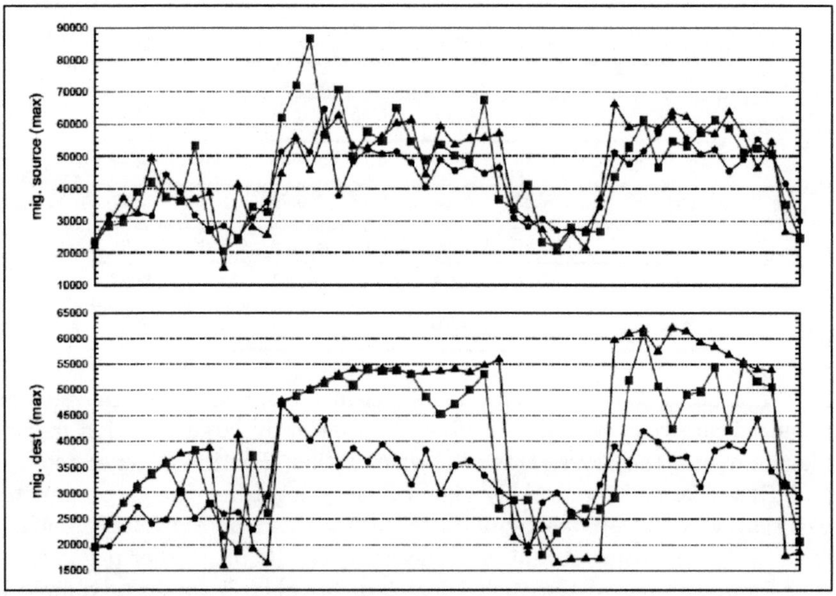

FIGURE 2. Number of migrating vertices (ℓ_∞-norm) in each frame of the biggertrace sequence for PDIBAP (circles), METIS (triangles), and JOSTLE (squares). Lower values are better.

obtained for static partitioning [**23**], which is due to the fact that parallel KL (re)partitioners often compute worse solutions than their serial counterparts for static partitioning.

The results for the migration volume are not consistent. All tools have a similar amount of best values. The fact that PARMETIS is competetitive is slightly surprising when compared to previous results [**22**], where it compared worse. Also unexpectedly, PDIBAP shows significantly higher migration costs for the instance biggerbubbles. Our experiments indicate that PDIBAP has a more constant migration volume, while the values for parallel JOSTLE and PARMETIS show a higher amplitude. It depends on the instance which strategy pays off. This behavior is shown in Figure 2. It displays the migration volumes in the ℓ_∞-norm for each frame within the benchmark sequence called *slowrot*, which is smaller but similar to the ones used in our main experimental study.

These results lead to the conclusion that PDIBAP's implicit optimization with the iterative algorithms BUBBLE-FOS/C and TRUNCCONS focusses more on good partitions than on small migration costs. In some cases the latter objective should receive more attention. As currently no explicit mechanisms for migration optimization are integrated, such mechanisms could be implemented if one finds in other experiments that the migration costs become too high with PDIBAP.

TABLE 3. Average running times in seconds for the benchmark problems. Lower values are better, best values per instance are written in bold. The values marked by * denote averaged times (or, in case of –, incomparable values) where parallel JOSTLE did not finish the whole sequence due to a premature crash.

Sequence	PARMETIS		Par. JOSTLE		PDIBAP	
	$k = 36$	$k = 60$	$k = 36$	$k = 60$	$k = 36$	$k = 60$
biggerslowtric	**0.27**	**0.22**	0.50	0.88	8.71	10.38
biggerbubbles	**0.38**	**0.30**	0.79	1.24	15.02	19.19
biggertrace	**0.33**	**0.26**	0.56	0.59	9.27	10.77
	$k = 60$	$k = 84$	$k = 60$	$k = 84$	$k = 60$	$k = 84$
hugetric	**0.68**	**0.64**	2.41*	4.68*	55.36	62.37
hugetrace	**0.85**	**0.76**	–	–	50.56	56.69

It is interesting to note that further experiments indicate a multilevel approach to be indeed necessary in order to produce sufficiently large partition movements that keep up with the movements of the simulation. Partitions generated by multilevel PDIBAP are of a noticeably higher quality regarding the graph partitioning metrics than those computed by TRUNCCONS without multilevel approach. Also, maybe surprisingly, using a multilevel hierarchy results in steadier migration costs.

The running time of the tools, depicted in Table 3, for the dynamic graph instances used in this study can be characterized as follows. PARMETIS is the fastest, taking from a fraction of a second up to a few seconds for each frame, with the average always being below one second. Parallel JOSTLE is approximately a factor of 2-4 slower than PARMETIS (without counting sequences where parallel JOSTLE crashed prematurely). PDIBAP, however, is significantly slower than both tools, with an average slowdown of about 28-97 compared to PARMETIS. It requires from a few seconds up to a few minutes for each frame, with the average being 10-20 seconds for the small benchmarks and about a minute for the large ones.

The scalability of PDIBAP is not good due to the linear dependence on k in the running time. PARMETIS is able to profit somewhat from more processors regarding execution time. PDIBAP and parallel JOSTLE, however, become slower with increasing k. Neglecting communication, the running time of PDIBAP should remain nearly constant for growing k when it computes a k-partitioning with k processors. However, in this parallel setting the communication overhead yields growing running times. Therefore, one can conclude that PDIBAP is more suitable for simulations with a small number of processors.

We would like to stress that a high repartitioning quality is often very important. Usually, the most time consuming parts of numerical simulations are the numerical solvers. Hence, a reduced communication volume provided by an excellent partitioning can pay off unless the repartitioning time is extremely high. Nevertheless, a further acceleration of shape-optimizing load balaincing is of utmost importance. Minutes for each repartitioning step might be problematic for some targeted applications.

6. Conclusions

With this work we have demonstrated that the shape-optimizing repartitioning algorithm DIBAP based on disturbed diffusion can be a good alternative to traditional KL-based methods for balancing the load in parallel adaptive numerical simulations. In particular, the parallel implementation PDIBAP is very suitable for simulations of small to medium scale, i. e., when the number of vertices and edges in the dynamic graphs are on the order of several millions. While PDIBAP is still significantly slower than the state-of-the-art, it usually computes considerably better solutions w. r. t. edge cut and communication volume. In situations where the quality of the load balancing phase is more important than its running time – e. g., when the computation time between the load balancing phases is relatively high – the use of PDIBAP is expected to pay off.

As part of future work, we aim at an improved multilevel process and faster partitioning methods. It would also be worthwhile to investigate if BUBBLE-FOS/C and TRUNCCONS can be further adapted algorithmically, for example to reduce the dependence on k in the running time.

References

[1] David Bader, Henning Meyerhenke, Peter Sanders, and Dorothea Wagner, *10th DIMACS implementation challenge*, http://www.cc.gatech.edu/dimacs10/, 2012.

[2] David Bader, Henning Meyerhenke, Peter Sanders, and Dorothea Wagner (eds.), *Proceedings of the 10th DIMACS implementation challenge*, Contemporary Mathematics, American Mathematical Society, 2012.

[3] N. A. Baker, D. Sept, M. J. Holst, and J. A. McCammon, *The adaptive multilevel finite element solution of the Poisson-Boltzmann equation on massively parallel computers*, IBM J. of Research and Development **45** (2001), no. 3.4, 427 –438.

[4] U. Catalyurek and C. Aykanat, *Hypergraph-partitioning-based decomposition for parallel sparse-matrix vector multiplication*, IEEE Transactions on Parallel and Distributed System **10** (1999), no. 7, 673–693.

[5] Umit V. Catalyurek, Erik G. Boman, Karen D. Devine, Doruk Bozdağ, Robert T. Heaphy, and Lee Ann Riesen, *A repartitioning hypergraph model for dynamic load balancing*, J. Parallel Distrib. Comput. **69** (2009), no. 8, 711–724.

[6] C. Chevalier and F. Pellegrini, *PT-Scotch: a tool for efficient parallel graph ordering*, Parallel Comput. **34** (2008), no. 6-8, 318–331, DOI 10.1016/j.parco.2007.12.001. MR2428880

[7] G. Cybenko, *Dynamic load balancing for distributed memory multiprocessors*, Parallel and Distributed Computing **7** (1989), 279–301.

[8] R. Diekmann, R. Preis, F. Schlimbach, and C. Walshaw, *Shape-optimized mesh partitioning and load balancing for parallel adaptive FEM*, Parallel Computing **26** (2000), 1555–1581.

[9] C. M. Fiduccia and R. M. Mattheyses, *A linear-time heuristic for improving network partitions*, Proceedings of the 19th Conference on Design automation (DAC'82), IEEE Press, 1982, pp. 175–181.

[10] G. Fox, R. Williams, and P. Messina, *Parallel computing works!*, Morgan Kaufmann, 1994.

[11] Leo Grady and Eric L. Schwartz, *Isoperimetric graph partitioning for image segmentation*, IEEE Trans. Pattern Anal. Mach. Intell. **28** (2006), no. 3, 469–475.

[12] B. Hendrickson and R. Leland, *A multi-level algorithm for partitioning graphs*, Proceedings Supercomputing '95, ACM Press, 1995, p. 28 (CD).

[13] Bruce Hendrickson and Tamara G. Kolda, *Graph partitioning models for parallel computing*, Parallel Comput. **26** (2000), no. 12, 1519–1534, DOI 10.1016/S0167-8191(00)00048-X. MR1786938

[14] Manuel Holtgrewe, Peter Sanders, and Christian Schulz, *Engineering a scalable high quality graph partitioner*, IPDPS, IEEE, 2010, pp. 1–12.

[15] Y. F. Hu and R. J. Blake, *An improved diffusion algorithm for dynamic load balancing*, Parallel Comput. **25** (1999), no. 4, 417–444, DOI 10.1016/S0167-8191(99)00002-2. MR1684706

[16] George Karypis and Vipin Kumar, *MeTiS: A Software Package for Partitioning Unstructured Graphs, Partitioning Meshes, and Computing Fill-Reducing Orderings of Sparse Matrices, Version 4.0*, Univ. of Minnesota, Minneapolis, MN, 1998.

[17] ———, *Multilevel k-way partitioning scheme for irregular graphs*, Journal of Parallel and Distributed Computing **48** (1998), no. 1, 96–129.

[18] B. W. Kernighan and S. Lin, *An efficient heuristic for partitioning graphs*, Bell Systems Technical Journal **49** (1970), 291–308.

[19] Stuart P. Lloyd, *Least squares quantization in PCM*, IEEE Trans. Inform. Theory **28** (1982), no. 2, 129–137, DOI 10.1109/TIT.1982.1056489. MR651807 (84a:94012)

[20] O. Marquardt and S. Schamberger, *Open benchmarks for load balancing heuristics in parallel adaptive finite element computations*, Proceedings of the International Conference on Parallel and Distributed Processing Techniques and Applications, (PDPTA'05), CSREA Press, 2005, pp. 685–691.

[21] H. Meyerhenke and S. Schamberger, *Balancing parallel adaptive FEM computations by solving systems of linear equations*, Proceedings of the 11th International Euro-Par Conference, Lecture Notes in Computer Science, vol. 3648, Springer-Verlag, 2005, pp. 209–219.

[22] Henning Meyerhenke, *Dynamic load balancing for parallel numerical simulations based on repartitioning with disturbed diffusion*, Proc. Internatl. Conference on Parallel and Distributed Systems (ICPADS'09), IEEE Computer Society, 2009, pp. 150–157.

[23] Henning Meyerhenke, Burkhard Monien, and Thomas Sauerwald, *A new diffusion-based multilevel algorithm for computing graph partitions*, Journal of Parallel and Distributed Computing **69** (2009), no. 9, 750–761, Best Paper Awards and Panel Summary: IPDPS 2008.

[24] Henning Meyerhenke, Burkhard Monien, and Stefan Schamberger, *Graph partitioning and disturbed diffusion*, Parallel Computing **35** (2009), no. 10–11, 544–569.

[25] Vitaly Osipov and Peter Sanders, *n-level graph partitioning*, Algorithms—ESA 2010. Part I, Lecture Notes in Comput. Sci., vol. 6346, Springer, Berlin, 2010, pp. 278–289, DOI 10.1007/978-3-642-15775-2_24. MR2762861

[26] François Pellegrini, *A parallelisable multi-level banded diffusion scheme for computing balanced partitions with smooth boundaries*, Proc. 13th International Euro-Par Conference, LNCS, vol. 4641, Springer-Verlag, 2007, pp. 195–204.

[27] Peter Sanders and Christian Schulz, *Engineering multilevel graph partitioning algorithms*, Algorithms—ESA 2011, Lecture Notes in Comput. Sci., vol. 6942, Springer, Heidelberg, 2011, pp. 469–480, DOI 10.1007/978-3-642-23719-5_40. MR2893224 (2012k:68259)

[28] Peter Sanders and Christian Schulz, *Distributed evolutionary graph partitioning*, Meeting on Algorithm Engineering & Experiments (ALENEX'12), SIAM, 2012.

[29] Stefan Schamberger, *Shape optimized graph partitioning*, Ph.D. thesis, Universität Paderborn, 2006.

[30] K. Schloegel, G. Karypis, and V. Kumar, *Wavefront diffusion and LMSR: Algorithms for dynamic repartitioning of adaptive meshes*, IEEE Transactions on Parallel and Distributed Systems **12** (2001), no. 5, 451–466.

[31] ———, *Graph partitioning for high performance scientific simulations*, The Sourcebook of Parallel Computing, Morgan Kaufmann, 2003, pp. 491–541.

[32] Kirk Schloegel, George Karypis, and Vipin Kumar, *Multilevel diffusion schemes for repartitioning of adaptive meshes*, Journal of Parallel and Distributed Computing **47** (1997), no. 2, 109–124.

[33] ———, *A unified algorithm for load-balancing adaptive scientific simulations*, Proceedings of Supercomputing 2000, IEEE Computer Society, 2000, p. 59 (CD).

[34] ———, *Parallel static and dynamic multi-constraint graph partitioning*, Concurrency and Computation: Practice and Experience **14** (2002), no. 3, 219–240.

[35] Klaus Stüben, *An introduction to algebraic multigrid*, Multigrid (U. Trottenberg, C. W. Oosterlee, and A. Schüller, eds.), Academic Press, 2000, Appendix A, pp. 413–532.

[36] Aleksandar Trifunović and William J. Knottenbelt, *Parallel multilevel algorithms for hypergraph partitioning*, J. Parallel Distrib. Comput. **68** (2008), no. 5, 563–581.

[37] Denis Vanderstraeten, R. Keunings, and Charbel Farhat, *Beyond conventional mesh partitioning algorithms and the minimum edge cut criterion: Impact on realistic applications*, Proceedings of the Seventh SIAM Conference on Parallel Processing for Scientific Computing (PPSC'95), SIAM, 1995, pp. 611–614.

[38] C. Walshaw and M. Cross, *Mesh partitioning: a multilevel balancing and refinement algorithm*, SIAM J. Sci. Comput. **22** (2000), no. 1, 63–80 (electronic), DOI 10.1137/S1064827598337373. MR1769526 (2001b:65153)

[39] C. Walshaw and M. Cross, *Parallel optimisation algorithms for multilevel mesh partitioning*, Parallel Comput. **26** (2000), no. 12, 1635–1660, DOI 10.1016/S0167-8191(00)00046-6. MR1786940

[40] C. Xu and F. C. M. Lau, *Load balancing in parallel computers*, Kluwer, 1997.

Institute of Theoretical Informatics, Karlsruhe Institute of Technology (KIT), Am Fasanengarten 5, 76131 Karlsruhe, Germany

E-mail address: meyerhenke@kit.edu

Graph partitioning for scalable distributed graph computations

Aydın Buluç and Kamesh Madduri

ABSTRACT. Inter-node communication time constitutes a significant fraction of the execution time of graph algorithms on distributed-memory systems. Global computations on large-scale sparse graphs with skewed degree distributions are particularly challenging to optimize for, as prior work shows that it is difficult to obtain balanced partitions with low edge cuts for these graphs. In this work, we attempt to determine the optimal partitioning and distribution of such graphs, for load-balanced parallel execution of communication-intensive graph algorithms. We use breadth-first search (BFS) as a representative example, and derive upper bounds on the communication costs incurred with a two-dimensional partitioning of the graph. We present empirical results for communication costs with various graph partitioning strategies, and also obtain parallel BFS execution times for several large-scale DIMACS Challenge instances on a supercomputing platform. Our performance results indicate that for several graph instances, reducing work and communication imbalance among partitions is more important than minimizing the total edge cut.

1. Introduction

Graph partitioning is an essential preprocessing step for distributed graph computations. The cost of fine-grained remote memory references is extremely high in case of distributed memory systems, and so one usually restructures both the graph layout and the algorithm in order to mitigate or avoid inter-node communication. The goal of this work is to characterize the impact of common graph partitioning strategies that minimize edge cut, on the parallel performance of graph algorithms on current supercomputers. We use breadth-first search (BFS) as our driving example, as it is representative of communication-intensive graph computations. It is also frequently used as a subroutine for more sophisticated algorithms such as finding connected components, spanning forests, testing for bipartiteness, maximum flows **[10]**, and computing betweenness centrality on unweighted graphs **[1]**. BFS has recently been chosen as the first representative benchmark for ranking supercomputers based on their performance on data intensive applications **[5]**.

2010 *Mathematics Subject Classification.* Primary 05C70; Secondary 05C85, 68W10.

Key words and phrases. graph partitioning, hypergraph partitioning, inter-node communication modeling, breadth-first search, 2D decomposition.

This work was supported by the Director, Office of Science, U.S. Department of Energy under Contract No. DE-AC02-05CH11231.

Given a distinguished *source vertex* s, breadth-first search (BFS) systematically explores the graph G to discover every vertex that is reachable from s. Let V and E refer to the vertex and edge sets of G, whose cardinalities are $n = |V|$ and $m = |E|$. We assume that the graph is unweighted; equivalently, each edge $e \in E$ is assigned a weight of unity. A *path* of *length* l from vertex s to t is defined as a sequence of edges $\langle u_i, u_{i+1} \rangle$ (edge directivity assumed to be $u_i \rightarrow u_{i+1}$ in case of directed graphs), $0 \leq i < l$, where $u_0 = s$ and $u_l = t$. We use $d(s,t)$ to denote the *distance* between vertices s and t: the length of the *shortest path* connecting s and t. BFS implies that all vertices at a distance k (or *level* k) from vertex s should be first *visited* before vertices at distance $k+1$. The distance from s to each reachable vertex is typically the final output. In applications based on a breadth-first graph traversal, one might optionally perform auxiliary computations when visiting a vertex for the first time. Additionally, a *breadth-first spanning tree* rooted at s containing all the reachable vertices can also be maintained.

Level-synchronous BFS implementations process all the vertices that are k hops away from the root (at the kth level), before processing any further vertices. For each level expansion, the algorithm maintains a *frontier*, which is the set of active vertices on that level. The kth frontier is denoted by F_k, which may also include any duplicates and previously-discovered vertices. The *pruned frontier* is the unique set of vertices that are discovered for the first time during that level expansion.

In Section 2, we review parallel BFS on distributed memory systems. Sections 3 and 4 provide an analysis of the communication costs of parallel BFS, and relate them to the metrics used by graph and hypergraph partitioning. We detail the experimental setup for our simulations and real large-scale runs in Section 5. Section 6 presents a microbenchmarking study of the collective communication primitives used in BFS, providing evidence that the 2D algorithm has lower communication costs. This is partly due to its better use of interconnection network resources, independent of the volume of data transmitted. We present performance results in Section 7 and summarize our findings in Section 8.

2. Parallel Breadth-first Search

Data distribution plays a critical role in parallelizing BFS on distributed-memory machines. The approach of partitioning vertices to processors (along with their outgoing edges) is the so-called 1D partitioning, and is the method employed by Parallel Boost Graph Library [**6**]. A two-dimensional edge partitioning is implemented by Yoo et al. [**11**] for the IBM BlueGene/L, and by us [**2**] for different generations of Cray machines. Our 2D approach is different in the sense that it does a *checkerboard* partitioning (see below) of the sparse adjacency matrix of the underlying graph, hence assigning contiguous submatrices to processors. Both 2D approaches achieved higher scalability than their 1D counterparts. One reason is that key collective communication phases of the algorithm are limited to at most $\sqrt{p}$ processors, avoiding the expensive all-to-all communication among p processors. Yoo et al.'s work focused on low-diameter graphs with uniform degree distribution, and ours primarily studied graphs with skewed degree distribution. A thorough study of the communication volume in 1D and 2D partitioning for BFS, which involves decoupling the performance and scaling of collective communication operations from the number of words moved, has not been done for a large set of graphs. This paper attempts to fill that gap.

The 1D row-wise partitioning (left) and 2D checkerboard partitioning (right) of the sparse adjacency matrix of the graph are as follows:

$$A^{1D} = \begin{pmatrix} A_1 \\ \hline \ldots \\ \hline A_p \end{pmatrix}, \qquad A^{2D} = \left(\begin{array}{c|c|c} A_{1,1} & \ldots & A_{1,p_c} \\ \hline \vdots & \ddots & \vdots \\ \hline A_{p_r,1} & \ldots & A_{p_r,p_c} \end{array}\right) \tag{2.1}$$

The nonzeros in the ith row of the sparse adjacency matrix A represent the outgoing edges of the ith vertex of G, and the nonzeros in the jth column of A represent the incoming edges of the jth vertex.

In 2D partitioning, processors are logically organized as a mesh with dimensions $p_r \times p_c$, indexed by their row and column indices. Submatrix $A_{i,j}$ is assigned to processor $P(i,j)$. The indices of the submatrices need not be contiguous, and the submatrices themselves need not be square in general. In 1D partitioning, sets of vertices are directly assigned to processors, whereas in 2D, sets of vertices are collectively owned by all the processors in one dimension. Without loss of generality, we will consider that dimension to be the row dimension. These sets of vertices are labeled as $V_1, V_2, ..., V_{p_r}$, and their outgoing edges are labeled as $\text{Adj}^+(V_1), \text{Adj}^+(V_2), ..., \text{Adj}^+(V_{p_r})$. Each of these adjacencies is distributed to processors that are members of a row dimension: $\text{Adj}^+(V_1)$ is distributed to $P(1,:)$, $\text{Adj}^+(V_2)$ is distributed to $P(2,:)$, and so on. The colon notation is used to index a slice of processors, e.g. processors in the ith processor row are denoted by $P(i,:)$.

Level-synchronous BFS with 1D graph partitioning comprises three main steps:

- **Local discovery:** Inspect outgoing edges of vertices in current frontier.
- **Fold:** Exchange discovered vertices via an *All-to-all* communication phase, so that each processor gets the vertices that it owns after this step.
- **Local update:** Update distances/parents locally for newly-visited vertices.

The parallel BFS algorithm with 2D partitioning has four steps:

- **Expand:** Construct the current frontier of vertices on each processor by a collective *All-gather* step along the processor column $P(:,j)$ for $1 \leq j \leq p_c$.
- **Local discovery:** Inspect outgoing edges of vertices in the current frontier.
- **Fold:** Exchange newly-discovered vertices via an collective *All-to-all* step along the processor row $P(i,:)$, for $1 \leq i \leq p_r$.
- **Local update:** Update distances/parents locally for newly-visited vertices.

In contrast to the 1D case, communication in the 2D algorithm happens only along one processor dimension. If *Expand* happens along one processor dimension, then *Fold* happens along the other processor dimension. Detailed pseudo-code for the 1D and 2D algorithms can be found in our earlier paper [**2**]. Detailed micro benchmarking results in Section 6 show that the 2D algorithm has a lower communication cost than the 1D approach due to the decreased number of communicating processors in collectives. The performance of both algorithms is heavily dependent on the performance and scaling of MPI collective `MPI_Alltoallv`. The 2D algorithm also depends on the `MPI_AllGatherv` collective.

3. Analysis of Communication Costs

In prior work [**2**], we study the performance of parallel BFS on synthetic Kronecker graphs used in the Graph 500 benchmark. We observe that the communication volume is $O(m)$ with a random ordering of vertices, and a random partitioning of the graph (i.e., assigning m/p edges to each processor). The edge cut is also $O(m)$ with random partitioning. While it can be shown that low-diameter real-world graphs do not have sparse separators [**8**], constants matter in practice, and any decrease in the communication volume, albeit not asymptotically, may translate into reduced execution times for graph problems that are typically communication-bound.

We outline the communication costs incurred in 2D-partitioned BFS in this section. 2D-partitioned BFS also captures 1D-partitioned BFS as a degenerate case. We first distinguish different ways of aggregating edges in the local discovery phase of the BFS approach:

(1) No aggregation at all, local duplicates are not pruned before fold.
(2) Local aggregation at the current frontier only. Our simulations in Section 7.1 follow this assumption.
(3) Local aggregation over all (current and past) locally discovered vertices by keeping a persistent bitmask. We implement this optimization for gathering parallel execution results in Section 7.2.

Consider the expand phase. If the adjacencies of a single vertex v are shared among $\lambda^+ \leq p_c$ processors, then its owner will need to send the vertex to $\lambda^+ - 1$ neighbors. Since each vertex is in the pruned frontier once, the total communication volume for the expand phases over all iterations is equal to the communication volume of the same phase in 2D sparse-matrix vector multiplication (SpMV) [**4**]. Each iteration of BFS is a sparse-matrix sparse-vector multiplication of the form $A^\mathsf{T} \times F_k$. Hence, the column-net hypergraph model of A^T accurately captures the cumulative communication volume of the BFS expand steps, when used with the *connectivity*-1 metric.

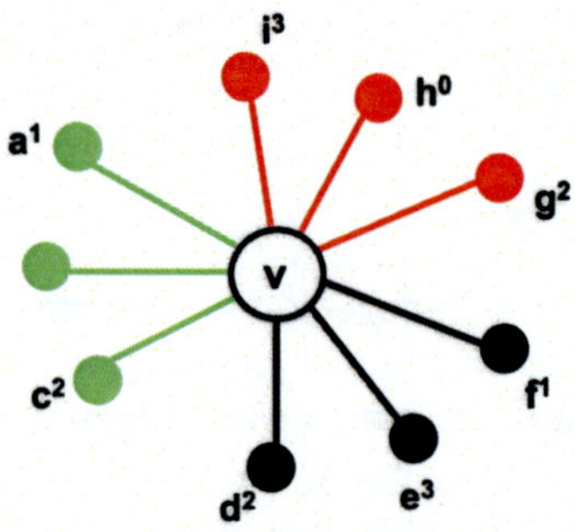

FIGURE 1. Example illustrating communication in fold phase of BFS: Partitioning of $\mathrm{Adj}^-(v)$.

Characterizing communication for the fold phase is more complicated. Consider a vertex v of in-degree 9, shown in Figure 1. In terms of the sparse matrix representation of the graph discussed above, this corresponds to a column with 9 nonzeros. We label the adjacencies $\mathrm{Adj}^-(v)$ with a superscript denoting the earliest BFS iteration in which they are discovered. Vertex h in the figure belongs to F_0, vertices a and f to F_1, and so on. Furthermore, assume that the adjacencies of v span three processors, with the color of the edges indicating the partitions they belong to. We denote non-local vertices with $\mathrm{RemoteAdj}^-(v)$. Since v belongs to the black partition, $\mathrm{RemoteAdj}^-(v)$ is $\mathrm{Adj}^-(v) \setminus \{d, e, f\}$ in this case.

The communication cost of the fold phase is complicated to analyze due to the *space-time partitioning* of edges in the graph in a BFS execution. We can annotate every edge in the graph using two integers: the partition the edge belongs to, and the BFS phase in which the edge is traversed (remember each edge is traversed exactly once).

The communication volume due to a vertex v in the fold phase is at most $|\text{RemoteAdj}^-(v)|$, which is realized when every $e \in \text{RemoteAdj}^-(v)$ has a distinct *space-time partitioning* label, i.e. no two edges are traversed by the same remote process during the same iteration. The *edgecut* of the partitioned graph is the set of all edges for which the end vertices belong to different partitions. The size of the *edgecut* is equal to $\sum_{v \in V} |\text{RemoteAdj}^-(v)|$, giving an upper bound for the overall communication volume due to fold phases.

Another upper bound is $O(\textit{diameter} \cdot (\lambda^- - 1))$, which might be lower than the *edgecut*. Here, $\lambda^- \leq p_r$ is the number of processors among which $\text{Adj}^-(v)$ is partitioned, and *diameter* gives the maximum number of BFS iterations. Consequently, the communication volume due to discovering vertex v, *comm*(v), obeys the following inequality: $\textit{comm}(v) \leq \min\left(\textit{diameter} \cdot (\lambda^- - 1), |\text{RemoteAdj}^-(v)|\right)$. In the above example, this value is $\min(8, 6) = 6$.

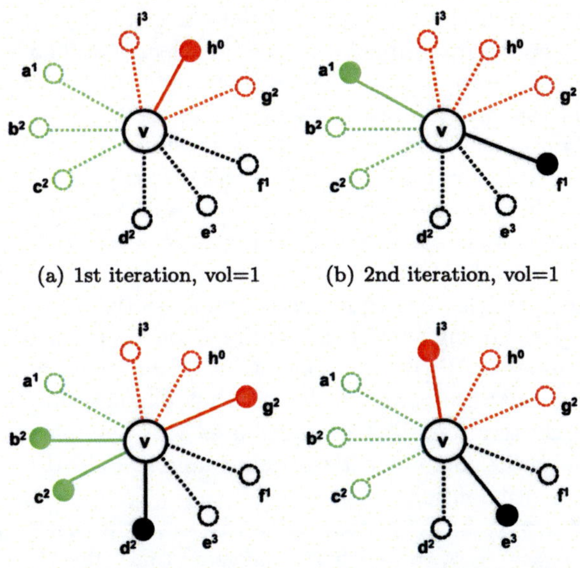

(a) 1st iteration, vol=1 (b) 2nd iteration, vol=1

(c) 3rd iteration, vol=2 (d) 4th iteration, vol=1

FIGURE 2. Partitioning of $\text{Adj}^-(v)$ per BFS iteration.

Figure 2 shows the space-time edge partitioning of $\text{Adj}^-(v)$ per BFS step. In the first step, the communication volume is 1, as the red processor discovers v through the edge (h, v) and sends it to the black processor for marking. In the second step, both green and black processors discover v and communication volume is 1 from green to black. Continuing this way, we see that the actual aggregate communication in the fold phase of v is 5.

The row-net hypergraph model of A^{T} is an optimistic lower-bound on the overall communication volume of the fold phases using the *connectivity* -1 metric. On the other hand, modeling the fold phase with the *edgecut* metric would be a pessimistic upper bound (in our example, the graph model would estimate communication due to v to be 6). It is currently unknown which bound is tighter in practice for different classes of graphs. If we implement global aggregation (global replication of discovered vertices), the total communication volume in the fold phase will decrease all the way down to the SpMV case of $(\lambda^{-}-1)$. However, this involves an additional communication step similar to the expand phase, in which processors in the column dimension exchange newly-visited vertices.

4. Graph and Hypergraph Partitioning Metrics

We consider several different orderings of vertices and edges and determine the incurred communication costs. Our baseline approach is to take the given ordering of vertices and edges as-is (i.e., the *natural* ordering), and to partition the graph into 1D or 2D (checkerboard) slices as shown in Equation 2.1. The second scenario is to randomly permute vertex identifiers, and then partition the graph via the baseline approach. These two scenarios do not explicitly optimize for an objective function. We assume the load-balanced *2D vector distribution* [**2**], which matches the 2D matrix distribution for natural and random orderings. Each processor row (except the last one) is responsible for $t = \lfloor n/p_r \rfloor$ elements. The last processor row gets the remaining $n - \lfloor n/p_r \rfloor (p_r - 1)$ elements. Within the processor row, each processor (except the last) is responsible for $\lfloor t/p_c \rfloor$ elements.

We use the graph partitioner METIS [**7**] to generate a 1D row-wise partitioning with balanced vertices per partition and simultaneously minimizing the number of cut edges. Lastly, we experiment with hypergraph partitioning, which exactly captures total communication costs of sparse matrix-dense vector multiplication in its objective function [**4**]. We use PaToH [**3**] and report results with its row-wise and checkerboard partitioning algorithms. Our objective is to study how graph and hypergraph partitioning affect computational load balance and communication costs. In both use cases of PaToH, we generate a symmetric permutation as output, since input and output vectors have to be distributed in the same way to avoid data shuffling after each iteration. PaToH distributes both the matrix and the vectors in order to optimize the communication volume, and so PaToH runs might have an unbalanced vector distribution.

We define $V(d,p)$ to be the number of words sent by processor p in the dth BFS communication phase, on a run with P processors that takes D level-synchronous iterations to finish. We compute the following machine-independent counts that give the incurred communication.

(1) Total communication over the course of BFS execution:

$$\mathit{TotalVolume} = \sum_{d=1}^{D} \sum_{p=1}^{P} V(d,p).$$

(2) Sum of maximum communication volumes for each BFS step:

$$\mathit{MaxVolume} = \sum_{d=1}^{D} \max_{p \in \{1 \ldots P\}} V_{expand}(d,p) + \sum_{d=1}^{D} \max_{p \in \{1 \ldots P\}} V_{fold}(d,p).$$

Although we report the total communication volume over the course of BFS iterations, we are most concerned with the *MaxVolume* metric. It is a better approximation for the time spent on remote communication, since the slowest processor in each phase determines the overall time spent in communication.

5. Experimental Setup

Our parallel BFS implementation is level-synchronous, and so it is primarily meant to be applied to low-diameter graphs. However, to quantify the impact of barrier synchronization and load balance on the overall execution time, we run our implementations on several graphs, both low- and high-diameter.

We categorize the following DIMACS Challenge instances as low diameter: the synthetic Kronecker graphs (**`kron_g500-simple-logn`** and **`kron_g500-logn`** families), Erdős-Rényi graphs (`er-fact1.5` family), web crawls (**`eu-2005`** and others), citation networks (`citationCiteseer` and others), and co-authorship networks (**`coAuthorsDBLP`** and others). Some of the high-diameter graphs that we report performance results on include **`hugebubbles-00020`**, graphs from the **`delaunay`** family, road networks (**`road_central`**), and random geometric graphs.

Most of the DIMACS test graphs are small enough to fit in the main memory of a single machine, and so we are able to get baseline serial performance numbers for comparison. We are currently using serial partitioning software to generate vertex partitions and vertex reorderings, and this has been a limitation for scaling to larger graphs. However, the performance trends with DIMACS graphs still provide some interesting insights.

We use the k-way multilevel partitioning scheme in METIS (v5.0.2) with the default command-line parameters to generate balanced vertex partitions (in terms of the number of vertices per partition) minimizing total edge cut. We relabel vertices and distribute edges to multiple processes based on these vertex partitions. Similarly, we use PaToH's column-wise and checkerboard partitioning schemes to partition the sparse adjacency matrix corresponding to the graph. While we report communication volume statistics related to checkerboard partitioning, we are still unable to use these partitions for reordering, since PaToH edge partitions are not necessarily aligned.

We report parallel execution times on Hopper, a 6392-node Cray XE6 system located at Lawrence Berkeley National Laboratory. Each node of this system contains two twelve-core 2.1 GHz AMD Opteron Magny-Cours processors. There are eight DDR3 1333-MHz memory channels per node, and the observed memory bandwidth with the STREAM [**9**] benchmark is 49.4 GB/s. The main memory capacity of each node is 32 GB, of which 30 GB is usable by applications. A pair of compute nodes share a Gemini network chip, and these network chips are connected to form a 3D torus (of dimensions $17 \times 8 \times 24$). The observed MPI point-to-point bandwidth for large messages between two nodes that do not share a network chip is 5.9 GB/s. Further, the measured MPI latency for point-to-point communication is 1.4 microseconds, and the cost of a global barrier is about 8 microseconds. The maximum injection bandwidth per node is 20 GB/s.

We use the GNU C compiler (v4.6.1) for compiling our BFS implementation. For inter-node communication, we use Cray's MPI implementation (v5.3.3), which is based on MPICH2. We report performance results up to 256-way MPI process/task concurrency in this study. In all experiments, we use four MPI tasks per node, with every task constrained to six cores to avoid any imbalances due to Non-Uniform Memory Access (NUMA) effects. We did not explore multithreading within a node in the current study. This may be another potential source of load imbalance, and we will quantify this in future work. More details on multithreading within a node can be found in our prior work on parallel BFS [**2**].

To compare performance across multiple systems using a rate analogous to the commonly-used floating point operations per second, we normalize the serial and parallel execution times by the number of edges visited in a BFS traversal and present a *Traversed Edges Per Second* (TEPS) rate. For an undirected graph with a single connected component, the BFS algorithm would visit every edge in the component twice. We only consider traversal execution times from vertices that appear in the largest connected component in the graph (all the DIMACS test instances we used have one large component), compute the mean search time (harmonic mean of TEPS) using at least 20 randomly-chosen sources vertices for each benchmark graph, and normalize the time by the cumulative number of edges visited to get the TEPS rate.

6. Microbenchmarking Collectives Performance

In our previous paper [**2**], we argue that the 2D algorithm has a lower communication cost because the inverse bandwidth is positively correlated with the communicator size in collective operations. In this section, we present a detailed microbenchmarking study that provides evidence to support our claim. A subcommunicator is a sub partition of the entire processor space. We consider the 2D partitioning scenario here. The 1D case can be realized by setting the column processor dimension to one. We have the freedom to perform either one of the communication phases (Allgatherv and Alltoallv) in contiguous ranks, where processes in the same subcommunicator map to sockets that are physically close to each other. The default mapping is to pack processes along the rows of the processor grid, as shown in Figure 3 (we refer to this ordering as *contiguous ranks*). The alternative method is to reorder ranks so that they are packed along the columns of the processor grid (referred to as *spread-out ranks*). The alternative remapping decreases the number of nodes spanned by each column subcommunicator. This increases contention, but can potentially increase available bandwidth.

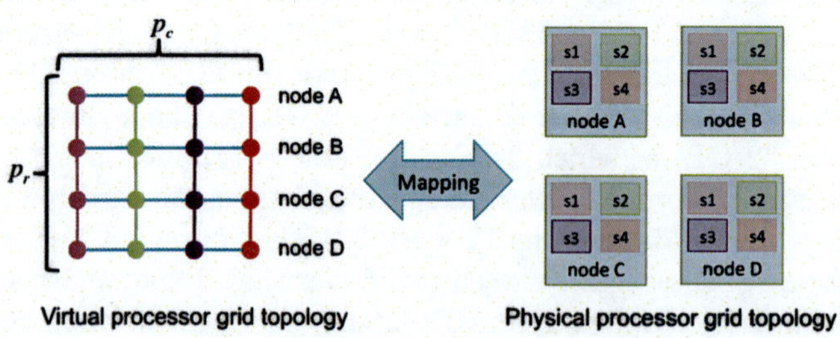

FIGURE 3. Mapping of column subcommunicators from a 4×4 virtual process grid to a physical network connecting 4 nodes, each having 4 sockets. Each column subcommunicator (shown with a different color) spans multiple physical nodes. One MPI process maps to one socket.

We consider both the cases of spread-out and contiguous ranks on Hopper, and microbenchmark Allgatherv and Alltoallv operations by varying processor grid configurations. We benchmark each collective at 400, 1600, and 6400 process counts. For each process count, we use a square $\sqrt{p} \times \sqrt{p}$ grid, a tall skinny $(2\sqrt{p}) \times (\sqrt{p}/2)$ grid, and a short fat $(\sqrt{p}/2) \times (2\sqrt{p})$ grid, making a total of nine different process configurations for each of the four cases: Allgatherv spread-out, Alltoallv spread-out, Allgatherv packed, Alltoallv packed. We perform linear regression on mean inverse bandwidth (measured as microseconds/MegaBytes) achieved among all subcommunicators when all subcommunicators work simultaneously. This mimics the actual BFS scenario. We report the mean as opposed to minimum, because the algorithm does not require explicit synchronization across subcommunicators.

In each run, we determine constants a, b, c that minimize the sum of squared errors ($SS_{res} = \sum(y_{obsd} - y_{est})^2$) between the observed inverse bandwidth and the inverse bandwidth estimated via the equation $\beta(p_r, p_c) = a\,p_r + b\,p_c + c$. The results are summarized in Table 1. If the observed t-value of any of the constants are below the critical t-value, we force its value to zero and rerun linear regression. We have considered other relationships that are linear in the coefficients, such as power series and logarithmic dependencies, but the observed t-values were significantly below the critical t-value for those hypotheses, hence not supporting them. We also list r^2, coefficient of determination, which shows the ratio (between 0.0 and 1.0) of total variation in β that can be explained by its linear dependence on p_r and p_c. Although one can get higher r^2 scores by using higher-order functions, we opt for linear regression in accordance to Occam's razor, because it adequately explains the underlying data in this case.

TABLE 1. Regression coefficients for $\beta(p_r, p_c) = a\,p_r + b\,p_c + c$. Alltoallv (a2a) happens along the rows and Allgatherv (ag) along the columns. Shaded columns show the runs with spread-out ranks. Dash ('–') denotes uncorrelated cases.

Regression coefficients	Pack along rows		Pack along columns	
	β_{ag}	β_{a2a}	β_{ag}	β_{a2a}
a	0.0700	0.0246	–	0.0428
b	0.0148	–	–	0.0475
c	2.1957	1.3644	2.3822	4.4861
SS_{res}	1.40	0.46	0.32	7.66
r^2	0.984	0.953	0.633	0.895

We see that both the subcommunicator size (the number of processes in each subcommunicator) and the total number of subcommunicators affect the performance in a statistically significant way. The linear regression analysis shows that the number of subcommunicators have a stronger effect on the performance than the subcommunicator size for the Allgatherv operation on spread-out ranks (0.0700 vs 0.0148). For Alltoallv operation on spread-out ranks, however, their effects are comparable (0.0428 vs 0.0475). Increasing the number of subcommunicators increases both the contention and the physical distance between participating processes. Subcommunicator size does not change the distance between each participant in a communicator and the contention, but it can potentially increase the available bandwidth by using a larger portion of the network. We argue that it is that extra available bandwidth that makes subcommunicator size important for the Alltoallv case, because it is more bandwidth-hungry than Allgatherv.

For Alltoallv runs with contiguous ranks, we find that the total number of subcommunicators does not affect the inverse bandwidth in a statistically significant way. We truncate the already-low coefficient to zero since its observed t-values are significantly below the critical t-value. The subcommunicator size is positively correlated with the inverse bandwidth. This supports our original argument that larger subcommunicators degrade performance due to sub-linear network bandwidth scaling. For Allgatherv runs with contiguous ranks, however, we see that neither parameter affects the performance in a statistically significant way.

We conclude that the number of processors inversely affect the achievable bandwidth on the Alltoallv collective used by both the 1D and 2D algorithms. Hence, the 2D algorithm uses available bandwidth more effectively by limiting the number of processors in each subcommunicator.

7. Performance Analysis and Results

7.1. Empirical modeling of communication. We first report machine-independent measures for communication costs. For this purpose, we simulate parallel BFS using a MATLAB script whose inner kernel, a single BFS step local to a processor, is written in `C++` using `mex` for speed. For each partition, the simulator does multiple BFS runs (in order) starting from different random vertices to report an accurate average, since BFS communication costs, especially the *MaxVolume* metric, depend on the starting vertex. When reporting the ratio of *TotalVolume* to the total number of edges in Table 2, the denominator counts each edge twice (since an adjacency is stored twice).

TABLE 2. Percentage of *TotalVolume* for 1D row-wise partitioning to the total number of edges (lower is better). N denotes the natural ordering, R denotes the ordering with randomly-permuted vertex identifiers, and P denotes reordering using PaToH.

Graph	$p = 4 \times 1$			$p = 16 \times 1$			$p = 64 \times 1$		
	N	R	P	N	R	P	N	R	P
coPapersCiteseer	4.7%	14.7%	1.9%	8.7%	47.9%	3.4%	**10.8%**	**102.5%**	4.8%
coAuthorsCiteseer	37.6%	79.9%	5.9%	59.3%	143.9%	11.3%	68.7%	180.3%	15.6%
citationCiteseer	64.8%	75.0%	7.8%	125.0%	139.0%	16.9%	164.9%	176.1%	29.0%
coPapersDBLP	7.6%	18.4%	3.7%	15.7%	58.2%	7.6%	21.0%	118.8%	11.7%
coAuthorsDBLP	45.2%	81.3%	10.9%	74.9%	148.9%	19.8%	90.1%	182.5%	27.2%
eu-2005	5.3%	23.2%	0.3%	8.7%	63.8%	1.9%	12.3%	107.4%	7.2%
kronecker-logn18	7.7%	7.6%	6.3%	22.7%	23.1%	19.5%	47.5%	53.4%	45.0%
delaunay_n20	52.4%	123.7%	0.2%	59.3%	178.0%	0.6%	60.6%	**194.4%**	**1.4%**
rgg_n_2_20_s0	0.2%	85.5%	0.1%	**0.6%**	**160.1%**	0.3%	2.5%	188.9%	0.6%

The reported communication volume for the expand phase is exact, in the sense that a processor receives a vertex v only if it owns one of the edges in $\mathrm{Adj}^+(v)$ and it is not the owner of v itself. We count a vertex as one word of communication. In contrast, in the fold phase, the discovered vertices are sent in $\langle parent, vertex\_id \rangle$ pairs, resulting in two words of communication per discovered edge. This is why values in Table 2 sometimes exceed 100% (i.e. more total communication than the number of edges), but are always less than 200%. For these simulations, we report numbers for both 1D row-wise and 2D checkerboard partitioning when partitioning with the natural ordering, partitioning after random vertex relabeling, and partitioning using PaToH. The performance trends obtained with 1D partitions generated using METIS (discussed in Section 7.2) are similar to the ones obtained with PaToH partitions, and we do not report the METIS simulation counts in current work.

For 1D row-wise partitioning, random relabeling increases the total communication volume (i.e., the edge cut), by a factor of up to 10× for low-diameter graphs (realized in `coPaperCiteseer` with 64 processors) and up to 250× for high-diameter graphs (realized in `rgg_n_2_20_s0` with 16 processors), compared to the natural ordering. Random relabeling never decreases the communication volume. PaToH can sometimes drastically reduce the total communication volume, as observed for the graph `delaunay_n20` (15× reduction compared to natural ordering and 45× reduction compared to random relabeling for 64 processors) in Table 2. However, it is of little use with synthetic Kronecker graphs.

TABLE 3. Ratio of *TotalVolume* with 2D checkerboard partitioning to the *TotalVolume* with 1D row-wise partitioning (less than 1 means 2D improves over 1D).

Graph	$p = 2 \times 2$			$p = 4 \times 4$			$p = 8 \times 8$		
	N	R	P	N	R	P	N	R	P
coPapersCiteseer	1.32	0.67	0.64	1.25	0.46	0.74	1.35	0.39	0.81
coAuthorsCiteseer	1.45	0.91	0.66	1.47	0.88	0.76	1.60	0.97	0.85
citationCiteseer	0.91	0.28	0.63	0.88	0.84	0.70	0.91	0.93	0.71
coPapersDBLP	1.13	0.68	0.64	1.01	0.48	0.66	1.07	0.42	0.72
coAuthorsDBLP	1.35	0.92	0.69	1.31	0.91	0.76	1.40	1.00	0.85
eu-2005	1.89	0.73	1.29	1.90	0.56	0.60	1.63	0.57	0.48
kronecker-logn18	0.71	0.73	0.52	0.51	0.51	0.42	0.43	0.39	0.34
delaunay_n20	1.79	0.95	0.60	2.16	1.09	0.59	2.45	1.24	0.60
rgg_n_2_20_s0	**135.54**	0.75	0.61	**60.23**	0.80	0.64	**18.35**	0.99	0.66

Table 3 shows that 2D checkerboard partitioning generally decreases total communication volume for random and PaToH orderings. However, when applied to the default natural ordering, 2D in general increases the communication volume.

TABLE 4. Ratio of $P \cdot \mathit{MaxVolume}$ to *TotalVolume* for 1D row-wise partitioning (lower is better).

Graph	$p = 4 \times 1$			$p = 16 \times 1$			$p = 64 \times 1$		
	N	R	P	N	R	P	N	R	P
coPapersCiteseer	1.46	1.01	1.23	1.81	1.02	1.76	2.36	1.07	2.44
coAuthorsCiteseer	1.77	1.02	1.55	2.41	1.06	2.06	2.99	1.21	2.86
citationCiteseer	1.16	1.02	1.39	1.33	1.07	2.17	1.53	1.21	2.93
coPapersDBLP	1.56	1.01	1.22	1.99	1.02	1.86	2.40	1.05	2.41
coAuthorsDBLP	1.84	1.01	1.39	2.58	1.05	1.85	3.27	1.13	2.43
eu-2005	1.37	1.10	1.05	3.22	1.28	3.77	**7.35**	1.73	**9.36**
kronecker-logn18	1.04	1.06	1.56	1.22	1.16	1.57	1.63	1.42	1.92
delaunay_n20	2.36	1.03	1.71	3.72	1.13	3.90	**6.72**	1.36	**8.42**
rgg_n_2_20_s0	2.03	1.03	2.11	4.70	1.13	6.00	**9.51**	1.49	**13.34**

The $(P \cdot \mathit{MaxVolume})/\mathit{TotalVolume}$ metric shown in Tables 4 and 5 show the expected slowdown due to load imbalance in per-processor communication. This is an understudied metric that is not directly optimized by partitioning tools. Random

TABLE 5. Ratio of $P \cdot MaxVolume$ to $TotalVolume$ for 2D checkerboard partitioning (lower is better).

Graph	$p = 2 \times 2$			$p = 4 \times 4$			$p = 8 \times 8$		
	N	R	P	N	R	P	N	R	P
coPapersCiteseer	1.38	1.29	1.20	2.66	1.07	1.59	4.82	1.04	2.12
coAuthorsCiteseer	1.46	1.29	1.56	2.57	1.08	1.95	4.76	1.08	2.52
citationCiteseer	1.29	1.29	1.40	1.35	1.08	2.08	1.71	1.07	2.63
coPapersDBLP	1.40	1.29	1.28	2.35	1.07	1.81	4.00	1.03	1.96
coAuthorsDBLP	1.51	1.28	1.28	2.51	1.08	1.81	4.57	1.12	1.97
eu-2005	1.70	1.32	1.78	3.38	1.15	3.25	8.55	1.19	8.58
kronecker-logn18	1.31	1.31	2.08	1.14	1.12	1.90	1.12	1.09	1.93
delaunay_n20	1.40	1.30	1.77	3.22	1.12	4.64	8.80	1.18	11.15
rgg_n_2_20_s0	3.44	1.31	2.38	8.25	1.13	6.83	53.73	1.18	17.07

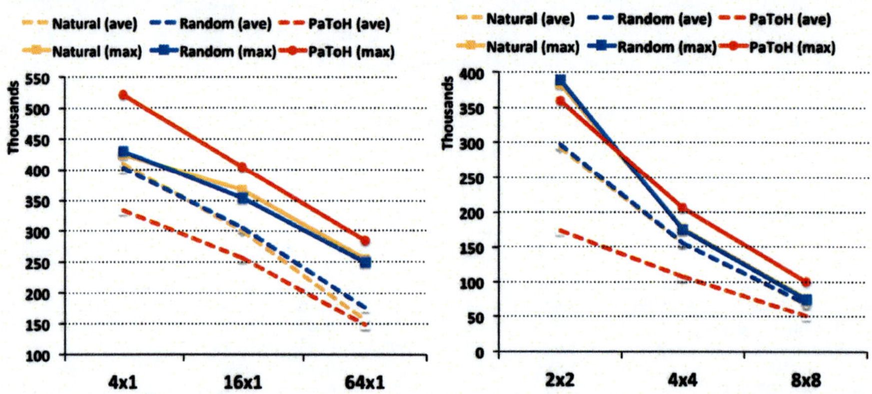

(a) 1D row-wise model (b) 2D checkerboard model

FIGURE 4. Maximum and average communication volume scaling for various partitioning strategies. y-axis is in thousands of words received.

relabeling of the vertices results in partitions that are load-balanced per iteration. The maximum occurs for the eu-2005 matrix on 64 processors with 1D partitioning, but even in this case, the maximum (1.73×) is less than twice the average. In contrast, both natural and PaToH orderings suffer from imbalances, especially for higher processor counts.

To highlight the problems with minimizing total (hence average) communication as opposed to the maximum, we plot communication volume scaling in Figure 4 for the Kronecker graph we study. The plots show that even though PaToH achieves the lowest average communication volume per processor, its maximum communication volume per processor is even higher than the random case. This partly explains the computation times reported in Section 7.2, since the maximum communication per processor is a better approximation for the overall execution time.

Edge count imbalances for different partitioning strategies can be found in the Appendix. Although they are typically low, they only represent the load imbalance due to the number of edges owned by each processor, and not the number of edges traversed per iteration.

7.2. Impact of Partitioning on parallel execution time. Next, we study parallel performance on Hopper for some of the DIMACS graphs. To understand the relative contribution of intra-node computation and inter-node communication to the overall execution time, consider the Hopper microbenchmark data illustrated in Figure 5. The figure plots the aggregate bandwidth (in GB/s) with multi-node parallel execution (and four MPI processes per node) and a fixed data/message size. The collective communication performance rates are given by the total number of bytes received divided by the total execution time. We also generate a *random memory references* throughput rate (to be representative of the local computational steps discussed in Section 2), and this assumes that we use only four bytes of every cache line fetched. This rate scales linearly with the number of sockets. Assigning appropriate weights to these throughput rates (based on the the communication costs reported in the previous section) would give us a lower bound on execution time, as this assumes perfect load balance.

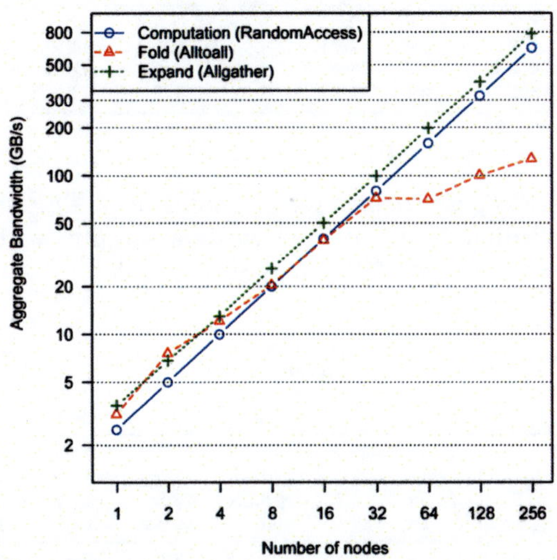

FIGURE 5. Strong-scaling performance of collective communication with large messages and intra-node random memory accesses on Hopper.

We report parallel execution time on Hopper for two different parallel concurrencies, $p = 16$ and $p = 256$. Tables 6 and 7 give the serial performance rates (with natural ordering) as well as the relative speedup with different reorderings, for several benchmark graphs. There is a 3.5× variation in serial performance rates, with the skewed-degree graphs showing the highest performance and the high diameter graphs `road_central` and `hugebubbles-00020` the lowest performance. For the parallel runs, we report speedup over the serial code with the natural ordering. Interestingly, the random-ordering variants perform best in all of the low-diameter graph cases. The performance is better than PaToH- and METIS-partitioned variants in all cases. The table also gives the impact of checkerboard partitioning on the running time. There is a moderate improvement for the random variant, but the checkerboard scheme is slower for the rest of the schemes. The variation in relative speedup across graphs is also surprising. The synthetic low-diameter graphs demonstrate the best speedup overall. However, the speedups for the real-world low-diameter graphs are 1.5× lower, and the relative speedups for the high-diameter graphs are extremely low.

TABLE 6. BFS performance (in millions of TEPS) for single-process execution, and observed relative speedup with 16 MPI processes (4 nodes, 4 MPI processes per node). The fastest variants are highlighted in each case. *M* denotes METIS partitions.

Graph	Perf Rate $p = 1 \times 1$	Relative Speedup $p = 16 \times 1$				Rel. Speedup over 1D $p = 4 \times 4$			
	N	N	R	M	P	N	R	M	P
coPapersCiteseer	24.9	5.6×	9.7×	8.0×	6.9×	0.4×	**1.0×**	0.4×	0.5×
eu-2005	23.5	6.1×	7.9×	5.0×	4.3×	0.5×	**1.1×**	0.5×	0.6×
kronecker-logn18	24.5	12.6×	12.6×	1.8×	4.4×	1.1×	**1.1×**	1.4×	0.8×
er-fact1.5-scale20	14.1	11.2×	11.2×	11.5×	10.0×	1.1×	**1.2×**	0.8×	1.1×
road_central	7.2	3.5×	2.2×	3.5×	3.6×	0.6×	0.9×	0.5×	0.5×
hugebubbles-00020	7.1	3.8×	2.7×	3.9×	2.1×	0.7×	0.9×	0.6×	0.6×
rgg_n_2_20_s0	14.1	2.5×	3.4×	2.6×	2.6×	0.6×	1.2×	0.6×	0.7×
delaunay_n18	15.0	1.9×	1.6×	1.9×	1.3×	0.9×	1.4×	0.7×	1.4×

TABLE 7. BFS performance rate (in millions of TEPS) for single-process execution, and observed relative speedup with 256 MPI processes (64 nodes, 4 MPI processes per node).

Graph	Perf Rate $p = 1 \times 1$	Relative Speedup $p = 256 \times 1$				Rel. Speedup over 1D $p = 16 \times 16$			
	N	N	R	M	P	N	R	M	P
coPapersCiteseer	24.9	10.8×	22.4×	12.9×	18.1×	0.5×	**2.5×**	0.7×	0.5×
eu-2005	23.5	12.9×	21.7×	8.8×	17.2×	0.6×	**2.7×**	0.6×	0.3×
kronecker-logn18	24.5	42.3×	41.9×	16.3×	23.9×	2.6×	**2.6×**	0.3×	1.1×
er-fact1.5-scale20	14.1	57.1×	58.0×	50.1×	50.4×	1.6×	**1.6×**	1.1×	1.2×
road_central	7.2	1.2×	0.9×	1.3×	1.7×	1.9×	2.1×	1.1×	0.9×
hugebubbles-00020	7.1	1.6×	1.5×	1.6×	2.0×	1.5×	2.2×	2.0×	0.8×
rgg_n_2_20_s0	14.1	1.5×	1.3×	1.6×	2.1×	1.2×	1.2×	1.3×	1.1×
delaunay_n18	15.0	0.6×	0.4×	0.5×	0.8×	1.8×	1.9×	2.1×	1.6×

Figure 6 gives a breakdown of the average parallel BFS execution and inter-node communication times for 16-processor parallel runs, and provides insight into the reason behind varying relative speedup numbers. For all the low-diameter graphs, at this parallel concurrency, execution time is dominated by local computation. The local discovery and local update steps account for up to 95% of the total time, and communication times are negligible. Comparing the computational time of random ordering vs. METIS reordering, we find that BFS on the METIS-reordered graph is significantly slower. The first reason is that METIS partitions are highly unbalanced in terms of the number of edges per partition for this graph, and so we can expect a certain amount of imbalance in local computation. The second reason is a bit more subtle. Partitioning the graph to minimize edge cut does not guarantee that the local computation steps will be balanced, even if the number of edges per process are balanced. The per-iteration work is dependent on the number of vertices in the current frontier and their distribution among processes. Randomizing

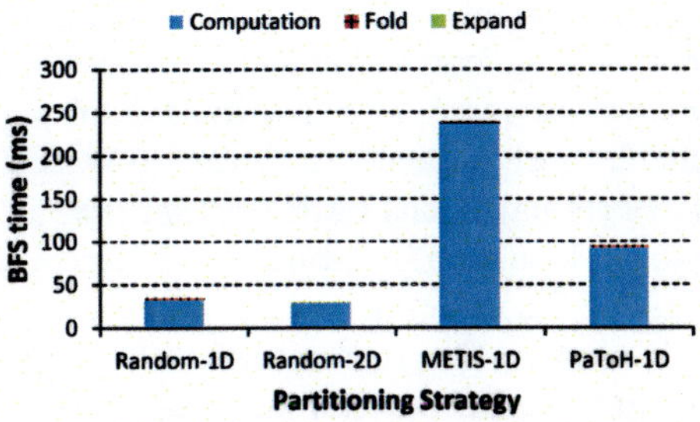

(a) kronecker-logn18 (total)

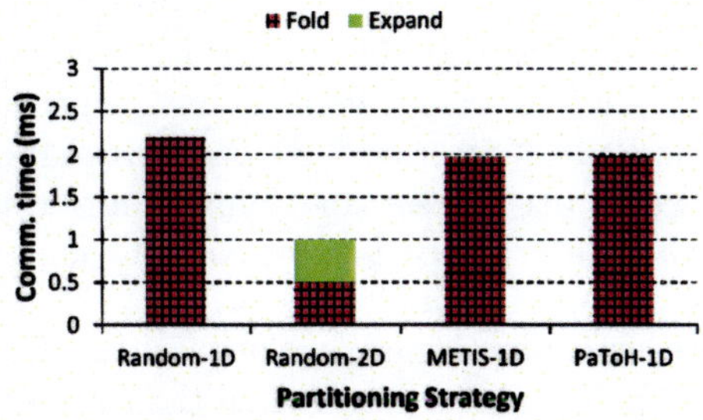

(b) kronecker-logn18 (comm)

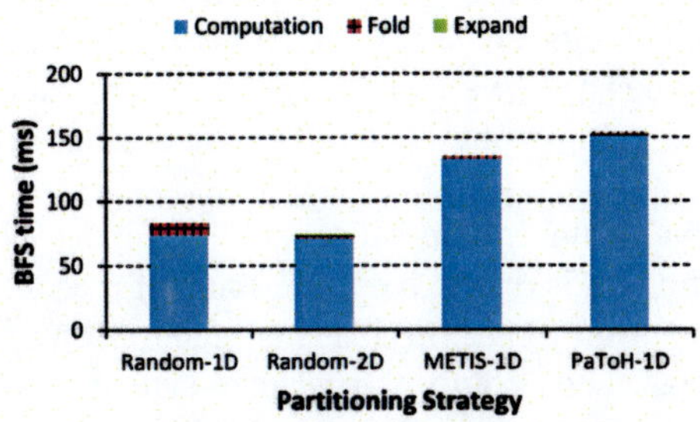

(c) eu-2005 (total)

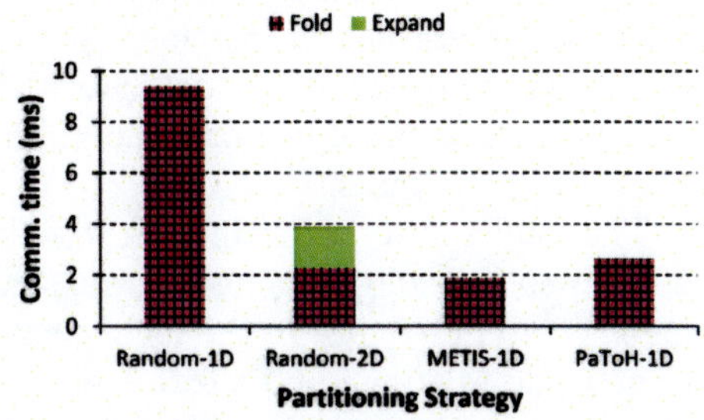

(d) eu-2005 (comm)

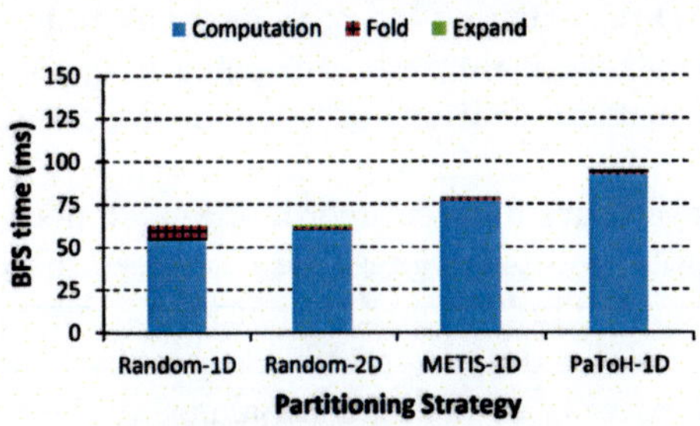

(e) coPapersCiteseer (total)

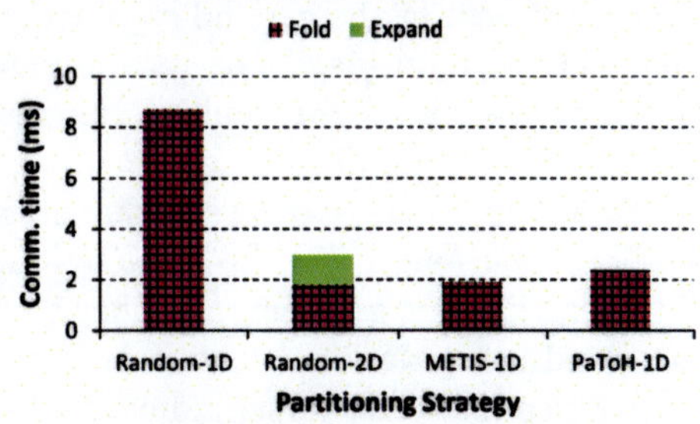

(f) coPapersCiteseer (comm)

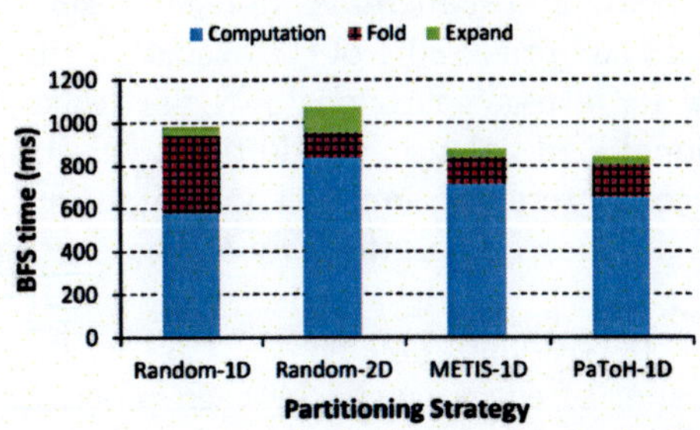

(g) road_central (total)

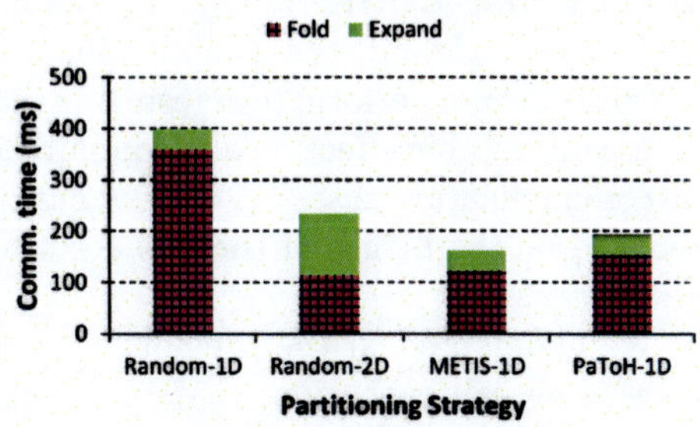

(h) road_central (comm)

FIGURE 6. Average BFS execution time for various test graphs with 16 MPI processes (4 nodes, 4 MPI processes per node).

vertex identifiers destroys any inherent locality, but also improves local computation load balance. The partitioning tools reduce edge cut and enhance locality, but also seem to worsen load balance, especially for skewed degree distribution graphs. The PaToH-generated 1D partitions are much more balanced in terms of number of edges per process (in comparison to the METIS partitions for Kronecker graphs), but the average BFS execution still suffers from local computation load imbalance. Next, consider the web crawl `eu-2005`. The local computation balance even after randomization is not as good as the synthetic graphs. One reason might be that the graph diameter is larger than the Kronecker graphs. 2D partitioning after randomization only worsens the load balance. The communication time for the fold step is somewhat lower for METIS and PaToH partitions compared to random partitions, but the times are not proportional to the savings projected in Table 4. This deserves further investigation. `coPapersCiteseer` shows trends similar to `eu-2005`. Note that the communication time savings going from 1D to 2D partitioning are different in both cases.

The tables also indicate that the level-synchronous approach performs extremely poorly on high-diameter graphs, and this is due to a combination of reasons. There is load imbalance in the local computation phase, and this is much more apparent after METIS and PaToH reorderings. For some of the level-synchronous phases, there may not be sufficient work per phase to keep all 16/256 processes busy. The barrier synchronization overhead is also extremely high. For instance, observe the cost of the expand step with 1D partitioning for `road_central` in Figure 6. This should ideally be zero, because there is no data exchanged in expand for 1D partitioning. Yet, multiple barrier synchronizations of a few microseconds turn out to be a significant cost.

Table 7 gives the parallel speedup achieved with different reorderings at 256-way parallel concurrency. The Erdős-Rényi graph gives the highest parallel speedup for all the partitioning schemes, and they serve as an indicator of the speedup achieved with good computational load balance. The speedup for real-world graphs is up to $5\times$ lower than this value, indicating the severity of the load imbalance problem. One more reason for the poor parallel speedup may be that these graphs are smaller than the Erdős-Rényi graph. The communication cost increases in comparison to the 16-node case, but the computational cost comprises 80% of the execution time. The gist of these performance results is that for level-synchronous BFS, partitioning has a considerable effect on the computational load balance, in addition to altering the communication cost. On current supercomputers, the computational imbalance seems to be the bigger of the two costs to account for, particularly at low process concurrencies.

As highlighted in the previous section, partitioners balance the load with respect to overall execution, that is the number of edges owned by each processor, not the number of edges traversed per BFS iteration. Figure 7 shows the actual imbalance that happens in practice due to the level-synchronous nature of the BFS algorithm. Even though PaToH limits the overall edge count imbalance to 3%, the actual per iteration load imbalances are severe. In contrast, random vertex numbering yields very good load balance across MPI processes and BFS steps.

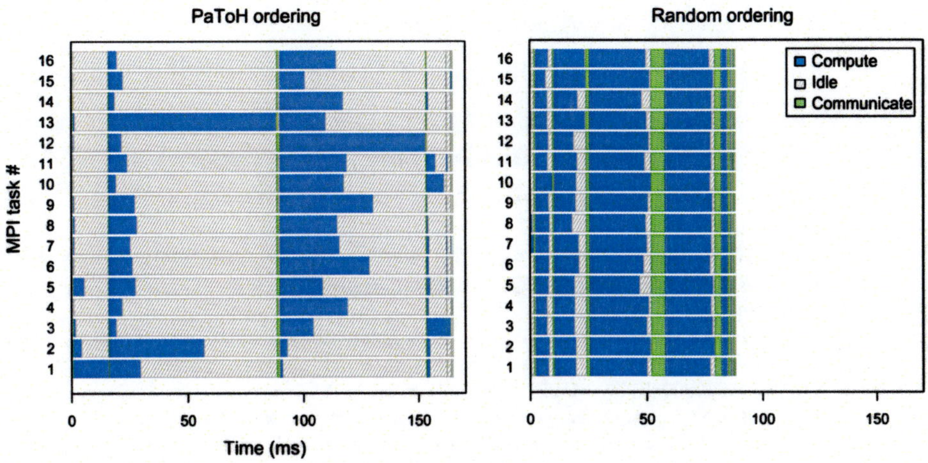

FIGURE 7. Parallel BFS execution timeline for the eu-2005 graph with PaToH and random vertex ordering (16 MPI processes, 4 nodes, 4 processes per node).

8. Conclusions and Future Work

Our study highlights limitations of current graph and hypergraph partitioners for the task of partitioning graphs for distributed computations. The crucial limitations are:

(1) The frequently-used partitioning objective function, total communication volume, is not representative of the execution time of graph problems such as breadth-first search, on current distributed memory systems.
(2) Even well-balanced vertex and edge partitions do not guarantee load-balanced execution, particularly for real-world graphs. We observe a range of relative speedups, between 8.8× to 50×, for low-diameter DIMACS graph instances.
(3) Although random vertex relabeling helps in terms of load-balanced parallel execution, it can dramatically reduce locality and increase the communication cost to worst-case bounds.
(4) Weighting the fold phase by a factor of two is not possible with two-phase partitioning strategies employed in current checkerboard method in PaToH, but it is possible with the single-phase fine grained partitioning. However, fine grained partitioning arbitrarily assigns edges to processors, resulting in communication among all processors instead of one processor grid dimension.

Although *MaxVolume* is a better metric than *TotalVolume* in predicting the running time, BFS communication structure heavily depends on run-time information. Therefore, a dynamic partitioning algorithm that captures the access patterns in the first few BFS iterations and repartitions the graph based on this feedback can be a more effective way of minimizing communication.

We plan to extend this study to consider additional distributed-memory graph algorithms. Likely candidates are algorithms whose running time is not so heavily dependent on the graph diameter. We are also working on a hybrid hypergraph-graph model for BFS where fold and expand phases are modeled differently.

Acknowledgments

We thank Bora Uçar for fruitful discussions and his insightful feedback on partitioning.

References

[1] U. Brandes, *A faster algorithm for betweenness centrality*, J. Mathematical Sociology **25** (2001), no. 2, 163–177.

[2] A. Buluç and K. Madduri, *Parallel breadth-first search on distributed memory systems*, Proc. ACM/IEEE Conference on Supercomputing, 2011.

[3] Ü.V. Çatalyürek and C. Aykanat, *PaToH: Partitioning tool for hypergraphs*, 2011.

[4] Ümit V. Çatalyürek, Cevdet Aykanat, and Bora Uçar, *On two-dimensional sparse matrix partitioning: models, methods, and a recipe*, SIAM J. Sci. Comput. **32** (2010), no. 2, 656–683, DOI 10.1137/080737770. MR2609335 (2011g:05176)

[5] *The Graph 500 List*, `http://www.graph500.org`, last accessed May 2012.

[6] D. Gregor and A. Lumsdaine, *The Parallel BGL: A Generic Library for Distributed Graph Computations*, Proc. Workshop on Parallel/High-Performance Object-Oriented Scientific Computing (POOSC'05), 2005.

[7] G. Karypis and V. Kumar, *Multilevel k-way partitioning scheme for irregular graphs*, Journal of Parallel and Distributed Computing **48** (1998), no. 1, 96–129.

[8] Richard J. Lipton, Donald J. Rose, and Robert Endre Tarjan, *Generalized nested dissection*, SIAM J. Numer. Anal. **16** (1979), no. 2, 346–358, DOI 10.1137/0716027. MR526496 (80d:65041)

[9] J.D. McCalpin, *Memory bandwidth and machine balance in current high performance computers*, IEEE Tech. Comm. Comput. Arch. Newslett, 1995.

[10] Yossi Shiloach and Uzi Vishkin, *An $O(n^2 \log n)$ parallel MAX-FLOW algorithm*, J. Algorithms **3** (1982), no. 2, 128–146, DOI 10.1016/0196-6774(82)90013-X. MR657270 (83e:68045)

[11] A. Yoo, E. Chow, K. Henderson, W. McLendon, B. Hendrickson, and Ü. V. Çatalyürek, *A scalable distributed parallel breadth-first search algorithm on BlueGene/L*, Proc. ACM/IEEE Conf. on High Performance Computing (SC2005), November 2005.

Appendix on edge count per processor

Tables 8 and 9 show the per-processor edge count (non-zero count in the graph's sparse adjacency matrix, denoted as $m_i, i \in P$ in the table) load imbalance for 1D and 2D checkerboard partitionings, respectively. The reported imbalances are for the storage of the graph itself, and exclude the imbalance among the frontier vertices. This measure affects memory footprint and local computation load balance. 1D row-wise partitioning gives very good edge balance for high-diameter graphs, which is understandable due to their local structure. This locality is not affected by any ordering either. For low-diameter graphs that lack locality, natural ordering can result in up to a 3.4× higher edge count on a single processor than the average. Both the random ordering and PaToH orderings seem to take care of this issue, though. On the other hand, 2D checkerboard partitioning exacerbates load imbalance in the natural ordering. For both low and high diameter graphs, a high imbalance, up to $10 - 16\times$, may result with natural ordering. Random ordering lowers it to at most 11% and PaToH further reduces it to approximately $3 - 5\%$.

TABLE 8. Edge count imbalance: $\max_{i \in P}(m_i)/\text{average}_{i \in P}(m_i)$ with 1D row-wise partitioning (lower is better, 1 is perfect balance).

Graph	$p = 4 \times 1$			$p = 16 \times 1$			$p = 64 \times 1$		
	N	R	P	N	R	P	N	R	P
coPapersCiteseer	2.11	1.00	1.00	2.72	1.02	1.00	3.14	1.06	1.00
coAuthorsDBLP	1.90	1.00	1.00	2.60	1.03	1.00	**3.40**	1.04	1.00
eu-2005	1.05	1.01	1.01	1.50	1.05	1.02	2.40	1.06	1.02
kronecker-logn18	1.03	1.02	1.01	1.10	1.08	1.02	1.29	1.21	1.02
rgg_n_2_20_s0	1.01	1.00	1.03	1.02	1.00	1.02	1.02	1.00	1.02
delaunay_n20	1.00	1.00	1.02	1.00	1.00	1.02	1.00	1.00	1.02

TABLE 9. Edge count imbalance: $\max_{i \in P}(m_i)/\text{average}_{i \in P}(m_i)$ with 2D checkerboard partitioning (lower is better, 1 is perfect balance).

Graph	$p = 2 \times 2$			$p = 4 \times 4$			$p = 8 \times 8$		
	N	R	P	N	R	P	N	R	P
coPapersCiteseer	3.03	1.01	1.02	**7.43**	1.00	1.03	**15.90**	1.02	1.02
coAuthorsDBLP	2.46	1.00	1.03	5.17	1.02	1.01	**10.33**	1.02	1.02
eu-2005	1.91	1.03	1.03	3.73	1.06	1.03	**9.20**	1.13	1.05
kronecker-logn18	1.03	1.01	1.01	1.06	1.04	1.03	1.15	1.11	1.03
rgg_n_2_20_s0	2.00	1.00	1.04	4.01	1.00	1.04	8.05	1.01	1.03
delaunay_n20	1.50	1.00	1.04	2.99	1.00	1.03	**5.99**	1.01	1.04

LAWRENCE BERKELEY NATIONAL LABORATORY

THE PENNSYLVANIA STATE UNIVERSITY

Using graph partitioning for efficient network modularity optimization

Hristo Djidjev and Melih Onus

ABSTRACT. The paper reviews an approach for finding the communities of a network developed by the authors [WAW'06, Lecture Notes in Computer Science, Volume 4936/2008, 117-128, IEEE TPDS vol. PP, issue 99, 2012], which is based on a reduction of the modularity optimization problem to the minimum weighted cut problem, and gives an experimental evaluation of an implementation based on that approach on graphs from the 10th DIMACS Implementation Challenge Testbed. Specifically, we describe a reduction from the problem of finding a partition of the nodes of a graph G that maximizes the modularity to the problem of finding a partition that minimizes the weight of the cut in a complete graph on the same node set as G, and weights dependent on a random graph model associated with G. The resulting minimum cut problem can then be solved by modifying existing codes for graph partitioning. We compare the performance of our implementation based on the Metis graph partitioning tool [SIAM J. Sci. Comp. 20, 359–392] against one of the best performing algorithms described in this volume.

1. Introduction

One way to extract information about the structure of a network or a graph and the relationships between its nodes is to divide it into *communities*, groups of nodes with denser links within the same group and sparser links between nodes in different groups. For instance, in a citation network, papers on related topics form communities and, in social networks, communities may define groups of people with similar interests.

The intuitive notion of communities given above is too vague as it is not specific about how dense the in-group links and how sparse the between group links should be. There are several formal definitions of communities, but the most popular currently is the one based on the modularity of a partition. Modularity [**31, 35**] is a measure of community quality of a partition of a network and measures the difference between the fraction of the links with endpoints in the same set of the partition and the expected fraction of that number in a network with a random

2010 *Mathematics Subject Classification.* Primary 05C85; Secondary 90C27, 90C35.

The work of the first author has been supported by the Department of Energy under contract W-705-ENG-36 and by the Los Alamos National Laboratory Directed Research and Development Program (LDRD), projects 20110093DR and 20130252ER.

Some of the results of this paper have been previously reported in IEEE Transactions on Parallel and Distributed Systems, vol. PP, issue: 99, 2012.

placement of the links. Formally, let $G = (V, E)$ be the graph representing the network and $\mathcal{P} = \{V_1, \ldots, V_k\}$, $k \geq 1$, be a *partition* of V, i.e., such that $V_1 \cup \cdots \cup V_k = V$ and $V_i \cap V_j = \emptyset$ for $i \neq j$. We refer to the sets V_i as communities. Let $\mathcal{G}$ be a random graph on the same set of nodes as G. Then the *modularity* of $\mathcal{P}$ with respect to $\mathcal{G}$ is defined as

$$\text{mod}(\mathcal{P}, G, \mathcal{G}) = \frac{1}{m}\sum_{i=1}^{k}(|E(V_i)| - \mathcal{E}(V_i, \mathcal{G})), \tag{1.1}$$

where m is the number of the edges of G, $E(V_i)$ denotes the set of all edges of G whose both endpoints are in V_i and $\mathcal{E}(V_i, \mathcal{G})$ denotes the expected number of edges in $\mathcal{G}$ with endpoints in V_i.

There are two choices that have been most often used for the random graph $\mathcal{G}$. The random graph model $G(n,p)$ of Erdös-Rényi [**17**] defines equal edge probabilities between all pairs of nodes. If n is the number of the nodes of G, m is the number of the edges, and p is chosen as m/n^2, then the expected number of edges of $G(n,p)$ is m. The alternative and more often used choice for a random graph in the definition of the modularity is based on the Chung and Lu model [**10**]. In that graph model, the expected node degrees match the node degrees of G. It defines an edge in $\mathcal{G}$ between nodes v and w of G with probability $\text{d}(v)\text{d}(w)/(2m)$, where by $\text{d}(x)$ we denote the degree of node x.

By the definition of modularity, a higher modularity indicates a larger fraction of in-community edges and, hence, a community structure of higher quality. Hence, the community detection problem can be formally defined as a *modularity optimization problem*, namely, given a graph G, find a partition of the nodes of G with maximum modularity. The minimum value of the modularity for a given $\mathcal{G}$ over the set of all partitions is called modularity of G, which we will denote by $\text{mod}(G, \mathcal{G})$. The modularity optimization problem has been shown to be NP-hard [**9**].

Hence, polynomial algorithms for finding an exact solution are unlikely, and various researchers have tried to construct heuristic algorithms for solving the modularity optimization problem. Clauset, Newman and Moore [**11**] construct an agglomerative algorithm that starts with a partition where each node represents a separate community and iteratively merge pairs of communities in order of maximum modularity gain, thereby building a dendrogram of the graph. They also consruct a data structure that makes the search of the best pair to merge very efficient. Guimerà and Amaral [**22**] use simulated annealing in a procedure that iteratively updates an initial partitioning aiming at increasing modularity. Simmulated annealing is used in order to try to avoid converging to a local minimum. Another physics-based approach is employed by Reichardt and Bornholdt [**38**], who simulate spin glass energy minimization for finding a community structure defined as the configuration of minimum energy. White and Smyth [**43**] and Newman [**33**] use a spectral approach by computing the eigenvector of the modularity matrix defined as the adjacency matrix of the input graph, appropriately updated to take into account the contribution of the random graph probabilities.

In this paper we describe a community detection method that reduces modularity optimization to the problem of finding a minimum weighted cut, which latter problem can be solved efficiently by using methods and tools developed for graph partitioning. Our approach was originally reported in [**12–14**], where we have compared our methods against the algorithms from [**11, 22, 33, 38**] on artificial graphs and showed that our algorithm is comparable in accuracy with the most accurate of these algorithms, while its scalability is significantly higher. In this paper we will first review the reduction from modularity optimization to minimum weighted cut, and then describe briefly how the resulting minimum cut problem can be solved by modifying the Metis graph partitioning code. Then we will compare the performance of the resulting algorithm against another algorithm described in this volume, using data sets from the 10th DIMACS Implementation Challenge collection.

2. Reduction of modularity optimization to minimum weighted cut

By the modularity definition (1.1) we have

$$m \cdot \mathrm{mod}(G, \mathcal{G}) = \max_{\mathcal{P}}\{ \sum_{i=1}^{k}(\,|E(V_i)| - \mathcal{E}(V_i, \mathcal{G}) \,)\}$$

$$= -\min_{\mathcal{P}}\{ -\sum_{i=1}^{k}(\,|E(V_i)| - \mathcal{E}(V_i, \mathcal{G}) \,)\}$$

$$= -\min_{\mathcal{P}}\{ \,(|E| - \sum_{i=1}^{k} |E(V_i)| \,) - (|E| - \sum_{i=1}^{k} \mathcal{E}(V_i, \mathcal{G}) \,)\}. \tag{2.1}$$

The first term $|E| - \sum_{i=1}^{k} |E(V_i)|$ of (2.1) is the number of the edges that connect all pairs of nodes from different sets of the partition. A *cut* of a graph is generally defined as a set C of edges whose removal divides the nodes of the graph into two or more sets such that all edges in C connect nodes from different sets. We extend this definition so that $C = \emptyset$ is also considered a valid cut, although it corresponds to a partition of a single set containing all the nodes of the graph. (The reason is that such partitions are also allowed in the definition of the modularity and, in fact, are essential as they correspond to a graph with a modularity structure of a single community.) We denote $\mathrm{cut}(\mathcal{P}, G) = E(G) - \cup_{i=1}^{k} E(V_i)$.

The second term $|E| - \sum_{i=1}^{k} \mathcal{E}(V_i, \mathcal{G})$ of (2.1), which we denote by $\mathcal{E}\mathrm{cut}(\mathcal{P}, G, \mathcal{G})$, corresponds to the expected value of the cut size of $\mathcal{P}$ in $\mathcal{G}$. The assumption that we make about the random graph model is that it preserves the expected number of the edges, hence $|E|$ is equal to the expected number of edges of $\mathcal{G}$. The two random graph models that we consider in this paper, the Erdös-Rényi and the Chung-Lu models, have this property, as we show below.

Hence,

$$m \cdot \mathrm{mod}(G, \mathcal{G}) = -\min_{\mathcal{P}}\{ \,|\mathrm{cut}(\mathcal{P}, G)| - \mathcal{E}\mathrm{cut}(\mathcal{P}, G, \mathcal{G})\},$$

which shows that the modularity optimization problem is equivalent to the problem of finding a partition that minimizes the difference of two cut. In order to merge the two cuts into a cut of a single graph, we define a new graph as follows.

We define a complete graph G' with the same vertices as G and a weight on each edge (v, w) defined by

$$\mathrm{wt}(v, w) = \begin{cases} 1 - p(v, w), & \text{if } (v, w) \in E(G) \\ -p(v, w), & \text{if } (v, w) \notin E(G), \end{cases} \tag{2.2}$$

where $p(v, w)$ is the probability of an edge between nodes v and w in $\mathcal{G}$. Since G' is complete, the cut set for $\mathcal{P}$ in G' is $\mathrm{cut}(\mathcal{P}, G') = \{(v, w) \mid v \in V_i, w \in V_j, i < j\}$ and its weight is

$$\begin{aligned}\mathrm{wt}(\mathrm{cut}(\mathcal{P}, G')) &= \sum_{i<j} \sum_{(v,w)\in V_i\times V_j} \mathrm{wt}(v, w) \\ &= \sum_{i<j} \sum_{(V_i\times V_j)\cap E} 1 - \sum_{i<j} \sum_{(v,w)\in V_i\times V_j} p(v, w) \\ &= |\mathrm{cut}(\mathcal{P}, G)| - \mathcal{E}\mathrm{cut}(\mathcal{P}, G, \mathcal{G}).\end{aligned}$$

Then

$$\mathrm{mod}(G, \mathcal{G}) = -\min_{\mathcal{P}} \mathrm{wt}(\mathrm{cut}(\mathcal{P}, G'))/m,$$

and, hence, the modularity optimization problem is equivalent to the problem of finding a minimum weighted cut for G'.

In order to complete the reduction, we just need to show that the values of $p(v, w)$ for the two random graph models we consider satisfy the assumption about the expected number of edges. For the Erdös-Rényi model, $p(v, w)$ is typically chosen as $2m/n^2$, which gives that the expected number of edges of $\mathcal{G}$ is

$$\frac{1}{2} \sum_{(v,w)\in V\times V} p(v, w) = \frac{n^2 p(v, w)}{2} = m.$$

For the Chung-Lu model we have $p(v, w) = d(v)d(w)/(2m)$, which gives for the expected number of edges of $\mathcal{G}$

$$\begin{aligned}\frac{1}{2} \sum_{(v,w)\in V\times V} p(v, w) &= \frac{1}{4m} \sum_{(v,w)\in V\times V} d(v)d(w) \\ &= \frac{1}{4m} \sum_{v\in V} d(v) \sum_{w\in V} d(w) = \frac{(2m)^2}{4m} = m.\end{aligned}$$

The above approach can be generalized in a straightforward manner to graphs with positively weighted edges. For this end, in definition (1.1), m is replaced with the sum M of all edge weights, $|E(V_i)|$ with the sum of the weights of all edges between nodes in V_i, and the expected number of edges in $\mathcal{G}$ corresponding to $E(V_i)$ with the expected weight of those edges. Finally, the random graph model $\mathcal{G}$ is replaced by a complete graph with weighted edges. For instance, for the Erdös-Rényi model, the probability $p(v, w)$ of an edge between nodes v and w is replaced by the weight $\mathrm{wt}(v, w)$, which is defined as $\mathrm{wt}(v, w) = 2M/n^2$ and in the Chung-Lu model the weight is defined as $D(v)D(w)/(2M)$, where $D(v)$ denotes the sum of the weights of all edges in G incident with v.

3. Implementation of the modularity optimization algorithm based on the Metis package

In the previous section we showed that finding a partition in G maximizing the modularity is equivalent to finding a minimum weighted cut in the complete graph G'. While the minimum cut problem is polynomial-time solvable in the case of nonnegative weights, this case does not apply to our problem as the weights of G' can be negative. The general decision version of the minimum cut problem is NP-complete as the maximum cut problem, which is NP-complete [**18**, problem

ND16, p.210], can be reduced to it. Hence one has to look for approximation or heuristic based algorithms for the modularity optimization problem.

While different versions of the minimum cut problem have been widely researched from theoretical point of view, much less attention has been paid to their implementation. The graph partitioning (GP) problem, which is related to the minimum cut problem, has received much greater attention from practitioners and very efficient implementations have been developed. The reason is that GP has important applications such as load balancing for high-performance computing and VLSI circuit design. For that reason, we are using a GP tool as a basis of our weighted minimum cut implementation, thereby solving the modularity optimization problem.

The GP problem asks, given a graph and an integer k, to find a partitioning of the vertices of the graph into equally sized (within difference at most one) sets such that the number (or weight) of the edges between different sets is minimized. Hence, GP is similar to the minimum cut problem, with the following differences: (i) in GP the sizes of the parts have to be balanced, while in minimum cut they can be arbitrary; (ii) in GP the number of the parts is an input variable given by the user, while in minimum cut and modularity optimization it is subject to optimization.

Any graph partitioning tool can be chosen as a basis for implementing (after appropriate modifications) the modularity optimization algorithm. The specific GP tool that we chose for our implementation is Metis [**23, 24**]. The reason is that Metis is considered an efficient and accurate tool for graph partitioning and that it is publicly available as a source code.

Metis is using multilevel strategy to find a good solution in a scalable manner. This type of multilevel strategy involves three phases: coarsening, partitioning, and uncoarsening. In the *coarsening* phase the size of the original graph is reduced in several stages, where at each stage connected subsets of nodes are contracted into single nodes, reducing as a result the number of the nodes of the graph roughly by half. The coarsening continues until the size of the resulting graph becomes reasonably small, say about 100 nodes. The final small graph is partitioned during the *partitioning* phase using some existing partitioning algorithms. In particular, Metis uses a graph-growing heuristic where one constructs one set of the partition by starting with a randomly selected node and then adding nodes to it in a breadth-first manner. The *uncoarsening* phase involves projecting the found solution from smaller graphs to larger ones, and refining the solution after each projection. This refinement step is one of the most important and sensitive for the quality of the final partition step. Metis implements it using the Kernighan-Lin algorithm. That algorithm computes for each node a quantity called *gain* that is equal to the change in the size (weight) of the cut if that node is moved from its current part to the other one. Then nodes with maximum gains are exchanged between partitions, making sure the balance between the sizes of the parts is maintained and also avoiding some local minima by allowing a certain number of negative-gain node swaps. See [**23, 24**] for more details about the implementation of Metis.

The modifications to Metis that need to be made are of two types: first, ones that take care of the above mentioned difference between GP and minimum cut problems and, second, ones aiming at improving the efficiency of the algorithm. Specifically, the minimum cut problem that we need to solve is on a complete graph G', whose number of edges is of order $\Omega(n^2)$, where n is the number of the nodes of G, while the number of the edges of the original graph G is typically of order $O(n)$. We will briefly discuss these two types of modifications below.

Removing the GP restriction of balanced part sizes is easy; in Metis we have just to omit checking the balance of the partition. Finding the right number of parts can be done in the following recursive way. We divide the original graph into two parts using the algorithm described above. If both parts are non-empty, we recurse on each part, and, if one of the part is empty (and the other contains all nodes), we are done. The latter case corresponds to the situation where the current graph (or subgraph) contains a single community.

The final issue we discuss is how to avoid the necessity of working explicitly with the edges of G' that are not in G and, as a result, to avoid the $\Omega(n^2)$ bound on the running time. The idea is to use explicitly in the algorithm only the edges of G, while handling implicitly the other ones by correcting the computed values in constant time. For instance, suppose that we have a partition $\mathcal{P}$ of the nodes of G' in two sets of sizes n_1 and n_2, respectively, and we have computed the weight of the corresponding cut in G, say w_G. Our goal is to evaluate the corresponding cut in G'. Assume that the random class model is $G(n,p)$. Then the weight of the cut corresponding to $\mathcal{P}$ in G' is $w_{G'} = w_G - n_1 n_2 p$ by formula (2.2). Hence it takes $O(1)$ time to compute $w_{G'}$ knowing $w_{G'}$. In a similar way one can compute the weight of the cut in G' in the case of the Chung-Li model, see **[14]** for details.

4. Comparison on DIMACS testbed graphs

We have tested our algorithm against the algorithm that was ranked at the top in the DIMACS Challenge with respect to its accuracy. Although the Challenge

TABLE 1. Comparison of our algorithm with the algorithm of Ovelgönne and Geyer-Schulz **[37]**

Network	Modularity (Our Paper)	Modularity (OG)	Run Time (Our Paper)	Run Time (OG)
as-22july06	0.6338	0.6776	3.26	25.28
astro-ph	0.7162	0.7424	1.14	36.48
caidaRouterLevel	0.8421	0.8719	8.67	324.36
celegans-metabolic	0.4224	0.4490	0.03	0.22
citationCiteseer	0.7732	0.8228	12.17	579.16
coAuthorsCiteseer	0.8844	0.9051	18.97	618.75
cond-mat-2005	0.7187	0.7460	3.54	111.34
email	0.5654	0.5802	0.08	0.81
G-n-pin-pout	0.3829	0.4998	6.12	680.77
kron-g500-logn16	0.0402	0.0533	12.42	4682.84
memplus	0.5913	0.7000	0.53	20.47
PGPgiantcompo	0.8750	0.8862	0.64	8.45
polblogs	0.4260	0.4269	0.14	0.64
power	0.9329	0.9397	0.19	1.92
rgg-n-2-7-s0	0.9732	0.9780	6.55	174.76
smallworld	0.7455	0.7930	17.61	267.86
coPapersDBLP	0.8309	0.8666	142.91	6388.64
in-2004	0.9717	0.9806	1321.52	1717.6

website already contains ranking results for several of the algorithms [**39**], including ours, those algorithms were not run on the same computer. Furthermore, the data on the website has not been converted into an easy to read format. Therefore, we believe it is worth including in this paper a direct comparison of our algorithm against a top performing algorithm from the challenge.

Ovelgonne and Geyer-Schulz's algorithm [**37**], ranked as number one in the Challenge, exploits the idea of ensemble learning. It learns weak graph clusterings and uses them to find a strong graph clustering. Table 1 compares the performance of our algorithm with that algorithm.

The test graphs in our experiments are the Co-author and Citation Networks and the Clustering Instances datasets of the DIMACS Challenge testbed [**40**], [**2**], [**4**], [**6**], [**8**], [**7**], [**42**], [**1**], [**32**], [**25**], [**29**], [**19**], [**26**], [**28**], [**16**], [**30**], [**41**], [**27**], [**36**], [**34**], [**5**], [**21**], [**15**], [**20**], and [**3**]. All experiments have been run on an Intel(R) Core(TM) i3 CPU M370 2.40 GHz processor notebook computer with 4G of memory.

For each experiment, the table shows the average running time and modularity of the partition for each of the algorithms. The results show modularity of the clusterings that our algorithm finds is 7% less on average, but our algorithm is 48 times faster on average. For one instance (kron-g500-logn16), our algorithm is 390 times faster.

One of the reasons that the modularities of our partitions are lower than the modularities produced by the code of [**37**] is that our algorithm is based on a version of Metis that is known to perform poorly on power law graphs. Hence our algorithm inherits the same weakness. Most of the networks in the testbed have power law or non-uniform degree distribution, which may explain some of the results. There is a newer version of Metis that is claimed to partition power law graphs successfully and it can be used for a new implementation of our algorithm.

5. Conclusion

We proved in this paper that the modularity optimization problem is equivalent to the problem of finding a minimum cut of a complete graph with real edge weights. We also showed that the resulting minimum cut problem can be solved based on existing software for graph partitioning. Our implementation was based on Metis, but we believe most other high-quality graph partitioning tools can be used for the same purpose. Of particular interest will be using a parallel partitioner as this will yield a parallel code for community detection.

References

[1] L. A. Adamic and N. Glance, *The political blogosphere and the 2004 us election*, WWW-2005 Workshop on the Weblogging Ecosystem (2005).

[2] Albert-László Barabási and Réka Albert, *Emergence of scaling in random networks*, Science **286** (1999), no. 5439, 509–512, DOI 10.1126/science.286.5439.509. MR2091634

[3] Alex Arenas, *http://deim.urv.cat/ aarenas/data/welcome.htm.*

[4] D. Baird and R.E. Ulanowicz, *The seasonal dynamics of the chesapeake bay ecosystem*, Ecol. Monogr. **59** (1989), 329–364.

[5] M. Boguñá, R. Pastor-Satorras, A. Diaz-Guilera, and A. Arenas, *Pgp network*, Physical Review E **70** (2004).

[6] Paolo Boldi, Bruno Codenotti, Massimo Santini, and Sebastiano Vigna, *Ubicrawler: A scalable fully distributed web crawler*, Software: Practice & Experience **34** (2004), no. 8, 711–726.

[7] Paolo Boldi, Marco Rosa, Massimo Santini, and Sebastiano Vigna, *Layered label propagation: A multiresolution coordinate-free ordering for compressing social networks*, Proceedings of the 20th international conference on World Wide Web, ACM Press, 2011.
[8] Paolo Boldi and Sebastiano Vigna, *The WebGraph framework I: Compression techniques*, Proc. of the Thirteenth International World Wide Web Conference (WWW 2004) (Manhattan, USA), ACM Press, 2004, pp. 595–601.
[9] Ulrik Brandes, Daniel Delling, Marco Gaertler, Robert Görke, Martin Hoefer, Zoran Nikoloski, and Dorothea Wagner, *On modularity clustering*, IEEE Trans. Knowl. Data Eng. **20** (2008), no. 2, 172–188.
[10] Fan Chung and Linyuan Lu, *Connected components in random graphs with given expected degree sequences*, Ann. Comb. **6** (2002), no. 2, 125–145, DOI 10.1007/PL00012580. MR1955514 (2003k:05123)
[11] Aaron Clauset, M. E. J. Newman, and Cristopher Moore, *Finding community structure in very large networks*, Physical Review E **70** (2004), 066111.
[12] H. Djidjev, *A fast multilevel algorithm for graph clustering and community detection*, Algorithms and Models for the Web-Graph, Lecture Notes in Computer Science, vol. 4936, 2008.
[13] Hristo Djidjev and Melih Onus, *A scalable multilevel algorithm for community structure detection*, WAW'06, Tech. Report LA-UR-06-6261, Los Alamos National Laboratory, September 2006.
[14] Hristo N. Djidjev and Melih Onus, *Scalable and accurate graph clustering and community structure detection*, IEEE Transactions on Parallel and Distributed Systems **99** (2012), no. PrePrints.
[15] J. Duch and A. Arenas, *C. elegans metabolic network*, Physical Review E **72** (2005).
[16] ———, *Condensed matter collaborations 2003*, Phys. Rev. E **72** (2005).
[17] P. Erdős and A. Rényi, *On random graphs. I*, Publ. Math. Debrecen **6** (1959), 290–297. MR0120167 (22 #10924)
[18] Michael R. Garey and David S. Johnson, *Computers and intractability*, W. H. Freeman and Co., San Francisco, Calif., 1979. A guide to the theory of NP-completeness; A Series of Books in the Mathematical Sciences. MR519066 (80g:68056)
[19] M. Girvan and M. E. J. Newman, *Community structure in social and biological networks*, Proc. Natl. Acad. Sci. USA **99** (2002), no. 12, 7821–7826 (electronic), DOI 10.1073/pnas.122653799. MR1908073
[20] P. Gleiser and L. Danon, *Jazz musicians network*, Adv. Complex Syst. **565** (2003).
[21] R. Guimerà, L. Danon, A. Diaz-Guilera, F. Giralt, and A. Arenas, *E-mail network urv*, Physical Review E **68** (2003).
[22] Roger Guimerà and Luis A. Nunes Amaral, *Functional cartography of complex metabolic networks*, Nature **433** (2005), 895.
[23] George Karypis and Vipin Kumar, *Multilevel graph partitioning schemes.*, International Conference on Parallel Processing, 1995, pp. 113–122.
[24] George Karypis and Vipin Kumar, *A fast and high quality multilevel scheme for partitioning irregular graphs*, SIAM J. Sci. Comput. **20** (1998), no. 1, 359–392 (electronic), DOI 10.1137/S1064827595287997. MR1639073 (99f:68158)
[25] D. E. Knuth, *Les miserables: coappearance network of characters in the novel les miserables*, The Stanford GraphBase: A Platform for Combinatorial Computing (1993).
[26] D. Lusseau, K. Schneider, O. J. Boisseau, P. Haase, E. Slooten, and S. M. Dawson, *Dolphin social network*, Behavioral Ecology and Sociobiology **54** (2003), 396–405.
[27] M. E. J. Newman, *The structure of scientific collaboration networks*, Proc. Natl. Acad. Sci. USA **98** (2001), no. 2, 404–409 (electronic), DOI 10.1073/pnas.021544898. MR1812610
[28] M. E. J. Newman, *Condensed matter collaborations 2005*, Proc. Natl. Acad. Sci. USA **98** (2001), 404–409.
[29] M. E. J. Newman, *High-energy theory collaborations*, Proc. Natl. Acad. Sci. USA **98** (2001), 404–409.
[30] M. E. J. Newman, *The structure of scientific collaboration networks*, Proc. Natl. Acad. Sci. USA **98** (2001), no. 2, 404–409 (electronic), DOI 10.1073/pnas.021544898. MR1812610
[31] M. E. J. Newman, *Mixing patterns in networks*, Phys. Rev. E (3) **67** (2003), no. 2, 026126, 13, DOI 10.1103/PhysRevE.67.026126. MR1975193 (2004f:91126)
[32] M. E. J. Newman, *Coauthorships in network science*, Phys. Rev. E **74** (2006).

[33] M. E. J. Newman, *Finding community structure in networks using the eigenvectors of matrices*, Phys. Rev. E (3) **74** (2006), no. 3, 036104, 19, DOI 10.1103/PhysRevE.74.036104. MR2282139 (2007j:82115)

[34] M. E. J. Newman, *Finding community structure in networks using the eigenvectors of matrices*, Phys. Rev. E (3) **74** (2006), no. 3, 036104, 19, DOI 10.1103/PhysRevE.74.036104. MR2282139 (2007j:82115)

[35] M. E. J. Newman and M. Girvan, *Finding and evaluating community structure in networks*, Physical Review E **69** (2004), 026113.

[36] Mark Newman, *Internet: a symmetrized snapshot of the structure of the internet at the level of autonomous systems*, The University of Oregon Route Views Project (2006).

[37] Michael Ovelgönne and Andreas Geyer-Schulz, *A divisive clustering technique for maximazing the modularity*, In: 10th DIMACS Implementation Challenge (Atlanta, Georgia), 2012.

[38] Jörg Reichardt and Stefan Bornholdt, *Statistical mechanics of community detection*, Phys. Rev. E (3) **74** (2006), no. 1, 016110, 14, DOI 10.1103/PhysRevE.74.016110. MR2276596 (2007h:82089)

[39] *Tenth DIMACS implementation challenge results*, `http://www.cc.gatech.edu/dimacs10/results`, accessed: 1/9/2012.

[40] D. Watts and S. Strogatz, *Collective dynamics of small-world networks*, Nature (1998).

[41] ———, *Neural network*, Nature **393** (1998), 440–442.

[42] ———, *Power grid*, Nature **393** (1998), 440–442.

[43] S. White and P. Smyth, *A spectral clustering approach to finding communities in graph*, Proceedings of the SIAM International Conference on Data Mining, 2005.

Information Sciences, Los Alamos National Labratory, Los Alamos, New Mexico 87545

Department of Computer Engineering, Bilkent University, Bilkent, Ankara, 06800, Turkey

Modularity maximization in networks by variable neighborhood search

Daniel Aloise, Gilles Caporossi, Pierre Hansen, Leo Liberti, Sylvain Perron, and Manuel Ruiz

ABSTRACT. Finding communities, or clusters, in networks, or graphs, has been the subject of intense studies in the last ten years. The most used criterion for that purpose, despite some recent criticism, is modularity maximization, proposed by Newman and Girvan. It consists in maximizing the sum for all clusters of the number of inner edges minus the expected number of inner edges assuming the same distribution of degrees. Numerous heuristics, as well as a few exact algorithms have been proposed to maximize modularity. We apply the Variable Neighborhood Search metaheuristic to that problem. Computational results are reported for the instances of the 10th DIMACS Implementation Challenge. The algorithm presented in this paper obtained the second prize in the quality modularity (sub)challenge of the referred competition, finding the best known solutions for 11 out of 30 instances.

1. Introduction

Clustering is an important chapter of data analysis and data mining with numerous applications in natural and social sciences as well as in engineering and medicine. It aims at solving the following general problem: given a set of entities, find subsets, or clusters, which are homogeneous and/or well-separated. As the concepts of homogeneity and of separation can be made precise in many ways, there are a large variety of clustering problems [**HJ**, **JMF**, **KR**, **M**]. These problems in turn are solved by exact algorithms or, more often and particularly for large data sets, by heuristics, of which there are frequently a large variety. An exact algorithm provides, hopefully in reasonable computing time, an optimal solution together with a

1991 *Mathematics Subject Classification.* Primary 90C27, 90C59, 91C20.

Key words and phrases. Modularity, community, clustering.

The first author was partially supported by the National Council for Scientific and Technological Development - CNPq/Brazil grant number 305070/2011-8.

The second author was partially supported by FQRNT (Fonds de recherche du Québec – Nature et technologies) team grant PR-131365 and NSERC(Natural Sciences and Engineering Research Council of Canada) grant 298138-2009.

The third author was partially supported by FQRNT team grant PR-131365, NSERC grant 105574-07 and Digiteo grant *Senior Chair* 2009-14D "RMNCCO".

The fourth author was partially supported by Digiteo grant *Senior Chair* 2009-14D "RMNCCO".

The fifth author was partially supported by FQRNT team grant PR-131365 and NSERC grant 327435-06.

proof of its optimality. A heuristic provides, usually in moderate computing time, a near optimal solution or sometimes an optimal solution but without proof of its optimality.

In the last decade, clustering on networks, or graphs, has been extensively studied, mostly in the physics and computer science communities, with recently a few forays from operations research. Rather than using the term cluster, the words *module* or *community* are often adopted in the physics literature. We use below the standard notation and terminology for graphs, i.e, a graph $G = (V, E, \omega)$ is composed of a set V of n vertices v_j and a set E of m edges $e_{ij} = \{v_i, v_j\}$. These edges may be weighted by the function $\omega(\{u,v\})$. If they are unweighted $\omega(\{u,v\}) = 1$. A subgraph $G_C = (C, E_C, \omega)$ of a graph $G = (V, E, \omega)$ induced by a set of vertices $C \subseteq V$ is a graph with vertex set C and edge set E_C equal to all edges with both vertices in C. Such a subgraph corresponds to a cluster (or module, or community) and many heuristics aim at finding a partition $\mathcal{C}$ of V into pairwise disjoint nonempty subsets $V_1, V_2, \ldots, V_N$ inducing subgraphs of G and covering V. Various objective functions have been proposed for evaluating such a partition. Among the best known are *multiway cut* [**GH**], *normalized cut* [**SM**], *ratio cut* [**AY**] and *modularity* [**NG**]. Initially proposed by Girvan and Newman in 2002 [**GN**] as a stopping rule for a hierarchical divisive heuristic, modularity was considered later as an independent criterion allowing determination of optimal partitions as well as comparison between partitions obtained by various methods.

Modularity aims at finding a partition of V which maximizes the sum, over all modules, of the number of inner edges minus the expected number of such edges assuming that they are drawn at random with the same distribution of degrees as in G. The following precise definition of *modularity* is given in [**NG**]:

$$Q = \sum_{C \in \mathcal{C}} \left[a_C - e_C\right],$$

where a_C is the fraction of all edges that lie within module C and e_C is the expected value of the same quantity in a graph in which the vertices have the same expected degrees but edges are placed at random. A maximum value of Q near to 0 indicates that the network considered is close to a random one (barring fluctuations), while a maximum value of Q near to 1 indicates strong community structure. Observe that maximizing modularity gives an optimal partition together with the optimal number of modules.

Let the weight vertex function be defined as:

$$\omega(v) = \begin{cases} \displaystyle\sum_{\{u,v\} \in E} \omega(\{u,v\}) & \text{if } \{v,v\} \notin E \\ \displaystyle\sum_{\{u,v\} \in E, u \neq v} \omega(\{u,v\}) + 2\omega(\{v,v\}) & \text{if } \{v,v\} \in E. \end{cases}$$

The modularity for a module C may be written as

$$Q(C) = \frac{\displaystyle\sum_{\{u,v\} \in E_C} \omega(\{u,v\})}{\displaystyle\sum_{e \in E} \omega(e)} - \frac{\left(\displaystyle\sum_{v \in V_C} \omega(v)\right)^2}{4\left(\displaystyle\sum_{e \in E} \omega(e)\right)^2}. \tag{1}$$

Let $\mathcal{C}$ be a partition of V. The sum over modules of their modularities can be written as

$$(2)\qquad Q = \frac{\sum_{C\in\mathcal{C}} \sum_{\{u,v\}\in E_C} \omega(\{u,v\})}{\sum_{e\in E} \omega(e)} - \frac{\sum_{C\in\mathcal{C}} \left(\sum_{v\in V_C} \omega(v)\right)^2}{4\left(\sum_{e\in E} \omega(e)\right)^2}.$$

Numerous heuristics have been proposed to maximize modularity. They are based on divisive hierarchical clustering, agglomerative hierarchical clustering, partitioning, and hybrids. They rely upon various criteria for agglomeration or division [**BGLL, CNM, DDA, N04, WT**], simulated annealing [**GA, MD, MAD**], mean field annealing [**LH**], genetic search [**THB**], extremal optimization [**DA**], label propagation [**BC, LM**], spectral clustering [**N06, RMP, SDJB**], linear programming followed by randomized rounding [**AK**], dynamical clustering [**BILPR**], multilevel partitioning [**D**], contraction-dilation [**MHSWL**], multistep greedy search [**SC**], quantum mechanics [**NHZW**] and other approaches [**BGLL, CFS, FLZWD, KSKK, RZ, SDJB**]. For a more detailed survey, see [**F**]. While other metaheuristics have been applied to modularity maximization, it is the first time, to the best of our knowledge, that the Variable Neighborhood Search (VNS) metaheuristic is used for that purpose. In particular, by using decomposition, we were able to tackle larger problems than the previous metaheuristic approaches, reducing the size of the problem in which VNS is applied.

The paper is organized as follows: in Section 2, after giving an outline of the VNS metaheuristic, we discuss its application to modularity maximization. In Section 3, we recall and extend to the weighted case an exact method for modularity maximization which is used to evaluate the quality of the solutions obtained by our variable neighborhood search metaheuristic. Experimental results are presented in Section 4 in two tables corresponding to the results for Pareto and Quality challenges respectively. Brief conclusions are drawn in the last section.

2. Description of the heuristic

2.1. Outline of the variable neighborhood search metaheuristic. Variable Neighborhood Search (VNS) is a metaheuristic, or framework for building heuristics, aimed at solving combinatorial and global optimization problems. Since its inception, VNS has undergone many developments and has been applied in numerous fields (see [**HMP**] for a recent survey).

Metaheuristics address the problem of escaping, as much as possible, from local optima. A local maximum x_L of an optimization problem is such that

$$(3)\qquad f(x_L) \geq f(x), \forall x \in N(x_L)$$

where $N(x)$ denotes the feasible *neighborhood* of x, which can be defined in many different ways each one yielding a different neighborhood structure. In discrete optimization problems, a neighborhood structure consists of all vectors obtained from x by some simple modification. For instance, for x binary, one neighborhood structure can be defined by the set of all vectors obtained from x by complementing one of its components. Another possible neighborhood structure can be defined as the set of all vectors obtained from x by complementing two complementary components of x (i.e., one component is set from 0 to 1 and the other goes from 1 to 0). A *local search* or *improving* heuristic consists of choosing an initial solution x, and then moving to the best neighbor $x' \in N(x)$ in the case $f(x') > f(x)$. If no such neighbor exists, the heuristic stops, otherwise it is iterated.

If many local maxima exist for a problem, the range of values they span may be large. Moreover, the globaly optimum value $f(x^*)$ may differ substantially from the average value of a local maximum, or even from the best such value among many, obtained by some simple randomized heuristic. In order to escape from local maxima and, more precisely, the mountains of which they are the top, VNS exploits the idea of *neighborhood* change. In fact, VNS relies upon the following observations:

Fact 1: *A local maximum with respect to one neighborhood structure is not necessarily so for another*;

Fact 2: *A global maximum is a local maximum with respect to all possible neighborhood structures*;

Fact 3: *For many problems local maxima with respect to one or several neighborhoods are relatively close to each other.*

Let us denote with $\mathcal{N}_t$, $(t = 1, \ldots, t_{max})$, a finite set of pre-selected neighborhood structures, and with $\mathcal{N}_t(x)$ the set of solutions in the tth neighborhood of x. We call x a *local maximum* with respect to $\mathcal{N}_t$ if there is no solution $x' \in \mathcal{N}_t(x)$ such that $f(x') > f(x)$.

In the VNS framework, the neighborhoods used correspond to various types of moves, or perturbations, of the current solution, and are problem specific. The current best solution x found is the center of the search. When looking for a better one, a solution x' is drawn at random in an increasingly far neighborhood and a local ascent is performed from x', leading to another local maximum x''. If $f(x'') \leq f(x)$, x'' is ignored and one chooses a new neighbor solution x' in a further neighborhood of x. If, otherwise, $f(x'') > f(x)$, the search is re-centered around x'' restarting with the closest neighborhood. If all neighborhoods of x have been explored without success, one begins again with the closest one to x, until a stopping condition (e.g. maximum CPU time) is satisfied.

As the size of neighborhoods tends to increase with their distance from the current best solution x, close-by neighborhoods are explored more thoroughly than far away ones. This strategy takes advantage of the three Facts 1–3 mentioned above. Indeed it is often observed that most or all local maxima of combinatorial problems are concentrated in a small part of the solution space. Thus, finding a first local maximum x implies that some important information has been obtained: to get a better, near-optimal solution, one should first explore its vicinity.

The algorithm proposed in this work has two main components: (i) an improvement heuristic, and (ii) exploration of different types of neighborhoods for getting out of local maxima. They are used within a variable neighborhood decomposition search framework [**HMP**] which explores the structure of the problem concentrating on small parts of it. The basic components as well as the decomposition framework are described in the next sections.

2.2. Improvement heuristic. The improvement heuristic we used is the LPAm+ algorithm proposed by Liu and Murata in [**LM**]. LPAm+ is composed of a label propagation algorithm proposed by Barber and Clark [**BC**] and a community merging routine. A strong feature of this heuristic is that label propagation executes in near linear time (in fact, each iteration of label propagation executes in time proportional to m), while one round of merging pairs of communities can execute in $O(m \log n)$ [**SC**].

Label propagation is a similarity-based technique in which the label of a vertex is propagated to adjacent vertices according to their proximity. Label propagation algorithms for clustering problems assume that the label of a node correspond to its incumbent cluster index. Then, at each label propagation step, each vertex is sequentially evaluated for label updating according to a propagation rule. In [**BC**], the Barber and Clark propose a label propagation algorithm, called LPA, for modularity maximization. Their label updating rule for vertex v is (see. [**BC**] for details):

$$\ell_v \leftarrow \underset{\ell}{\operatorname{argmax}} \left(\sum_{u=1}^{n} B_{uv} \delta(\ell_u, \ell) \right) \tag{4}$$

where ℓ_v is the label of vertex v, $B_{uv} = \omega(\{u, v\}) - (\omega(u)\omega(v))/2m$, and $\delta(i, j)$ is the Kronecker's delta.

Moreover, the authors prove that the candidate labels for ℓ_v in eq.(4) can be confined to the labels of the vertices adjacent to v and an unused label. We decided to use this fact in order to speedup LPA. Let us consider a vertex $v^* \in C$, where C is a module of the current partition, and let us suppose that the modules to which its adjacent vertices belong have not changed since the last evaluation of v^*. In this case, v^* can be discarded for evaluation since no value has changed from the last instantiation of eq.(4) since the last evaluation of v^*. With that in mind, we decided to iterate over "labels" instead of over the vertices of the graph.

We used LPAm+ modified as follows. A list L of all labels is initialized with the clusters indices of the current partition. Then, from L, we proceed by picking a label $\ell \in L$ until L is empty. Each time a label ℓ is removed from L, we evaluate by means of eq.(4) all its vertices for label updating. If the label of a vertex is updated, yielding an improvement in the modularity value of the current partition, the old and the new labels of that vertex, denoted ℓ^{old} and ℓ^{new}, are inserted in L. Moreover, the labels of vertices which are connected to a node with label equal to either ℓ^{old} or ℓ^{new} are also inserted in L. This modification induces a considerable algorithmic speedup since only a few labels need to be evaluated as the algorithm proceeds.

We then tested this modified LPAm+, and proceeded to improve it based on empirical observations. In the final version, whenever a vertex relabeling yields an improvement, the old and the new labels of that vertex are added to L but only together with the labels of vertices which are adjacent to the relabeled vertex. This version was selected to be used in our experiments due to its benefits in terms of computing times and modularity maximization.

2.3. Neighborhoods for perturbations. In order to escape from local maxima, our algorithm uses five distinct neighborhoods for perturbing a solution. They are:

(1) SINGLETON: all the vertices in a cluster are made singleton clusters.
(2) DIVISION: splits a community into two equal parts. Vertices are assigned to each part randomly.

(3) NEIGHBOR: relabels each vertex of a cluster to one of the labels of its neighbors or to an unused label.
(4) EDGE: puts two linked vertices assigned to different clusters into one neighboring cluster randomly chosen.
(5) FUSION: merges two or more clusters into a single one.
(6) REDISTRIBUTION: destroys a cluster and spreads each one of its vertices to a neighboring cluster randomly chosen.

2.4. Variable Neighborhood Decomposition Search. Given the size of the instances proposed in the 10th DIMACS Implementation Challenge, a decomposition framework was used. It allows the algorithm to explore the search space more quickly since just a small part of the solution is searched for improvement at a time. This subproblem is isolated for improvement through selecting a subset of the clusters in the incumbent solution.

The decomposition proposed here is combined with the five neighborhoods presented in the previous section within a variable neighborhood schema. Thus, the decomposition executes over five distinct neighborhood topologies, with subproblems varying their size according to the VNS paradigm. The pseudo-code of the variable neighborhood decomposition search heuristic is given in Figure 1.

```
1  Algorithm VNDS(P)
2  Construct a random solution x ;
3  x ← LPAm+(x, P) ;
4  s ← 1;
5  while stopping condition not satisfied do
6      Construct a subproblem S from x with a randomly selected
       cluster and s − 1 neighboring clusters ;
7      Select randomly
       α ∈ {singleton, division, neighbor, fusion, redistribution} ;
8      x' ← shaking(x, α, S);
9      x' ← LPAm+(x', S) ;
10     if cost(x') > cost(x) then
11         x ← LPAm(x', P) ;
12         s ← 1;
13     else
14         s ← s + 1;
15         if s > min{MAX_SIZE, #clusters(x)} then
16             s ← 1;
17         end
18     end
19 end
20 return x
```

Algorithm 1: Pseudo-code of the decomposition heuristic.

The algorithm VNDS starts with a random solution for an input problem P in line 2. Then, in line 3 this solution is improved by applying our implementation of LPAm+. Note that LPAm+ receives two input parameters, they are: (i) the solution to be improved, and (ii) the space on which an improvement will be searched. In line 3, the local search is applied in the whole problem space P, which means that all vertices are tested for label updating, and all clusters are considered for merging. In line 4, the variable s which controls the current decomposition size is set to 1.

The central part of the algorithm VNDS consists of the loop executed in lines 5-19 until a stopping criterion is met (this can be the number of non improving iterations for the Pareto Challenge or maximum allowed CPU time for the Quality Challenge). This loop starts in line 6 by constructing a subproblem from a randomly selected cluster and $s-1$ neighboring clusters. Then, in line 7 a neighborhood α is randomly selected for perturbing the incumbent solution x. Our algorithm allows choosing α by specifying a probability distribution on the neighborhoods. Thus, the most successful neighborhoods are more often selected. The shaking routine is actually performed in line 8 in the chosen neigborhood α and in the search space defined by subproblem S. In the following, the improving heuristic LPAm+ is applied over x' in line 9 only in the current subproblem S. If the new solution x' is better than x in line 10, a faster version of the improving heuristic, denoted LPAm, is applied in line 11 over x' in the whole problem P. In this version, the improving heuristic does not evaluate merging clusters. The resulting solution of LPAm application is assigned to x in line 11 and s is reset to 1 in line 12. Otherwise, if x' is not better than x, the size of the decomposition is increased by one in line 14. This value is reset to 1 in line 16 if it exceeds the minimum between a given parameter MAX_SIZE and the number of clusters (i.e., #clusters(x)) in the current solution x (line 15). Finally, a solution x is returned by the algorithm in line 20.

3. Description of the exact method

Column generation together with branch-and-bound can be used to obtain an optimal partition. Column generation algorithms for clustering implicitly take into account all possible communities (or, in other words, all subsets of the set of entities under study). They replace the problem of finding simultaneously all communities in an optimal partition by a sequence of optimization problems for finding one community at a time, or more precisely and for the problem under study a community which improves the modularity of the current solution. In [**ACCHLP**], several stabilized column generation algorithms have been proposed for modularity maximization and compared on a series of well-known problems from the literature. The column generation algorithm based on extending the mixed integer formulation of Xu et al. [**XTP**] appears to be the most efficient. We summarize below an adaptation of this algorithm for the case of weighted networks.

Column generation is a powerful technique of linear programming which allows the exact solution of linear programs with a number of columns exponential in the size of the input. To this effect, it follows the usual steps of the simplex algorithm, apart from finding an entering column with a positive reduced cost in case of maximization which is done by solving an auxiliary problem. The precise form of this last problem depends on the type of problem considered. It is often a combinatorial optimization or a global optimization problem. It can be solved heuristically as long as a column with a reduced cost of the required sign can be found. When this is no longer the case, an exact algorithm for the auxiliary problem must be applied either to find a column with the adequate reduced cost sign, undetected by the heuristic, or to prove that there remains no such column and hence the linear programming relaxation is solved.

For modularity maximization clustering, as for other clustering problems with an objective function additive over the clusters, the columns correspond to the set T of all subsets of V, i.e., to all nonempty modules, or in practice to a subset T' of T. To express this problem, define $a_{it} = 1$ if vertex i belongs to module t and $a_{it} = 0$ otherwise. One can then write the model as

$$\max \sum_{t \in T} c_t z_t \tag{5}$$

$$\text{s.t.} \sum_{t \in T} a_{it} z_t = 1 \qquad \forall i = 1, \ldots, n \tag{6}$$

$$z_t \in \{0, 1\} \qquad \forall t \in T, \tag{7}$$

where c_t corresponds to the modularity value of the module indexed by t with $t = 1 \ldots 2^n - 1$. The problem (5)-(7) is too large to be written explicitly. A reduced problem with few columns, i.e., those with index $t \in T'$, is solved instead. One first relaxes the integrality constraints and uses column generation for solving the resulting linear relaxation.

The auxiliary problem, for the weighted case, can be written as follows:

$$\begin{aligned} \max_{x \in \mathbb{B}^m, D \in \mathbb{R}} \quad & \sum_{e} \in E \frac{x_e}{M} - \left(\frac{D}{2M}\right)^2 - \sum_{u \in V} \lambda_u y_u \\ \text{s.t.} \quad & D = \sum_{u \in V} \omega(u) y_u \\ & x_e \leq y_u \qquad \forall e = \{u, v\} \in E \\ & x_e \leq y_v \qquad \forall e = \{u, v\} \in E \end{aligned}$$

where $M = \sum_{e \in E} \omega(e)$. Variable x_e is equal to 1 if edge e belongs to the community which maximizes the objective function and to 0 otherwise. Similarly, y_u is equal to 1 if the vertex u belongs to the community and 0 otherwise. The objective function is equal to the modularity of the community to be determined minus the scalar product of the current value λ_u of the dual variables times the indicator variables y_u. As in [**ACCHLP**], the auxiliary problem is first solved with a VNS heuristic as long as a column with a positive reduced cost can be found. When this is no more the case, CPLEX is called to find such a column or prove that none remain. If the optimal solution of the linear relaxation is not integer, one proceeds to branching on the condition that two selected entities belong to the same community or to two different ones.

4. Experimental Results

The algorithms were implemented in C++ and compiled by gcc 4.5.2. Limited computational experiments allowed to set the parameters of the VNDS algorithm as follows:

- MAX_SIZE = 15
- Probability distribution for selecting α is drawn with:
  - 30% of chances of selecting SINGLETON
  - 30% of chances of selecting DIVISION
  - 28% of chances of selecting NEIGHBOR
  - 5% of chances of selecting EDGE
  - 4% of chances of selecting FUSION

– 3% of chances of selecting REDISTRIBUTION

The stopping condition in algorithm VNDS was defined depending on the challenge, Pareto or Quality, in which VNDS is used. Thus, the same algorithm is able to compete in both categories by just modifying how it is halted.

4.1. Results for exactly solved instances. Exact algorithms provide a benchmark of exactly solved instances which can be used to fine tune heuristics. More precisely, the comparison of the symmetric differences between the optimal solution and the heuristically obtained ones may suggest additional moves which improve the heuristic under study.

In general, a sophisticated heuristic should be able to find quickly an optimal solution for most or possibly all practical instances which can be solved exactly with a proof of optimality. Our first set of experiments aims to verify the effectiveness of the VNDS algorithm in the instances for which the optimal solution is proved by the exact method of Section 3. The instances tested here are taken from the *Clustering* chapter of the 10th DIMACS Implementation Challenge (http://www.cc.gatech.edu/dimacs10/archive/clustering.shtml).

Table 1 presents average solution values and CPU times (in seconds) obtained from five independent runs of the VNDS algorithm in a Intel X3353 with a 2.66 Ghz clock and 24Gb of RAM memory. The first column refers to the instance. The second and third columns refer to the number of nodes (n) and edges (m) of each instance. Fourth and fifth column show the VNDS average results. Finally, the sixth and seventh column present the optimal solution values proved by the exact algorithm.

TABLE 1. VNDS average results and optimal modularity values obtained by the exact algorithm of Section 3.

instance	n	m	Q_{avg}	t_{avg}	Q_{opt}	$\|C\|_{opt}$
karate	34	78	0.419790	0.00	0.419790	4
chesapeake	39	170	0.265796	0.00	0.265796	3
dolphins	62	159	0.528519	0.00	0.528519	5
lesmis	77	254	0.566688	0.00	0.566688	6
polbooks	105	441	0.527237	0.00	0.527237	5
adjnoun	112	425	0.312268	0.09	0.313367	7
football	115	613	0.604570	0.00	0.604570	10
jazz	198	2742	0.445144	0.01	0.445144	4

We note that VNDS finds the optimal solutions of instances `karate`, `chesapeake`, `dolphins`, `lesmis`, `polbooks`, `adjnoun`, `football`, and `jazz`. Except for instance `adjnoun`, where the optimal solution is found in 2 out of 5 runs, the optimal solutions of the aforementioned instances are obtained in all runs.

4.2. Results for Pareto Challenge. The results presented in this section and in the following one refers to the final modularity instances of the 10th DIMACS Implementation Challenge. Particularly for this section, results are presented both in terms of modularity values and CPU times, which are the two performance dimensions evaluated in the Pareto challenge. Computational experiments were performed on a Intel X3353 with a 2.66 Ghz clock and 24Gb of RAM memory.

Instances `kron_g500-simple-logn20` ($n = 1048576, m = 44619402$), `cage15` ($n = 5154859, m = 47022346$), `uk-2002` ($n = 18520486, m = 261787258$), `uk-2007-05` ($n = 105896555, m = 3301876564$), and `Er-fact1.5-scale25` ($\log_2 n = 25$) were not executed due to memory limitations.

In this challenge, VNDS stops whenever it attains either N iterations without improving the incumbent solution or after 3 hours of CPU. For this challenge we divided the instances into two categories. For the instances in category $P1$, VNDS uses $N = 1000$, while for those in category $P2$, the algorithm uses $N = 100$.

Table 2 shows average computational results obtained in five independent runs of algorithm VNDS. The results for one of these runs was sent to the DIMACS Pareto challenge. The first column in the table refers to the category of the instance indicated in the second column. The third and fourth columns refer to the number of nodes (n) and edges (m) of each instance. The fifth and sixth columns refer to average modularity values and computing times (in seconds), respectively. VNDS is stopped due to CPU time limit in the instances for which the average computing time $t_{avg} = 10800.00$.

We remark from Table 2:

- Instances `coPapersDBLP`, `audikw1`, and `ldoor` are stopped due to our CPU time limit in each one of the five independent runs.
- Considering all the instances presented in the table, VNDS was Pareto dominated (see http://www.cc.gatech.edu/dimacs10/data/dimacs10-rules.pdf) in the DIMACS challenge by at most 2 other algorithms in each instance. It is worthy mentioning that the organizing committee left open to the Pareto challenge participants the task of defining their own stopping condition for their algorithms. Consequently, Pareto challenge scores were sensitive to the strategy used by each team. For instance, considering only the instances we categorized in $P1$, which uses $N = 1000$ as stopping condition, VNDS was dominated by at most 1 other algorithm.
- In 9 out of 25 instances in the table, VNDS was not Pareto dominated by any other algorithm in the Pareto challenge.

4.3. Results for Quality challenge. Since the amount of work to compute a solution is not taken into consideration for this challenge, the VNDS algorithm was allowed to run for a longer period of time than before, the CPU time limit being the unique stopping condition. In our set of experiments, the instances were split into two different categories. The algorithm was allowed to run for 1800 seconds (30 minutes) for instances in category Qu1, and 18000 seconds (5 hours) for instances in category Qu2. Furthermore, in order to overcome memory limitations, VNDS was executed in a Intel Westmere-EP X5650 with a 2.66 Ghz clock and 48Gb of RAM memory for the largest instances. This allowed the algorithm to obtain solutions for instances `kron_g500-simple-logn20`, `cage15` and `uk-2002`.

Table 3 presents the computational results obtained in 10 independent runs of algorithm VNDS. We chose to present here the same results submitted to the DIMACS Implementation Challenge. The first column refers to the category of the instance indicated in the second column. Third and fourth columns refer to the best obtained modularity value and its corresponding number of clusters. Finally, the last column shows the rank position of the referred solution among 15 participating teams.

TABLE 2. Average modularity results for the Pareto Challenge of the 10th DIMACS Implementation Challenge.

category	instance	n	m	Q_{avg}	t_{avg}
P1	celegans_metabolic	453	2025	0.452897	1.97
	e-mail	1133	5451	0.427105	1.69
	polblogs	1490	16715	0.582510	4.96
	power	4941	6594	0.940460	11.87
	PGPgiantcompo	10680	24316	0.885451	28.03
P2	astro-ph	16726	47594	0.738435	10.53
	memplus	17758	54196	0.686220	78.44
	as-22july06	22963	48436	0.676091	31.01
	cond-mat-2005	40421	175691	0.739293	106.54
	kron_g500-simple-logn16	65536	2456071	0.061885	12.91
	preferentialAttachment	100000	499985	0.314690	1972.50
	smallworld	100000	499998	0.791905	33.82
	G_n_pin_pout	100000	501198	0.488209	2625.18
	luxembourg.osm	114599	119666	0.989307	90.30
	rgg_n_2_17_s0	131072	728753	0.977417	214.53
	caidaRouterLevel	192244	609066	0.867259	2037.24
	coAuthorsCiteseer	227320	814134	0.901432	1754.33
	citationCiteseer	268495	1156647	0.820164	9048.09
	coPapersDBLP	540486	15245729	0.862878	10800.00
	eu-2005	862664	16138468	0.941157	9914.73
	audikw1	943695	38354076	0.918034	10800.00
	ldoor	952203	22785136	0.968844	10800.00
	in-2004	1382908	13591473	0.980509	9583.98
	belgium.osm	1441295	1549970	0.994569	6549.13
	333SP	3712815	11108633	0.988247	10019.93

A few remarks are in order regarding the results shown in Table 3:

- The VNDS algorithm obtained the best solutions for 11 out of 30 instances of the modularity challenge (i.e., approx. 37% of the instances). Moreover, the algorithm figured in the first two positions in 25 out of 30 instances (i.e., approx. 83%).
- The algorithm does not appear to have its effectiveness influenced by the number of clusters. For example, the algorithm found the best solution for instance celegans_metabolic using 9 clusters as well as it obtained the best result for instance kron_g500-simple-logn16 using 10027 clusters.
- The algorithm was particularly bad for instance cage15. Actually, the result submitted to the challenge corresponded to a simple LPAm+ application. This was due to two combined reasons: (i) the instance was large with $n = 5154859$ nodes and $m = 47022346$, and (ii) LPAm+ did not get to decrease much the number of clusters (648819). This led to algorithm abortion before executing its main computing part.

5. Conclusion

Several integer programming approaches and numerous heuristics have been applied to modularity maximization. They are due mostly to the physics and computer sciences research communities. We have applied the variable neighborhood

TABLE 3. Modularity results for the Quality Challenge of the 10th DIMACS Implementation Challenge.

category	instance	Q_{best}	$\|\mathcal{C}\|_{best}$	rank
Qu1	celegans_metabolic	0.453248	9	#1
	e-mail	0.582828	10	#1
	polblogs	0.427105	278	#1
	power	0.940850	43	#1
	PGPgiantcompo	0.886081	111	#2
	astro-ph	0.74462	1083	#1
	memplus	0.695284	64	#3
	as-22july06	0.677575	42	#2
	cond-mat-2005	0.745064	1902	#2
	kron_g500-simple-logn16	0.065055	10027	#1
Qu2	preferentialAttachment	0.315993	9	#1
	smallworld	0.793041	242	#1
	G_n_pin_pout	0.499344	171	#2
	luxembourg.osm	0.98962	275	#1
	rgg_n_2_17_s0	0.978323	133	#1
	caidaRouterLevel	0.870905	477	#2
	coAuthorsCiteseer	0.9039	297	#2
	citationCiteseer	0.821744	201	#2
	coPapersDBLP	0.865039	326	#2
	eu-2005	0.941324	389	#2
	audikw1	0.917983	34	#1
	ldoor	0.969098	88	#2
	in-2004	0.980537	1637	#2
	belgium.osm	0.994761	580	#4
	333SP	0.988365	229	#4
	kron_g500-simple-logn20	0.049376	253416	#2
	cage15	0.343823	648819	#10
	uk-2002	0.990087	45165	#3

search metaheuristic to that problem and it proves to be very effective. For problems with known optimum values, the heuristic always found an optimal solution at least once. For the DIMACS Implementation Challenge, the best know solution was provided for 11 out of 30 instances. Overall, the proposed algorithm obtained the second prize in the modularity Quality challenge and the fifth place in the Pareto challenge.

References

[ACCHLP] D. Aloise, S. Cafieri, G. Caporossi, P. Hansen, L. Liberti, and S. Perron, *Column Generation Algorithms for Exact Modularity Maximization in Networks*, Physical Review E, vol. 82 (2010), no. 046112.

[AK] G. Agarwal and D. Kempe, *Modularity-maximizing graph communities via mathematical programming*, Eur. Phys. J. B **66** (2008), no. 3, 409–418, DOI 10.1140/epjb/e2008-00425-1. MR2465245 (2009k:91130)

[AY] C.J. Alpert and S.-Z. Yao, *Spectral Partitioning: The More Eigenvectors, The Better*, Proc. 32nd ACM/IEEE Design Automation Conf., 1995, 195–200.

[BC] M.J. Barber and J.W. Clark, *Detecting network communities by propagating labels under constraints*, Physical Review E, vol. 80 (2009), no. 026129.

[BGLL] V. Blondel, J.-L. Guillaume, R. Lambiotte, and E. Lefebvre, *Fast unfolding of communities in large networks*, Journal of Statistical Mechanics, (2008), p. P10008.

[BILPR] S. Boccaletti, M. Ivanchenko, V. Latora, A. Pluchino, and A. Rapisarda, *Detecting complex network modularity by dynamical clustering*, Physical Review E, vol. 75 (2007), 045102(R).

[CFS] D. Chen, R. Fu, and M. Shang, *A fast and efficient heuristic algorithm for detecting community structures in complex networks*, Physica A, vol. 388 (2009), 2741–2749.

[CHL] S. Cafieri, P. Hansen, and L. Liberti, *A locally optimal heuristic for modularity maximization of networks*, Physical Review E, vol. 83 (2011), 056105(1-8).

[CNM] A. Clauset, M. Newman, M., and C. Moore, *Finding community structure in very large networks*, Physical Review E, vol. 70 (2004), no. 066111.

[D] H. Djidjev, *A scalable multilevel algorithm for graph clustering and community structure detection*, LNCS, vol. 4936, (2008).

[DA] J. Duch, and A. Arenas, *Community identification using extremal optimization*, Physical Review E 72 (2005) 027104(2).

[DDA] L. Danon, A. Diaz-Guilera, and A. Arenas, *The effect of size heterogeneity on community identification in complex networks*, Journal of Statistical Mechanics, vol. P11010, 2006.

[DGSW] D. Delling, R. Görke, C. Schulz, D. Wagner. *ORCA reduction and contraction graph clustering*. In 5th Int. Conf. on Algorithmic Aspects in Information and Management (AAIM), 2009, 152–165.

[F] Santo Fortunato, *Community detection in graphs*, Phys. Rep. **486** (2010), no. 3-5, 75–174, DOI 10.1016/j.physrep.2009.11.002. MR2580414 (2011d:05337)

[FLZWD] Y. Fan, M. Li, P. Zhang, J. Wu, Z. Di, *Accuracy and precision of methods for community identification in weighted networks*, Physica A, vol. 377 (2007), 363–372.

[GA] R. Guimerà and A. Amaral, *Functional cartography of complex metabolic networks*, Nature, vol. 433 (2005), pp. 895–900.

[GH] Olivier Goldschmidt and Dorit S. Hochbaum, *A polynomial algorithm for the k-cut problem for fixed k*, Math. Oper. Res. **19** (1994), no. 1, 24–37, DOI 10.1287/moor.19.1.24. MR1290008 (95h:90154)

[GN] M. Girvan and M. E. J. Newman, *Community structure in social and biological networks*, Proc. Natl. Acad. Sci. USA **99** (2002), no. 12, 7821–7826 (electronic), DOI 10.1073/pnas.122653799. MR1908073

[HJ] Pierre Hansen and Brigitte Jaumard, *Cluster analysis and mathematical programming*, Math. Programming **79** (1997), no. 1-3, Ser. B, 191–215, DOI 10.1016/S0025-5610(97)00059-2. Lectures on mathematical programming (ismp97) (Lausanne, 1997). MR1464767 (98g:90043)

[HMP] Pierre Hansen, Nenad Mladenović, and José A. Moreno Pérez, *Variable neighbourhood search: methods and applications*, 4OR **6** (2008), no. 4, 319–360, DOI 10.1007/s10288-008-0089-1. MR2461646 (2009m:90198)

[JMF] A. Jain, M. Murty, and P. Flynn, *Data clustering: A review*, ACM Computing Surveys, vol. 31 (1999), 264–323.

[KR] Leonard Kaufman and Peter J. Rousseeuw, *Finding groups in data*, Wiley Series in Probability and Mathematical Statistics: Applied Probability and Statistics, John Wiley & Sons Inc., New York, 1990. An introduction to cluster analysis; A Wiley-Interscience Publication. MR1044997 (91e:62159)

[KSKK] J. Kumpula, J. Saramaki, K. Kaski, and J. Kertesz, *Limited resolution and multiresolution methods in complex network community detection*, Fluctuation and Noise Letters, vol. 7 (2007), L209–L214.

[LH] S. Lehmann and L. Hansen, *Deterministic modularity optimization*, European Physical Journal B, vol. 60 (2007), 83–88.

[LM] X. Liu and T. Murata, *Advanced modularity-specialized label propagation algorithm for detecting communities in networks*, Physica A, vol. 389 (2010), 1493–1500.

[LS] W. Li and D. Schuurmans, *Modular Community Detection in Networks*, In Proc. of the 22nd. Intl Joint Conf. on Artificial Intelligence, 2011, 1366–1371.

[M] Boris Mirkin, *Clustering for data mining*, Computer Science and Data Analysis Series, Chapman & Hall/CRC, Boca Raton, FL, 2005. A data recovery approach. MR2158827 (2006d:62002)

[MAD] A. Medus, G. Acuna, and C. Dorso, *Detection of community structures in networks via global optimization*, Physica A, vol. 358 (2005), 593–604.

[MD] Jonathan P. K. Doye and Claire P. Massen, *Self-similar disk packings as model spatial scale-free networks*, Phys. Rev. E (3) **71** (2005), no. 1, 016128, 12, DOI 10.1103/PhysRevE.71.016128. MR2139325 (2005m:82056)

[MHSWL] J. Mei, S. He, G. Shi, Z. Wang, and W. Li, *Revealing network communities through modularity maximization by a contraction-dilation method*, New Journal of Physics, vol. 11 (2009), 043025.

[N04] M. Newman, *Fast algorithm for detecting community structure in networks*, Physical Review E, vol. 69 (2004), 066133.

[N06] M. Newman, *Modularity and community structure in networks*, In Proc. of the National Academy of Sciences, 2006, 8577–8582.

[NG] M. Newman and M. Girvan, *Finding and evaluating community structure in networks*, Physical Review E, vol. 69 (2004), 026133.

[NHZW] Yan Qing Niu, Bao Qing Hu, Wen Zhang, and Min Wang, *Detecting the community structure in complex networks based on quantum mechanics*, Phys. A. **387** (2008), no. 24, 6215–6224, DOI 10.1016/j.physa.2008.07.008. MR2591580 (2010k:81089)

[NR] A. Noack and R. Rotta, *Multi-level algorithms for modularity clustering*, LNCS vol. 5526 (2009), 257–268.

[RMP] T. Richardson, P. Mucha, and M. Porter, *Spectral tripartitioning of networks*, Physical Review E, vol. 80 (2009), 036111.

[RZ] J. Ruan and W. Zhang, *Identifying network communities with a high resolution*, Physical Review E, vol. 77 (2008), 016104.

[SC] P. Schuetz and A. Caflisch, *Efficient modularity optimization by multistep greedy algorithm and vertex mover refinement*, Physical Review E, vol. 77 (2008), 046112.

[SDJB] Y. Sun, B. Danila, K. Josic, and K.E. Bassler, *Improved community structure detection using a modified fine-tuning strategy*, Europhysics Letters, vol. 86 (2009), 28004.

[SM] J. Shi and J. Malik, *Normalized cuts and image segmentation*, IEEE TPAMI, vol. 22 (2000), 888–905.

[THB] M. Tasgin, A. Herdagdelen, and H. Bingol, *Community detection in complex networks using genetic algorithms*, arXiv:0711.0491, (2007).

[WT] K. Wakita and T. Tsurumi, *Finding community structure in mega-scale social networks*, Tech. Rep. 0702048v1, arXiv, 2007.

[XTP] G. Xu, S. Tsoka, and L. Papageorgiou, *Finding community structures in complex networks using mixed integer optimization*, The European Physical Journal B, vol. 60 (2007), 231–239.

Universidade Federal do Rio Grande do Norte, Campus Universitário s/n, Natal-RN, Brazil, 59072-970
E-mail address: aloise@dca.ufrn.br

GERAD & HEC Montréal, 3000, chemin de la Côte-Sainte-Catherine, Montréal (Québec) Canada, H3T 2A7
E-mail address: gilles.caporossi@hec.ca

GERAD & HEC Montréal, 3000, chemin de la Côte-Sainte-Catherine, Montréal (Québec) Canada, H3T 2A7
E-mail address: pierre.hansen@hec.ca

LIX, École Polytechnique, F-91128 Palaiseau, France
E-mail address: liberti@lix.polytechnique.fr

GERAD & HEC Montréal, 3000, chemin de la Côte-Sainte-Catherine, Montréal (Québec) Canada, H3T 2A7
E-mail address: sylvain.perron@hec.ca

INP-Grenoble, 46, avenue Félix Viallet, 38031 Grenoble Cedex 1, France
E-mail address: manuel.ruiz@grenoble-inp.fr

Network clustering via clique relaxations: A community based approach

Anurag Verma and Sergiy Butenko

ABSTRACT. In this paper, we present a general purpose network clustering algorithm based on a novel clique relaxation concept of *k-community*, which is defined as a connected subgraph such that endpoints of every edge have at least k common neighbors within the subgraph. A salient feature of this approach is that it does not use any prior information about the structure of the network. By defining a cluster as a k-community, the proposed algorithm aims to provide a clustering of a network into k-communities with varying values of k. Even though the algorithm is not designed to optimize any particular performance measure, the computational results suggest that it performs well on a number of criteria that are used in literature to evaluate the quality of a clustering.

1. Introduction

Network (graph) based data mining is an emerging field that studies network representations of data sets generated by an underlying complex system in order to draw meaningful conclusions regarding the system's properties. In a network representation of a complex system, the network's nodes typically denote the system's entities, while the edges between nodes represent a certain kind of similarity or relationship between the entities. Network clustering, aiming to partition a network into clusters of similar elements, is an important task frequently arising within this context. The form of each cluster in the partitioning is commonly specified through a predefined graph structure. Since a cluster is typically understood as a "tightly knit" group of elements, the graph theoretic concept of a *clique*, which is a subset of nodes inducing a complete subgraph, is a natural formalization of a cluster that has been used in many applications. This results in partitioning into clusters with the highest possible level of cohesiveness one can hope for.

However, in many applications modeling clusters as cliques is excessively restrictive, since a highly cohesive structure might not get identified as a cluster by the mere absence of a few edges. In real life data sets, this is of critical importance because some edges could be missing either naturally or due to erroneous data collection. Moreover, given that networks arising in many important applications tend to be very large with respect to the number of nodes and very sparse in terms of edge density, the clique clustering usually results in meaninglessly large number of clusters in such situations. In addition, computing large cliques and good clique partitions are computationally challenging problems, as

2010 *Mathematics Subject Classification.* Primary 05C85; Secondary 68R10.

This work was partially supported by NSF award OISE-0853804 and AFOSR Award FA8651-12-2-0011.

finding a maximum clique and a minimum clique partition in a graph are classical NP-hard problems [**8**].

To circumvent these issues, researchers in several applied fields, such as social network analysis and computational biology, have defined and studied structures that relax some of the properties of cliques, and hence are called clique relaxations. Some of the popular clique relaxations include s-plexes, which require each vertex to be connected to all but s other vertices [**11**]; s-clubs, which require the diameter of the induced subgraph to be at most s [**2**]; and γ-quasi-cliques, which require the density of the induced subgraph to be at least γ [**1**]. It should be noted that each of 1-plex, 1-club and 1-clique trivially represent a clique. By relaxing the properties of a clique, namely the degree, diameter, and density, these clique relaxations capture clusters that are strongly but not completely connected. However, like the clique model, these clique relaxations still suffer from the drawback of being computationally expensive.

In 1983, Seidman [**10**] introduced the concept of a *k-core* that restricts the minimum number k of direct links a node must have with the rest of the cluster. Using k-cores to model clusters in a graph has considerable computational advantages over the other clique clique relaxation models mentioned above. Indeed, the problem of finding the largest k-core can be easily solved in polynomial time by recursively removing vertices of degree less than k. As a result, the k-core model has gained significant popularity as a network clustering tool in a wide range of applications. In particular, k-core clustering has been used as a tool to visualize very large scale networks [**4**], to identify highly interconnected subsystems of the stock market [**9**], and to detect molecular complexes and predict protein functions[**3, 5**]. On the downside, the size of a k-core may be much larger than k, creating a possibility of a low level of cohesion within the resulting cluster. Because of this, a k-core itself may not be a good model of a cluster, however, it has been observed that k-cores tend to contain other, more cohesive, clique relaxation structures, such as s-plexes, and hence computing a k-core can be used as a scale-reduction step while detecting other structures [**6**].

Most recently, the authors of the current paper proposed yet another clique relaxation model of a cluster, referred to as *k-community*, that aims to benefit from the positive properties of k-cores while ensuring a higher level of cohesion [**12**]. More specifically, a k-community is a connected subgraph such that endpoints of every edge have at least k common neighbors within the subgraph. The k-community structure has proven to be extremely effective in reducing the scale of very large, sparse instances of the maximum clique problem [**12**]. This paper explores the potential of using the k-community structure as a network clustering tool. Even though the proposed clustering algorithm does not aim to optimize any of the quantitative measures of clustering quality, the results of numerical experiments show that, with some exceptions, it performs quite well with respect to most of such measures available in the literature.

The remainder of this paper is organized as follows. Section 2 provides the necessary background information. Section 3 outlines the proposed network clustering algorithm. Section 4 reports the results of numerical experiments on several benchmark instances, and Section 5 concludes the paper.

2. Background

In this paper, a network is described by a simple (i.e., with no self-loops or multi-edges) undirected graph $G = (V, E)$ with the set $V = \{1, 2, \ldots, n\}$ of nodes and the set E of edges. We call an unordered pair of nodes u and v such that $\{u, v\} \in E$ *adjacent*

Algorithm 1 $k-$ `Community`(G): Algorithm to find the k-Communities of G

Input: G, k
Output: maximum k-community of G

```
repeat
    for every {i, j} ∈ E do
        if |N_G(i) ∩ N_G(j)| < k then
            E ← E \ {{i, j}}
        end if
    end for
until No edge is removed in the current iteration
G(V_k, E_k) ← G_e[E]      /* Edge induced subgraph */
return 𝒱_k ← Connected components of G(V_k, E_k). /* Each set of connected vertices
    forms a k-community*/
```

or *neighbors*. For a node u, let $N_G(u) = \{v : \{u,v\} \in E\}$ denote the neighborhood of u in G. Then the degree $deg_G(u)$ of u in G is given by the number of elements in $N_G(u)$. Let $\delta(G)$ denote the minimum degree of a node in G. For a subset C of nodes, $G[C] = (C, E \cap (C \times C))$ denotes the subgraph induced by C. Next we define two clique relaxation concepts, namely k-core and k-community, that play a key role in this paper.

DEFINITION 2.1 (k-core). A subset of nodes C is called a k-core if $G[C]$ is a connected graph and $\delta(G[C]) \geq k$.

Before defining a k-community, we need the following two preliminary definitions.

DEFINITION 2.2 (Neighbor of an edge). A node $t \in V$ is a neighbor of an edge $\{u,v\} \in E$ if it is connected to both u and v, i.e., $\{v,t\} \in E$ and $\{u,t\} \in E$.

DEFINITION 2.3 (Edge induced subgraph). An edge induced subgraph, denoted by $G_e[F]$ is given by $G_e[F] = (V(F), F)$, where $V(F) = \{u \in V : \exists\{u,v\} \in F\}$, i.e., $G_e[F]$ is a subset of edges F of a graph G together with all the incident vertices.

We are now ready to define a k-community, which can be seen as an edge analogue of the k-core as follows.

DEFINITION 2.4 (k-Community). A k-Community of a graph G is the set of nodes in the connected edge induced subgraph $G_e[E_k]$ with each edge in E_k having at least k neighboring vertices in the subgraph $G_e[E_k]$. If $G_e[E_k]$ is disconnected, then each component forms a k-community by itself.

Given a positive integer k, both of these structures find a cluster of vertices that satisfies some minimum node degree requirements. In the case of k-core, the presence of each node has to be supported by the presence of at least k neighbors, while in the case of k-community, the presence of each edge has to be supported by the presence of at least k alternative edge-disjoint paths of length two. It is instructive to note that every k-community is also a $(k+1)$-core, but the converse is not true.

Given a positive integer k, all the maximal k-communities of a graph G can be easily computed as outlined in Algorithm 1.

3. Clustering Algorithm

The algorithm described in this section is based on the idea of finding k-communities for large k and placing them in different clusters. To this end, we identify the largest k' such

Algorithm 2 `Basic` k-`Community Clustering`(G): Basic algorithm to find clusters in G

Input: G, k, l
Output: k-community clustering $\mathscr{C}$ of G

$G' \leftarrow G$
$\mathscr{C} \leftarrow \emptyset$
repeat
  $k \leftarrow$ highest integer such that k-`community`(G') is non-empty.
  Find all the k-Communities in G' and add them to $\mathscr{C}$.
  Find the set of vertices L that are not yet clustered.
  $G' \leftarrow G[L]$.
until $k \leq l$ or G' is empty
for every $v \in L$ **do**
  Add v to the cluster $C \in \mathscr{C}$ which maximizes $|N(v) \cap C|$.
end for
return $\mathscr{C}$

Algorithm 3 `Enhanced` k-`Community Clustering`(G): Enhanced algorithm to find clusters in G

Input: G, k, l, u
Output: k-community clustering $\mathscr{C}$ of G

$G' \leftarrow G$
$\mathscr{C} \leftarrow \emptyset$, best_mod $\leftarrow -1/2$
repeat
  $k \leftarrow$ highest integer such that k-`community`(G') is non-empty.
  Find all the k-Communities in G' and add them to $\mathscr{C}$.
  Find the set of vertices L that are not yet clustered.
  $G' \leftarrow G[L]$.
  if $k \leq u$ **then**
    $\mathscr{C}^k \leftarrow \mathscr{C}$
    for every $v \in L$ **do**
      Add v to the cluster $C^k \in \mathscr{C}^k$ which maximizes $|N(v) \cap C^k|$.
    end for
    if `Modularity`($\mathscr{C}^k$) < best_mod **then**
      $\mathscr{C} \leftarrow \mathscr{C}^{k-1}$
      break
    else
      best_mod $\leftarrow$ `Modularity`($\mathscr{C}^k$)
    end if
  end if
until k=1 or G' is empty
for every $v \in L$ **do**
  Add v to the cluster $C \in \mathscr{C}$ which maximizes the increase in `Modularity`($\mathscr{C}$).
end for
`LocalSearch`($\mathscr{C}$)
return $\mathscr{C}$

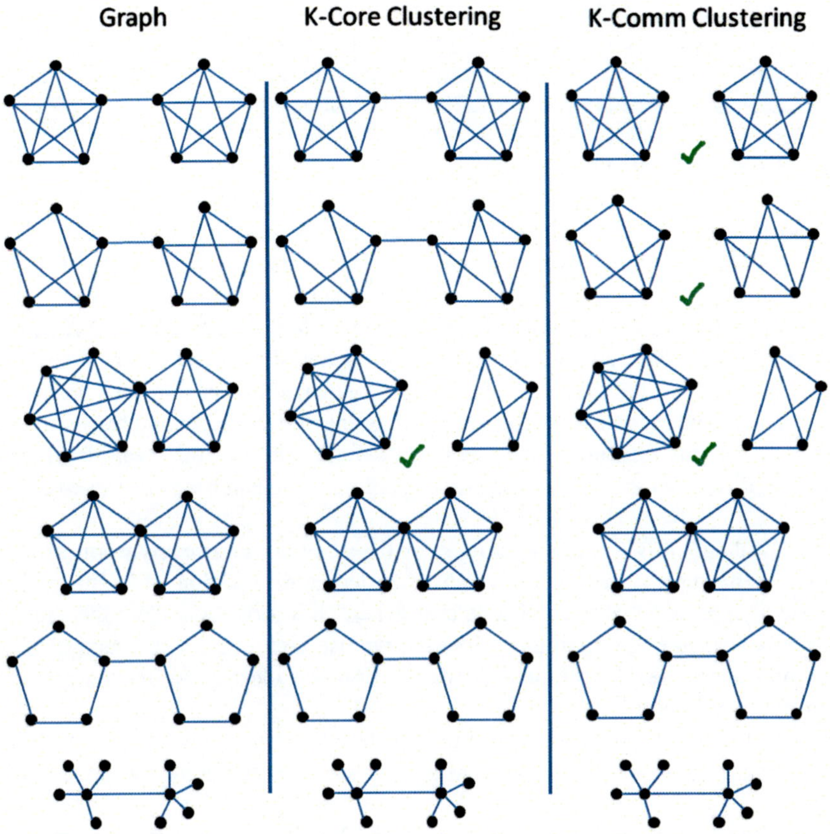

FIGURE 1. Clustering found by Algorithm 2 using the k-core and k-community-based approaches on some illustrative graphs. The diagram highlights the cases where community based approach is better than the core-based approach, and also when none of them perform well.

that the k'-community of G is non-empty, and place all k'-communities formed in distinct clusters. Once this has been done, all the nodes that have been placed in clusters are removed from G and the whole procedure is repeated till either k becomes small (reaches a lower bound l provided by the user) or no vertices are left to cluster. If any vertex is left to cluster, we attach it to the cluster that contains the most neighbors of that vertex. This basic procedure is described in Algorithm 2.

In this algorithm, we stop when k becomes small enough so that a k-community becomes meaningless. For example, any set of vertices that induce a tree will form a 0-community. While in some cases this might be the best possible option (the original graph is a forest), for most clustering instances we would like the vertices in a cluster to share more than just one edge with the remaining nodes. For this paper, the lower bound l was set to 1 in Algorithm 2.

It should be noted that the clustering provided by Algorithm 2 does not aim to optimize any criteria provided such as modularity, performance, average isolated inter-cluster conductance (aixc), average isolated inter-cluster expansion (aixe), and minimum intra-cluster density (mid) as described in the DIMACS 2011 challenge [7].

3.1. Enhancements. If optimizing a given measure is indeed the aim, an enhanced version of the basic algorithm is provided in Algorithm 3. The description of the enhanced Algorithm 3 uses modularity as a measure, but can as well have any other measure. A major improvement in Algorithm 3 over Algorithm 2 is that the decision of the what k is too small to be used for finding k-communities as clusters is made dynamically. Given a range $[l,u]$, the algorithm checks the modularity of the clustering found at each $k \leq u$ and stops as soon as reducing k also reduces modularity. In this manner, the formation of k-communities for small k that don't contribute to increasing modularity can be avoided. Furthermore, local search is done to increase modularity by moving vertices from one cluster to another cluster such that the increase in modularity is maximized. For the results in this paper, the range $[l,u]$ was set to $[1,6]$, and the time spent in local search was restricted to $10,000$ seconds.

An advantage of both these algorithms is that they do not use any prior information about the graph such as the number of clusters, degree distribution, etc. This makes it a very general approach that is applicable even when no information about the structure of the graph is available. Furthermore, although we use k-core and k-community to define clusters, new structures that fit the users description of a cluster can be incorporated into the algorithm fairly easily.

In both the Algorithms 2 & 3, we can replace k-community in steps 4-5 with k-core, with the remaining steps of the algorithm as they are, to obtain a k-core-based clustering algorithm.

Some illustrations of clusterings found by the k-core and k-community approach described in this section are provided in Figure 1. It should be noted that, although k-communities are strictly stronger relaxations, the clustering formed by the core-based approach can in some cases be better than that obtained using the community-based approach.

4. Computational Results

In this section we provide computational results obtained by using the k-community and k-core clustering on the graph sets provided in the DIMACS 2011 challenge [7]. The computational results were obtained on a desktop machine (Intel Xeon E5620@2.40GHz, 16 cores, 12GB RAM). All computations except for the final steps of attaching leftover vertices to already formed clusters and the local search used only one core. The local search and attaching leftover vertices were parallelized using OpenMP with 16 threads.

Table 1 presents the modularity and number of clusters found by Algorithm 2 using the k-core and k-community clustering for 27 graphs. For each graph, the higher of the two modularities as found be the two methods is highlighted in bold. It can be seen that k-community clustering is better on about half of the instances (14 of the 27 graphs tested). However, a closer look suggests that when the k-community based clustering significantly outperforms (difference in modularity more than 0.2) k-core clustering in 5 of those 14 instances, while k-community based clustering is significantly outperformed by k-core clustering only once out of the remaining 13 instances. Some noteworthy examples are the *preferentialAttachment, smallworld, luxembourg.osm* and *belgium.osm* graphs, where the almost all nodes in the graph are identified as 4-, 6-, 1- & 1-cores respectively and placed

TABLE 1. Modularity of clustering found by the basic Algorithm 2 using the k-community based and k-core based approaches. The modularity that is higher between the two methods is highlighted in bold.

Graphs	n	m	Method	Mod	Clusters	Time (s)
celegans_metabolic	453	2025	core	0.267	19	0.00
			comm	**0.331**	30	0.02
email	1133	5451	core	0.342	15	0.03
			comm	**0.394**	72	0.03
polblogs	1490	16715	core	**0.243**	8	0.03
			comm	0.219	32	0.06
power	4941	6594	core	0.295	24	0.05
			comm	**0.851**	189	0.09
PGPgiantcompo	10680	24316	core	**0.755**	398	0.47
			comm	0.732	655	0.64
astro-ph	16706	121251	core	**0.539**	918	1.70
			comm	0.538	1480	1.95
memplus	17758	54196	core	**0.555**	1238	0.56
			comm	0.554	1256	0.58
as-22july06	22963	48436	core	0.473	33	0.41
			comm	**0.519**	162	0.59
cond-mat-2005	40421	175691	core	**0.509**	2469	3.85
			comm	0.508	4016	4.99
kron_g500-simple-logn16	65536	2456071	core	-0.018	6	15.31
			comm	**-0.013**	28	38.60
preferentialAttachment	100000	499985	core	0.000	1	1.01
			comm	**0.145**	299	14.85
G_n_pin_pout	100000	501198	core	0.065	2	4.23
			comm	**0.136**	4479	33.93
smallworld	100000	499998	core	0.000	4	0.48
			comm	**0.570**	11129	9.64
luxembourg.osm	114599	119666	core	0.000	1	10.50
			comm	**0.955**	68	95.47
rgg_n_2_17_s0	131072	728753	core	**0.752**	7539	9.91
			comm	0.612	15572	13.64
caidaRouterLevel	192244	609066	core	**0.625**	5436	55.83
			comm	0.605	6005	78.57
coAuthorsCiteseer	227320	814134	core	**0.701**	17185	102.99
			comm	0.690	23562	127.65
citationCiteseer	268495	1156647	core	**0.481**	2145	91.69
			comm	0.433	11499	194.66
coPapersDBLP	540486	15245729	core	**0.670**	31213	1429.25
			comm	0.669	34267	1557.58
eu-2005	862664	16138468	core	0.304	18403	1965.33
			comm	**0.404**	30380	2570.01
audikw1	943695	38354076	core	0.241	10190	550.23
			comm	**0.389**	22076	1151.01
ldoor	952203	22785136	core	0.091	361	20.23
			comm	**0.392**	2	42.06
kron_g500-simple-logn20	1048576	44619402	core	-0.026	5	1554.64
			comm	**-0.025**	1788	3155.71
in-2004	1382908	13591473	core	**0.632**	29528	2774.93
			comm	0.625	43454	3416.69
belgium.osm	1441295	1549970	core	0.000	2	889.65
			comm	**0.983**	2326	7118.33
cage15	5154859	47022346	core	**0.813**	4958	14451.30
			comm	0.544	174163	259.33

in one huge cluster by the k-core clustering. On the other hand, the k-community clustering is able to identify a more meaningful clustering. The examples provided in Figure 1 point to some potential reasons why k-cores are not able to cluster these graphs as well as k-communities do.

TABLE 2. Modularity of clustering found by the enhanced Algorithm 3 using the k-community based and k-core based approaches. The modularity that is higher between the two methods is highlighted in bold. The improvement in modularity when compared to the basic Algorithm 2 and the time taken are also provided.

Graphs	n	m	Method	Mod	Improv	Time (s)
celegans_metabolic	453	2025	core	0.360	0.092	0.16
			comm	**0.402**	0.071	0.17
email	1133	5451	core	0.477	0.134	0.98
			comm	**0.542**	0.148	0.62
polblogs	1490	16715	core	0.419	0.176	2.75
			comm	**0.426**	0.206	0.16
power	4941	6594	core	0.759	0.464	0.55
			comm	**0.860**	0.009	0.50
PGPgiantcompo	10680	24316	core	0.835	0.080	1.54
			comm	**0.848**	0.116	1.59
astro-ph	16706	121251	core	**0.651**	0.112	25.93
			comm	0.646	0.108	6.94
memplus	17758	54196	core	**0.537**	-0.017	4.62
			comm	**0.537**	-0.017	4.45
as-22july06	22963	48436	core	0.513	0.041	113.67
			comm	**0.603**	0.084	43.85
cond-mat-2005	40421	175691	core	**0.625**	0.116	273.29
			comm	0.620	0.112	16.13
kron_g500-simple-logn16	65536	2456071	core	**0.023**	0.040	10019.40
			comm	0.014	0.027	1700.88
preferentialAttachment	100000	499985	core	0.000	0.000	22.00
			comm	**0.243**	0.097	10041.40
G_n_pin_pout	100000	501198	core	0.065	0.000	131.02
			comm	**0.212**	0.076	10047.00
smallworld	100000	499998	core	0.000	0.000	19.63
			comm	**0.753**	0.184	43.99
luxembourg.osm	114599	119666	core	0.000	0.000	29.00
			comm	**0.958**	0.003	233.72
rgg_n_2_17_s0	131072	728753	core	**0.871**	0.119	35.77
			comm	0.800	0.188	72.29
caidaRouterLevel	192244	609066	core	0.776	0.151	5447.02
			comm	**0.821**	0.216	340.88
coAuthorsCiteseer	227320	814134	core	**0.823**	0.122	397.66
			comm	0.817	0.127	211.66
citationCiteseer	268495	1156647	core	0.639	0.157	10142.10
			comm	**0.709**	0.276	483.39
coPapersDBLP	540486	15245729	core	**0.716**	0.046	2581.11
			comm	0.715	0.046	2720.36
eu-2005	862664	16138468	core	0.671	0.367	15205.00
			comm	**0.757**	0.353	11874.90
audikw1	943695	38354076	core	0.325	0.084	10826.60
			comm	**0.637**	0.248	11231.10
ldoor	952203	22785136	core	0.092	0.001	6130.62
			comm	**0.392**	0.000	847.84
kron_g500-simple-logn20	1048576	44619402	core	-0.024	0.002	11626.20
			comm	**0.010**	0.036	13737.80
in-2004	1382908	13591473	core	0.924	0.292	6033.33
			comm	**0.926**	0.302	5887.41
belgium.osm	1441295	1549970	core	0.000	0.000	55142.10
			comm	**0.983**	0.000	7112.92
cage15	5154859	47022346	core	**0.816**	0.004	25787.80
			comm	0.709	0.165	71808.90

TABLE 3. The modularity (Mod), coverage (Cov), mirror coverage (MCov), performance (Perf), average isolated inter-cluster conductance (Aixc), average isolated inter-cluster expansion (Aixe), and minimum intra-cluster density (Mid) found by the basic Algorithm 2 and enhanced Algorithm 3 using k-community. The higher Mod, Perf, Aixc, Aixe, and Mid entries amongst the two algorithms are highlighted in bold.

Graph	Method	Mod	Cov	Mcov	Perf	Aixc	Aixe	Mid
celegans_metabolic	Alg. 2	0.33	0.57	0.86	0.85	**0.50**	**3.25**	0.05
	Alg. 3	**0.40**	0.58	0.88	**0.87**	0.34	2.17	**0.06**
email	Alg. 2	0.39	0.44	0.96	**0.96**	**0.58**	**5.26**	0.02
	Alg. 3	**0.54**	0.62	0.93	0.93	0.38	3.15	**0.03**
polblogs	Alg. 2	0.22	0.39	0.91	**0.91**	**0.09**	**1.79**	0.02
	Alg. 3	**0.43**	0.93	0.68	0.68	0.01	0.04	**0.04**
power	Alg. 2	0.85	0.86	0.99	**0.99**	**0.16**	**0.48**	**0.02**
	Alg. 3	**0.86**	0.87	0.99	**0.99**	0.15	0.44	**0.02**
PGPgiantcompo	Alg. 2	0.73	0.74	1.00	**1.00**	**0.21**	**0.96**	**0.01**
	Alg. 3	**0.85**	0.89	0.95	0.95	0.11	0.52	0.00
astro-ph	Alg. 2	0.54	0.54	1.00	**1.00**	0.39	**2.85**	**0.04**
	Alg. 3	**0.65**	0.66	1.00	**1.00**	**0.58**	1.89	0.01
memplus	Alg. 2	**0.55**	0.63	0.99	**0.99**	**0.24**	1.09	**0.01**
	Alg. 3	0.54	0.83	0.76	0.76	0.21	**1.21**	0.00
as-22july06	Alg. 2	0.52	0.72	0.86	0.86	**0.33**	**1.17**	**0.00**
	Alg. 3	**0.60**	0.73	0.90	**0.90**	0.32	1.08	**0.00**
cond-mat-2005	Alg. 2	0.51	0.51	1.00	**1.00**	0.45	**2.40**	**0.01**
	Alg. 3	**0.62**	0.62	1.00	**1.00**	**0.71**	1.92	**0.01**
kron_g500-simple-logn16	Alg. 2	-0.01	0.33	0.73	**0.72**	**0.00**	**0.17**	**0.00**
	Alg. 3	**0.01**	0.67	0.47	0.47	**0.00**	0.06	**0.00**
preferentialAttachment	Alg. 2	0.15	0.47	0.56	0.56	**0.90**	**24.23**	**0.00**
	Alg. 3	**0.24**	0.38	0.88	**0.87**	0.77	7.30	**0.00**
G_n_pin_pout	Alg. 2	0.14	0.52	0.60	0.60	**0.80**	**8.79**	**0.00**
	Alg. 3	**0.21**	0.47	0.74	**0.74**	0.75	7.72	**0.00**
smallworld	Alg. 2	0.57	0.57	1.00	**1.00**	**0.49**	**4.91**	**0.13**
	Alg. 3	**0.75**	0.75	1.00	**1.00**	0.28	2.81	0.02
luxembourg.osm	Alg. 2	**0.96**	0.99	0.96	**0.96**	**0.03**	**0.07**	**0.00**
	Alg. 3	**0.96**	0.99	0.96	**0.96**	0.02	0.06	**0.00**
rgg_n_2_17_s0	Alg. 2	0.61	0.61	1.00	**1.00**	**0.45**	**4.71**	**0.20**
	Alg. 3	**0.80**	0.80	1.00	**1.00**	0.22	2.50	0.06
caidaRouterLevel	Alg. 2	0.61	0.62	0.99	**0.99**	**0.38**	1.81	**0.00**
	Alg. 3	**0.82**	0.85	0.97	0.97	0.96	**2.13**	**0.00**
coAuthorsCiteseer	Alg. 2	0.69	0.69	1.00	**1.00**	**0.31**	**1.83**	**0.01**
	Alg. 3	**0.82**	0.82	1.00	**1.00**	0.17	1.38	0.00
citationCiteseer	Alg. 2	0.43	0.45	0.98	0.98	**0.48**	**3.69**	**0.00**
	Alg. 3	**0.71**	0.72	0.99	**0.99**	0.29	2.49	**0.00**
coPapersDBLP	Alg. 2	0.67	0.67	1.00	**1.00**	**0.44**	**9.65**	**0.15**
	Alg. 3	**0.72**	0.72	1.00	**1.00**	0.30	8.75	0.10
eu-2005	Alg. 2	0.40	0.41	0.99	**0.99**	**0.67**	**21.32**	**0.00**
	Alg. 3	**0.76**	0.81	0.98	0.98	0.23	6.69	**0.00**
audikw1	Alg. 2	0.39	0.51	0.90	**0.90**	**0.83**	**51.18**	**0.00**
	Alg. 3	**0.64**	0.77	0.87	0.87	0.04	2.96	**0.00**
ldoor	Alg. 2	**0.39**	1.00	0.39	**0.39**	**0.00**	**0.11**	**0.00**
	Alg. 3	**0.39**	1.00	0.39	**0.39**	**0.00**	**0.11**	**0.00**
kron_g500-simple-logn20	Alg. 2	-0.03	0.37	0.79	**0.79**	**0.01**	**0.53**	**0.00**
	Alg. 3	**0.01**	0.67	0.59	0.59	0.00	0.01	**0.00**
in-2004	Alg. 2	0.62	0.63	1.00	**1.00**	**0.40**	**12.63**	**0.00**
	Alg. 3	**0.93**	0.94	0.99	0.99	0.19	2.33	**0.00**
belgium.osm	Alg. 2	**0.98**	0.98	1.00	**1.00**	**0.04**	**0.11**	**0.00**
	Alg. 3	**0.98**	0.98	1.00	**1.00**	**0.04**	**0.11**	**0.00**
cage15	Alg. 2	0.54	0.55	1.00	**1.00**	**0.67**	**10.87**	**0.00**
	Alg. 3	**0.71**	0.71	1.00	**1.00**	0.48	9.58	**0.00**

In addition, Table 1 also reports the time taken by the two approaches on each of the graphs. It can be seen that our approach scales well for large graphs, with graphs with up to 5 million vertices solved in reasonable time on a desktop machine.

Table 2 presents the modularity and number of clusters found by Algorithm 3 using the k-core and k-community clustering for the same 27 graphs. It can be seen that k-community based clustering outperforms k-core based clustering in 19 of the 27 instances. On an average, the improvement in the modularity was 0.099 for the k-core based clustering and 0.122 for the k-community based clustering. The time required for clustering increases, but is still within reasonable limit. A user can decide for or against using enhancements depending on the trade-off between the extra time required and the increase in modularity.

Table 3 presents the modularity, coverage, mirror coverage, performance, average isolated inter-cluster conductance, average isolated inter-cluster expansion, and minimum intra-cluster density for the clusterings found by the basic Algorithm 2 and the enhanced Algorithm 3 using the k-community based approach. For each graph, the table highlights the higher modularity, performance, average isolated inter-cluster conductance, average isolated inter-cluster expansion, and minimum intra-cluster density entries amongst the respective columns. It can be noted that while the enhanced Algorithm 3 increases the modularity, it has an adverse effect on the other clustering measures. This is an important observation that suggests that modularity maximization should not be used as the sole measure of good clustering.

5. Conclusion

This paper introduces k-community clustering, which can be thought of as something between k-core clustering and clique partitioning. The use of polynomially computable k-community not only provides a faster approach, but also provides a more effective clustering method by being able to identify cohesive structures that might not be cliques. k-Community clustering also provides advantages over k-core clustering due to the more cohesive nature of a k-community. As our computational results show, both the k-core and k-communities perform well for certain graphs, but k-community approach outperforms the k-core approach in general.

Acknowledgements

This work was partially supported by NSF award OISE-0853804 and AFOSR Award FA8651-12-2-0011.

References

[1] James Abello, Mauricio G. C. Resende, and Sandra Sudarsky, *Massive quasi-clique detection*, LATIN 2002: Theoretical informatics (Cancun), Lecture Notes in Comput. Sci., vol. 2286, Springer, Berlin, 2002, pp. 598–612, DOI 10.1007/3-540-45995-2_51. MR1966153

[2] Richard D. Alba, *A graph-theoretic definition of a sociometric clique*, J. Mathematical Sociology **3** (1973), 113–126. MR0395938 (52 #16729)

[3] M. Altaf-Ul-Amin, K. Nishikata, T. Koma, T. Miyasato, Y. Shinbo, M. Arifuzzaman, C. Wada, and M. Maeda et al., *Prediction of protein functions based on k-cores of protein-protein interaction networks and amino acid sequences*, Genome Informatics **14** (2003), 498–499.

[4] J. I. Alvarez-Hamelin, L. Dall'Asta, A. Barrat, and A. Vespignani, *k-core decomposition: a tool for the visualization of large scale networks*, CoRR **abs/cs/0504107** (2005).

[5] G. D. Bader and C. W. V. Hogue, *An automated method for finding molecular complexes in large protein interaction networks*, BMC Bioinformatics **4** (2003), no. 2.

[6] Balabhaskar Balasundaram, Sergiy Butenko, and Illya V. Hicks, *Clique relaxations in social network analysis: the maximum k-plex problem*, Oper. Res. **59** (2011), no. 1, 133–142, DOI 10.1287/opre.1100.0851. Electronic companion available online. MR2814224 (2012e:91241)

[7] DIMACS, *10th DIMACS Implementation Challenge: Graph Partitioning and Graph Clustering*, http://www.cc.gatech.edu/dimacs10/, 2011.

[8] Michael R. Garey and David S. Johnson, *Computers and intractability*, W. H. Freeman and Co., San Francisco, Calif., 1979. A guide to the theory of NP-completeness; A Series of Books in the Mathematical Sciences. MR519066 (80g:68056)

[9] J. Idicula, *Highly interconnected subsystems of the stock market*, NET Institute Working Paper No. 04-17, 2004, Available at SSRN: http://ssrn.com/abstract=634681.

[10] Stephen B. Seidman, *Network structure and minimum degree*, Social Networks **5** (1983), no. 3, 269–287, DOI 10.1016/0378-8733(83)90028-X. MR721295 (85e:05115)

[11] Stephen B. Seidman and Brian L. Foster, *A graph-theoretic generalization of the clique concept*, J. Math. Sociol. **6** (1978/79), no. 1, 139–154, DOI 10.1080/0022250X.1978.9989883. MR506325 (80j:92014)

[12] A. Verma and S. Butenko, *Maximum clique problem on very large scale sparse networks*.

INDUSTRIAL & SYSTEMS ENGINEERING, 3131 TAMU, TEXAS A&M UNIVERSITY, COLLEGE STATION, TEXAS
E-mail address: anuragverma@tamu.edu

INDUSTRIAL & SYSTEMS ENGINEERING, 3131 TAMU, TEXAS A&M UNIVERSITY, COLLEGE STATION, TEXAS
E-mail address: butenko@tamu.edu

Identifying base clusters and their application to maximizing modularity

Sriram Srinivasan, Tanmoy Chakraborty, and Sanjukta Bhowmick

ABSTRACT. Modularity maximization is an effective technique for identifying communities in networks that exhibit a natural division into tightly connected groups of vertices. However, not all networks possess a strong enough community structure to justify the use of modularity maximization. We introduce the concept of base clusters—that is group of vertices that form the kernel of each community and are always assigned together independent of the community detection algorithm used or the permutation of the vertices. If the number of vertices in the base clusters is high then the network is likely to have distinct communities and is suitable for the modularity maximization approach. We develop an algorithm for obtaining these base clusters and show that identifying base clusters as a preprocessing step can help in improving the modularity values for agglomerative methods.

1. Introduction

Many complex networks, such as those arising in biology [**V**], social sciences [**B1**] and epidemiology [**B3**] exhibit community structure, that is, there exists a natural division of groups of vertices that are tightly connected within themselves and sparsely connected across the groups. Identifying such naturally occurring communities is an important operation in analyzing complex networks. A popular method for obtaining good communities is by optimizing the modularity of the network. The higher the modularity, generally the better the distribution into communities. Therefore, many community detection algorithms are designed with the objective function of improving the modularity.

There exists several issues in using modularity as a metric for community detection. Finding the maximum modularity is a NP-complete problem [**B5**] and therefore like other combinatorial optimization problems, the ordering of the vertices in the network can significantly affect the results. Although high modularity values often indicate good divisions into communities, the highest modularity value need not reflect the best community division, as in examples exhibiting the resolution limit [**G1**]. Similarly a near-optimal modularity does not necessarily mean

---

2010 *Mathematics Subject Classification.* Primary 68R10.

Key words and phrases. Modularity maximization, agglomerative methods, communities in complex networks.

This work has been supported by the College of Information Science and Technology, University of Nebraska at Omaha and the FIRE grant from UNO Sponsor Office.

the division is also near-optimal. However, this metric has been effective in finding communities in networks, where there exists an inherent and strong community structure–the key proposition being that the network can be naturally divided into distinct communities. Most community detection algorithms, that are based on modularity optimization, however, do not contain any mechanism to ascertain whether the network indeed has a modularity structure. This is a "chicken-and-egg" problem because in order to discover communities, we first have to make sure that they exist.

In this paper, we propose a solution to this problem by introducing the concept of *base clusters* in communities. Base clusters consist of sets of vertices that form the kernel of each community in the network, and are groups of vertices that are always assigned to the same community, independent of the modularity maximization algorithm employed or the order in which the vertices are processed.

A naive, but effective method for identifying these base clusters of vertices would be to execute different community detection methods, and different vertex orderings and then comparing the groupings to find vertices that are always assigned to the same cluster. This approach has been implemented in [**O**] as part of their ensemble learning and recently in [**L1**] where they are called consensus clusters. However this technique is expensive because it requires executing *all* the algorithms in the set, and the effect of a bad permutation may persist over several of the methods. We propose an orthogonal method of finding base clusters that is based only on the topology of the network.

In addition to indicating whether a network indeed posseses community structure, base clusters can also be used as a preprocessing step to modularity maximization. First the base clusters are identified and assigned to the same community, because they are guaranteed to be in the same group, and then modularity maximization is applied to the smaller network. Combining base clusters as an initial step helps bias the network to move towards the correct community division and generally also increases modularity. In this paper, we study the effect of preprocessing using base clustering on two agglomerative modularity maximization methods–(i) proposed by Clauset *et. al.* in [**C**] (henceforth referred to as the CNM method) and (ii) proposed by Blondel *et. al.* in [**B2**] (henceforth referred to as the Louvain method). These two methods are both based on a greedy approach of combining pairs of vertices at each step that lead to the most increase in modularity.

The remainder of this paper is arranged as follows. In Section 2, we provide a brief overview of the network terminology used in this paper, short descriptions and a comparison of the CNM and Louvain methods and discussion on a few other preprocessing algorithms for modularity maximization. In Section 3, we present our main contribution–an algorithm to find base clusters in networks. In Section 4, we present experimental results of using base clusters as a preprocessing step to modularity maximization and discuss the effectiveness of this technique. We conclude in Section 5 with a discussion of future research.

2. Background

Terminology: A network (or graph) $G = (V, E)$ consists of a set of vertices V and a set of edges E. An edge $e \in E$ is defined by two vertices $\{u, v\}$ which are called its *endpoints*. A vertex u is a *neighbor* of v if they are share an edge. The *degree* of a vertex u is the number of its neighbors. A *path*, of length l, in a graph

G is an alternating sequence of $v_0, e_1, v_1, e_2, \ldots, e_l, v_l$ vertices and edges, such that for $j = 1, \ldots, l$; v_{j-1} and v_j are the endpoints of edge e_j, with no edges or internal vertices repeated. The *distance* of a vertex to another is the length of the shortest path between these vertices.

The *clustering coefficient* of a vertex indicates whether a vertex is part of a dense module. This value is computed as the ratio of the edges between the neighbors of a vertex to the total possible connections between the neighbors. If a vertex has a large clustering coefficient then all its neighbors are connected, therefore the vertex is part of a clique. Another metric related to the clustering coefficient is the *fill-in*. The *fill-in* of a vertex is the number of extra edges required such that the vertex and its neighbors form a clique. The fill-in of a vertex is computed as the total possible connections between the neighbors - the edges between the neighbors.

Clustering algorithms for networks are based on identifying tightly connected groups of vertices. However, mere comparison of edges within and outside groups is not always an appropriate measure for communities. This is because certain areas real-world complex networks, particularly those based on social sciences also exhibit random connections, and the effect of these random subnetworks have to be taken into account. The metric of modularity was proposed by Newman and Girvan [**N2**] and is based on the idea that there are no strong communities within random networks.

Modularity Maximization: Modularity on undirected graphs is computed as follows; Given a partition of a network into M groups, let C_{ij} represent the fraction of total links starting at a node in group i and ending at a node in group j. Let $a_i = \sum_j C_{ij}$ correspond to the fraction of links connected to subgroup i. Under random connections, the probability of links that begin at a node in i, is a_i, and the probability of links that end at a node in j, a_j. Thus, the expected number of within-community links, between nodes with group i, is a_i^2. The actual fraction of links within each group is C_{ii}. So, a comparison of the actual and expected values, summed over all groups of the partition gives us the modularity, which is the deviation of the partition from random connections: $Q = \sum(C_{ii} - a_i^2)$. Maximizing modularity is a popular method for finding good communities in networks. However finding the optimal modularity is an NP-hard problem [**B5**]. There exist many heuristics for maximizing modularity including spectral partitioning, divisive and agglomerative methods [**P**]. We now discuss the two agglomerative modularity maximization algorithms used in this paper;

The CNM method is a greedy agglomerative algorithm developed by Clauset *et. al.* [**C**]. This method initially considers every vertex in the network as an individual community. At each iteration, the pair of communities with the highest increase in modularity are merged. The process is repeated until there exists no combination of vertices that increase modularity. The runtime of the CNM method is improved by using heaps to store the edges and their associated increase in modularity.

The Louvain method, developed by Blondel et. al. [**B2**], also uses a greedy approach and initially assigns each vertex to an individual community. However, instead of a search over all edges, the Louvain method executes a local search over the edges of each vertex. Each vertex is combined with the neighbor that most increases its modularity, although in subsequent steps of the iteration, the neighbor itself can be detached from its original community to join a new one. Thus in one pass through the network the algorithm can identify multiple local communities and reduce the number of iterations. Additionally, by allowing vertices to be removed from earlier communities, the Louvain method provides a mechanism for correcting initial bad choices. This process of reassigning communities is repeated over several iterations (called the inner iteration), until modularity is increased. Once this first-phase allocation of communities is determined, the Louvain method joins the vertices within a community into supervertices (in the outer iteration). The inner iteration is then repeated over these supervertices. The steps of the inner and outer iterations are executed repeatedly until the number of communities is suitably small and the modularity cannot be increased any further.

The CNM method is generally the slower of the two, because it finds the maximum over all edges per iteration as opposed to the Louvain method which executes a combination for each vertex if possible. However, if the increase in modularity is equal over most of the edges in the network, the inner iterations in the Louvain method can spend many steps needlessly moving from one community to another. An advantage of the Louvain method over the CNM method is the opportunity to backtrack from a community if necessary, so long as it is within the same inner iteration. Despite these apparent differences, both the CNM and Louvain method are based on the same principle—a greedy combination of communities to maximize modularity. We posit that the difference is in the implementation of the methods. For example if the number of loops per inner iteration is set to 1, then the Louvain method would be exactly like the CNM method—combining pairs of community and not backtracking. Because of their similarity, we use two methods to compare the effectiveness of base clustering.

Comparsion Between Two Community Partitions Although modularity maximization is the designated objective function of the algorithms, comparing the values may not always give a clear picture about the community structure. This is because in certain networks two different partition schemes can give identical modularity values. One method of comparing across two partitions (obtained from different algorithms) is by using the Rand Index [**R1**]. Given two different partitions of a network, the Rand Index is computed as follows; Let a be the pair of vertices that are present in the same community over both the partitions, let b be the pair of nodes that were in different communities for both the partitions, then the Rand Index is computed as the ratio of the sum of a and b over all possible pairs of vertices. A high Rand Index (maximum value 1) indicates that the two partitions are equal and a low Rand Index indicates that they are very dissimilar.

Preprocessing For Modularity Maximization: Since the value of modularity is affected by implementation factors such as the vertex orderings, there exist several preprocessing techniques to improve the results, including simple but effective methods such as pruning the outlying (generally degree 1) vertices. The methods most similar to the base clustering approach include a seeded community method by Reidy *et. al.* [**R2**], an iterated random walk strategy by Liu *et. al.* [**L2**] and an ensemble learning based approach by Ovelgonne *et. al.* [**O**]. In the seeded community method, an initial set of seed vertices are given as input and the communities are grown outwardly from these seeds. The random walk method is based on the observation that given a network with community structure, a random walk generally remains within a community. In the preprocessing step several random walks are performed to obtain a better estimate of the vertices in the same community. The ensemble learning based approach executes a base algorithm several times to get a consensus of the smaller clusters. Once these clusters are obtained, then the base algorithm is executed over the entire graph. Note that all these three preprocessing steps are variations of finding a kernel community, like our proposed method of base clusters. Our method differs in that we try to estimate to base clusters based only on the topology of the network and instead of presupposing the existence of a communities, we use base clusters to estimate wether a network indeed has good community structure. Our clusters should ideally be invariant for a given network because they are not based on random selections such as the seeded method and the random walk nor on the effect of an underlying algorithm as in the case of the ensemble learning method. However, this is not always possible practically, and the benefits and issues of the base clustering method are discussed in Section 3.

Other works, not on preprocessing, but dealing with core communities include a study statistically significant communities by perturbing the connectivity of the network and then comparing change in community structures by Karrer *et. al.* [**K1**] and a recent work by Fortunato *et. al.* [**L1**] that looks at the consensus communities over different community detection algorithms on synthetically generated networks of varying degrees of community structure.

3. Finding Base Clusters in Complex Networks

Given a network, our objective is to estimate whether the network indeed posseses distinct communities. It has been observed that the permutation of the vertex order can change the partition to communities, and if the network does not have a strong community structure these partitions can significantly vary. We conducted a preliminary experiment for finding consensus communities—that is groups of vertices that are always grouped together over different permutations. As shown in Figure 1, the number of consensus communities, by using the CNM method, keep on increasing with the number of permutations of the vertices for the networks, Jazz (network of jazz musicians) [**G2**] and Power (network of a power grid) [**W**]. However, in spite of the relatively large number of different consensus groups, the bulk of the nodes (172 out of total of 198)of Jazz are concentrated in three large communities and the rest communities are composed of 2-3 vertices each. This result highlights that Jazz has a strong community structure and also that a small percentage of its nodes are not strictly attached to any of the major communities. In contrast only 72 of the 4941 nodes of the Power graph are in the three largest communities and the rest are scattered in groups of 3-4 over the rest of the smaller communities. Clearly the community structure of Power is not as strong as that of Jazz. However, both these networks are common benchmarks for evaluating modularity maximization algorithms.

We consider communities to have a kernel, that grows outwards to form the community. The vertices at the edge of the community are most likely to be the ones that migrate over different permutations, and the inner vertices form the consensus communities. We will focus on finding these kernels, which we call base clusters to distinguish that they do not represent the entire consensus community. We conjecture that the base clusters are the most tightly connected groups of vertices in the network which facilitates larger communities to be built around them.

A naive method for identifying these base clusters might be to search for densely connected set of vertices, preferably large cliques. However as shown in Figure 2a,

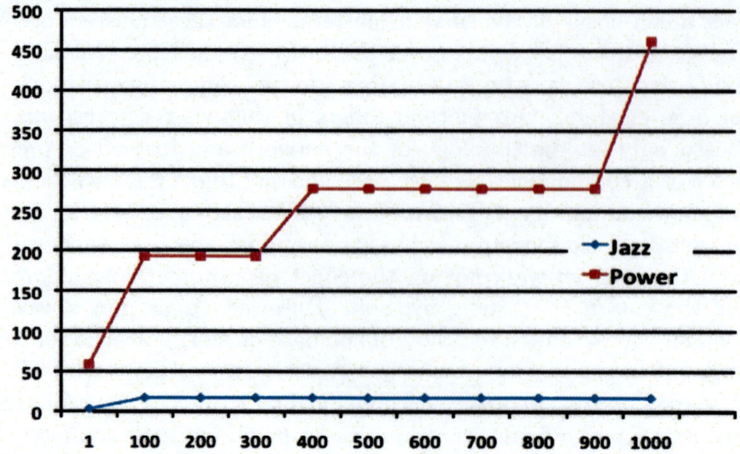

FIGURE 1. **Increase of Consensus Communities with Vertex Permutations.** The X-axis is the number of different permutations applied to the network and the Y-axis is the number of consensus communities under the CNM method. Jazz shows a relatively slow increase, indicating good community structure as compared to Power.

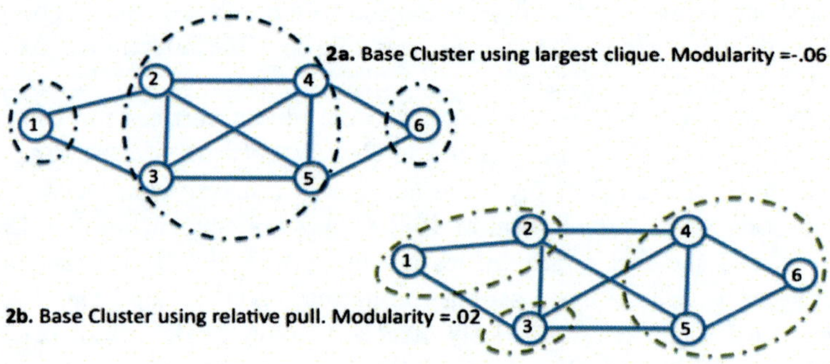

FIGURE 2. **Division of network into communities.** Figure 2a. combines a large clique, but with greater external pull. Figure 2b. distributes the pull amongst the communities.

members of cliques may not always fall in the same community. In the example vertices (2,3,4,5) form a clique. But a partition of the six vertices as ({1}, {2,3,4,5}, {6}) gives a negative modularity of -.06 This is because the even though the vertices in the clique are tightly connected amongst themselves, each subgroup (2,3) and (4,5) also have a strong connection to an external community. For example (2,3) has two edges to the external vertex (1) and also two edges to the internal vertex (4). Thus after (2,3) is combined it is equally likely that it can combine with (1) or with (4) or with (5).

Ideally, each subgroup within a base community should have more internal connections than external ones, to resist the pull from vertices outside the group. But it is expensive to find groups of vertices that satisfy this condition. We therefore temporize and look for clusters where *the number of internal connections is considerable greater than the external connections.* In the results presented in this paper we set the parameters such that the number of external connections is less than half the number of internal connections. However unless the network has extremely well-defined communities, even this condition is not always prevalent.

To accommodate base clusters with more external edges, we note that having more external edges is not necessarily bad so long as the external connections are to different communities. This way the "pull" from other communities is reduced, even though there are more outside connections. Figure 2b gives an example where the network is partitioned such that despite having more external edges, the "pull" is dissipated amongst different communities. The problem however, is that we have not yet grouped the vertices into communities. Therefore, we do not know which of edges point to the same community.

We use the community diameter to estimate the kernels. We define a community to have diameter d, if the shortest path between two vertices in that community is d. We assume that consensus communities have diameters of at least 2. Then, if a base cluster is composed of a core vertex and its distance-1 neighbors, the neighbors of neighbors, i.e. vertices at distance 2 from the core vertex are first ones that can be on the edge of the community. We identify base clusters such that these distance-2 vertices exert less pull on the distance-1 neighbors as follows;

We compute the fill-in of the vertices in the network and identify ones with low fill-in (generally 0-2). We form a temporary base community C composed of the vertex v and its neighbors. If the number of internal connections of each vertex in C is more than twice the number of external (to the core) connections then C is designated as a base community.

Otherwise, we consider set N of the distance-2 neighbors of v, that are not elements of C. The edges in N can be classified as follows; (i) one endpoint connected to a vertex in community C (type X); (ii) both endpoints connected to vertices in community N (type Y) and (iii) one end point connected to a vertex that is niether in C nor N (type Z). A vertex in C is considered to be suitable for the base cluster, if that vertex; (a) has fewer edges of type X than of type Y and (b)has fewer edges of type X and Y together than of type Z. Condition (a) ensures that the distance-2 neighbors do not have significantly more connections to the vertices in the base cluster to pull them out and condition (b) ensures that the set of external vertices has a larger "pull" from external communities other than C and therefore it is likely that they will not exert as much "pull" on the vertices within C.

It is possible that a vertex can be designated to be in multiple base clusters. If a vertex has multiple affiliations to several communities, we remove them. A side effect of removing these vertices is that the size of the base clusters now depends on the vertex ordering and the base clusters also become smaller. However this procedure reduces the ambiguity of the clusters, so we apply it for the current version of the algorithm. The pseudocode of our heuristic is shown in Algorithm 1.

Algorithm 1 Identifying Base Clusters in Networks.

```
    Input: A graph G = (V, E). Output: Set of base clusters C_1, C_2, ... C_n.
procedure FINDING BASE CLUSTERS
    Set max_fill to the Fill-In threshold                          ▷ Generally set from 0-2
    for all v ∈ V do
        Compute Fill-In of v
        if Fill-In of v ≤ max_fill then
            Create cluster C_v of v and its neighbors
            In_Edge=Internal Edges of C_v                          ▷ Both endpoints are in C_v
            Ex_Edge=External Edges of C_v                          ▷ Only one endpoint in C_v
            if Ex_Edge ≤ In_Edge/2 then
                Associate cluster id v for each vertex in C_v
                Mark C_v as base cluster
            else
                Create set N of n where, n is a distance-2 neighbor of v
                Y_Edge=Edges with both endpoints in N
                for all u ∈ C_v do
                    X_Edge=Edges with one endpoint in N and other in u
                    Z_Edge=Edges with one endpoint in N and other not in u
                    if X_Edge ≤ Y_Edge AND (X_Edge + Y_Edge) ≤ Z_Edge then
                        if Vertex u does not have a cluster id then
                            Associate cluster id v with u
                            Mark u as a vertex in base cluster
```

Our algorithm focuses on finding the innermost kernel of the consensus communities, and as such the size of the base clusters is likely to be considerably smaller than the ones found by the other preprocessing methods discussed in Section 2. However, recall that the primary objective of our algorithm is to check whether community structure at all exists in the network. In this respect, we are more successful than the other methods because our algorithm will not return any base community if there is no community in the network of diameter larger than two. For example, our method returns zero base clusters for the Delaunay meshes, which ideally do not have community structure. Our method also returns zero base clusters for the Football graph (network of American college football) [**G1**]. This is interesting because Football is known to have distinct communities. However, the diameters of the communities are in most cases at most two and the lowest fill-in of the vertices is more than 10. Due to the absence of tight kernels our algorithm cannot find any base clusters. The ratio of the number of vertices in base clusters to the total vertices provides an estimate of the strength of the communities in the network.

4. Modularity Maximization Using Base Clusters

Base clusters can also be used as a preprocessing step to improve the results of modularity maximization. The vertices with the same base cluster id are assigned to the same community and then a modularity maximization algorithm is applied to the transformed network. In this section we demonstrate the results of using this preprocessing technique combined with the CNM and Louvain methods.

Test Suites. Our test-suite consists of eight unweighted and undirected networks obtained from the clustering instances in the DIMACs website [**D1**]. These are;(i) Karate (network of member in a Zachary's karate club [**Z**] (V=34, E=78), (ii) Jazz (network of jazz musicians) [**G2**](V=198, E=2742), (iii) PolBooks (network of books about USA politics) [**K2**](V=105, E=441)), (iv) Celegans (metabolic network of C. elegans) [**D2**] (V=453, E=2025), (v) (social network of dolphins) [**L3**](V=62, E=159), (vi) Email (the network of e-mail interchanges between members of the Univeristy Rovira i Virgili) [**G3**](V=1133, E=5451), (vii) Power(topology of power grid in the western states of USA) [**W**](V=4941, E=6594) and (vii) PGP (component of the network of users of the Pretty-Good-Privacy algorithm) [**B4**] (V=10680, E=24316).

Algorithm Implementation. Although our underlying modularity maximization methods CNM and Louvain are extensively used in the network community, the available codes do not include provisions for preprocessing. We also could not find any easy to modify open source code that implements both the methods. Therefore to include the preprocessing step and to ensure a fair comparison we implemented the methods (in STL C++) along with the additional preprocessing for finding base clusters. The primary purpose of the code was to understand how using base clusters affect modularity maximization. Therefore although the results match the original versions, performance issues, such as execution time, are not optimized in our implementation. We anticipate in future to develop a faster version of the algorithm. Here we highlight some of the special characteristics of our implementation.

Unlike, most other implementations which uses adjacency lists, we use a compressed sparse row (CSR) structure to store the network. CSR is a standard format for storing sparse matrices. We used this storage because in future versions we plan to use matrix operations on the network. Additionally, even though the networks are undirected we store both directions of the edges (i.e. {v,w} as well as {w,v}). This is done to accommodate the code for directed networks when required. These features make the implementation slower than other versions of the algorithm. However, we are building towards a general software, not just an algorithm for base clusters. In these set of experiments time was used only to compare the different methods against each other in the same environment.

In the CNM code we implemented a heap, as is popularly used, to find the highest change in modularity. However, as the iterations progressed the heap continued to collect obsolete values associated with edges whose endpoints have merged to the same or different communities. The solution was either to recreate the heap after each iteration or to verify that the highest value in the heap with the value stored in the network, and continue until a valid value was obtained. Both these options are computationally expensive. We implemented a compromise where the heap is recreated only if a certain number of misses (top of the heap not being a valid value) is encountered. We set this value to 2.

In the Louvain implementation provided by the authors, there is a function for generating a random permutation of the vertices. The random permutation is not an essential part of the algorithm itself as it is described in [**B2**], but rather, we think, is included to ameliorate the effect of vertex perturbations. However, in our experiments we specifically want to see the effect of vertex permutations and compare its effects across the CNM and Louvain methods and their variations using base clusters. Therefore we did not include the random permutation within the Louvain implementation. The Louvain method also recreates a compressed

TABLE 1. **Comparison of the Modularity Values Obtained by Using the CNM Method and Base Cluster Preprocessing.** The CNM columns tabulate results obtained by using only the CNM method and the Base columns tabulate results obtained through base cluster preprocessing and then applying CNM. The Std Deviation column gives the standard deviation of the values over 60 permutations. The last column gives the percentage of vertices found to be in base clusters.

Name	Avg Modularity		Max Modularity		Std Deviation		Base
	CNM	Base	CNM	Base	CNM	Base	Cluster %
Karate	.3938	.4022	.4156	.4197	.006	.015	29%
Jazz	.43877	.4234	.4388	.4442	2e-05	.015	26%
PolBooks	.5019	.5140	.5019	.5260	3e-04	.011	27%
Celegans	.4046	.4231	.4149	.4327	.005	.004	30%
Dolphin	.4802	.4904	.5094	.5242	.012	.021	22%
Email	.4715	.4908	.5201	.5462	.118	.135	27%
Power	.8997	.9148	.9221	.9200	.117	.003	9%
PGP	.8628	.8616	.8696	.8716	.003	.003	≈ 40%

TABLE 2. **Comparison of the Modularity Values Obtained by Using the Louvain Method and Base Cluster Preprocessing.** The LVN columns tabulate results obtained by using only the LVN method and the Base columns tabulate results obtained through base cluster preprocessing and then applying Louvain. The Std Deviation column gives the standard deviation of the values over 60 permutations. The last column gives the best results of modularity as reported by [**N1**](O), [**L2**](R) and [**O**] (E).

Name	Avg Modularity		Max Modularity		Std Deviation		Best
	LVN	Base	LVN	Base	LVN	Base	Value
Karate	.4156	.4170	.4198	.4198	.007	.005	.4198(R)
Jazz	.4427	.4435	.445	.445	.002	.002	.445 (O)
Polbooks	.5258	.5266	.5268	.5268	.002	.002	.527 (O)
Celegans	.4355	.4320	.4421	.4447	.005	.006	.4501 (E)
Dolphins	.5202	.5200	. 5233	.5241	.002	.002	.529 (O)
Email	.5671	.5664	.5555	.5745	.003	.005	.5801 (E)
Power	.9360	.9359	.9365	.9370	.0003	.0004	.9396 (E)
PGP	.8776	.8775	.8807	.8796	.001	.002	.8861 (E)

network at the end of each outer loop. This process reduces the performance time significantly as the subsequent operations are executed on a much smaller network. In our code, we keep use the final community allocation of the vertices to identify which are compressed into a supernode, but retain the original network. Consequently, our execution times for the larger networks are substantially slower than compared to the code provided by the authors.

Empirical Results We applied 60 permutations to each of the networks in the test suite. The permutation orders were created using a random number generator.

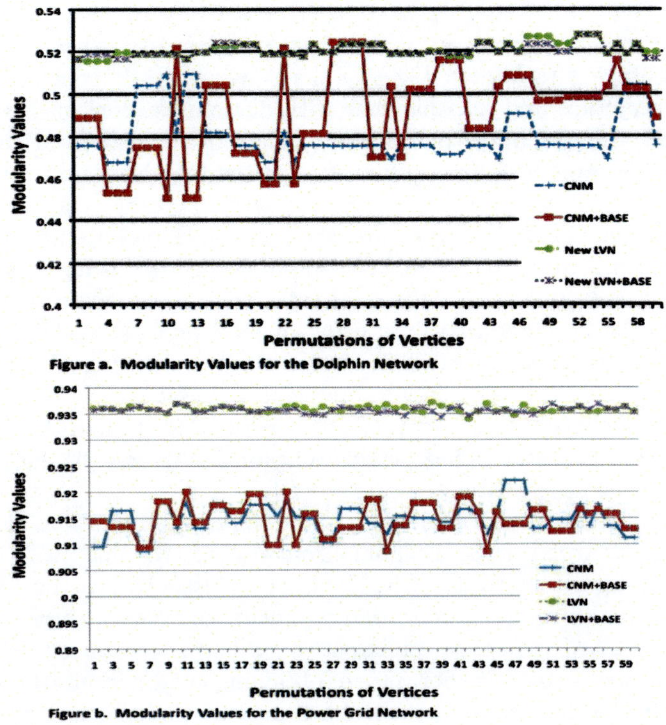

FIGURE 3. **Effect of Vertex Permutation on Computing Modularity**. Top Figure: The Dolphin Network. Bottom Figure: The Power Grid. CNM methods generally produce a lower value than Louvain methods. However, for networks with stronger community structure in certain permutations can produce equivalent modularity to the Louvain method.

For each permutation we applied the CNM and the Louvain method as well as the methods after finding and combining the base clusters. The statistics of the modularity obtained by these four methods is given in Tables 1 and 2.

We see that in general using base clustering increases the average modularity value as well as the highest one. There are a few exceptions, such as in average for Jazz and maximum for power in CNM and average for Email and Celegans and max for PGP in Louvain. In general, the improvement is higher for CNM, than for the Louvain methods. We believe that this is due to the backtracking feature of the Louvain algorithm. We also compare the standard deviations of the values across the different perturbations. The range of values in Louvain is not as affected by using base clusters as those of CNM. This phenomena once again points to the backtracking feature of the Louvain method, which automatically the process to adjust from any initial position to a good community partition. This leads us to conclude that the base clustering preprocessing would be most effective when the underlying algorithm does not contain self-adjusting steps.

The last column in Table 1 gives the percentage of vertices in the base clusters to the total number of vertices. We see that compared to other networks in the set the percentage is rather low (9%) for the Power network, which indicates poor community structure and also matches with our observations in Figure 1. The PGP network also has a low percentage (4%) of base cluster vertices, but since the network was large we only sampled 10% of the total vertices for fill-in. If the sample percentage is adjusted, the percentage of base clusters can go upto (40%).

The last column in Table 2 compares the best known modularity value obtained using other preprocessing methods. The ensemble strategy is denoted by (E), the random walk strategy by (R) and for networks where preprocessing was not used we tabulated the best known values listed in [**N1**] and denoted these as (O for other). For networks with well defined community structures (karate and jazz) base clustering can come very close to the highest modularity, but not so much for the others. We believe this is because (i) base clusters try to find the kernels of the communities and is therefore independent of modularity and (ii) due to the much smaller size of the base clusters.

Figure 3 plots the change in modularity over all the permutations of the Dolphin and the Power networks. In the Dolphin network we can see that using base clusters gives a significant boost to the CNM method. Also observe that although, in general, the Louvain methods can produce higher modularity, there exists certain case where the CNM with base communities method is equivalent to the Louvain method. This points to the importance of obtaining good permutations for a given algorithm and also indicates that the Dolphin network posses community structure. In contrast, the values in the Power network are well separated. As we know, Power network does not have as strong a community structure so perhaps separation of values by two algorithms is an indication of that. We plan to further investigate this phenomena in future.

Table 3 compares the difference in community structure across the original and the method with base cluster preprocessing using the Rand Index. Most of the values (with an exception in Email) are quite high (over 77%). However the values are generally higher for the Louvain method, once again reflecting the effect of self adjustment. Table 4 gives the average time (in seconds) to compute the original methods, the original methods with preprocessing and the time for only preprocessing. The codes were compiled with GNU-g++ and the experiments were run on a Xeon dual-core processor with 2.7GHz speed and a 32 GB RAM. We see that in some cases preprocessing helps reduce the overall agglomeration time, however finding the base clusters is generally as expensive as is our current implementation of modularity maximization. But that since the base clusters depend only on the network topology, finding them can be a one time operation. After that we can reuse the clusters for any underlying algorithm. Although, not implemented in this paper, this technique can help make base cluster preprocessing more cost effective.

It would also be instructive to compare how good our base cluster algorithm is in finding kernels of the consensus communities. However to analyze this we would have to compute the consensus communities themselves, such as by comparing the common groups over multiple perturbations. This is possible for small networks, but not for large ones like PGP–because as the number of vertices grows it is important to check out large number of perturbations (as close to n! as possible) to cover as much of the search space as possible. In this paper we have computed the

TABLE 3. **Comparison Between the Communities formed by using the Original Method and the Ones using Preprocessing.** The values of the Rand Index is generally high indicating similarity except at a few points. The similarity between the clusters obtained using the Louvain based methods is higher.

Name	CNM vs Base			LVN vs Base		
	Avg	Max	Min	Avg	Max	Min
Karate	.8873	.9643	.8146	.9599	1	.8823
Jazz	.9217	.9702	.8891	.9851	1	.9368
Polbooks	.8572	.9153	.9945	.9139	1	.9175
Celegans	.7966	.8236	.7707	.9068	1	.8829
Dolphins	.8344	.8773	.7911	.9810	1	.9074
Email	.8163	.8993	.6500	.9363	.9636	.8992
Power	.9804	.9839	.9759	.9889	.9926	.9834
PGP	.9682	.9757	.9567	.9976	1	.9947

TABLE 4. **Comparison of the Execution Time (In Seconds) of the Different Methods and the Time to Identify Base Clusters.** Using base cluster preprocessing can sometimes reduce the execution time. However, in some cases, obtaining base clusters can be as expensive as the agglomeration method. The performance of the algorithms can be improved by sampling selected vertices or using base clusters as a one time preprocessing operation for multiple methods.

Name	CNM	CNM+Base	LVN	LVN+Base	Base Only
Jazz	1.50	1.51	.57	.68	.45
Polbooks	.085	.067	.06	.05	.04
Celegans	3.67	1.80	1.35	1.50	.86
Dolphins	.01	.018	.003	.005	8e-04
Email	32.31	18.6	11.84	10.31	3.15
Power	52.59	50.19	24.12	24.68	31.4
PGP	760.78	757.25	579.88	577.87	25.79

consensus communities for Jazz, Dolphin and Power. Jazz has 86% of its vertices in the three highest consensus communities (our base cluster found 26%) and Dolphin has 74% of its vertices in the three highest consensus communities (our base cluster found 22%). These numbers are encouraging because we are only looking at the kernel —not the entire community and on inspecting the base clusters obtained, that in most cases they indeed belong to the same consensus cluster. However there are some false positives in that if nodes of two clusters are closely attached–they can appear as a base cluster. This happens for some permutations in Jazz and Dolphins, and those are the ones where the modularity is not as high. For example out of 53 of vertices, in Jazz, denoted to be in the base communities 5 were false positives. We found that the Louvain method is less forgiving of the false positives than the CNM method. In order to reduce the chances of false positives, for the

Louvain method, we only used cluster sizes ranging from 2-4. In future we plan to further modify the base cluster identification algorithm to reduce these false positives.

The Power network has just 1% of its vertices in the largest three consensus communities, yet by our method we were able to find 9% of the nodes. On inspection we found that this happened because the base cluster method picked up many of the smaller communities, that were built around a vertex with low fill-in. Once again, we need more stringent conditions in our algorithm to avoid picking up very small communities.

5. Discussion and Future Work

In this paper, we have attempted to answer the question–"how can we know whether a network indeed posses community structure ?". As an initial step in this investigation we proposed finding core kernels of communities, which we call base clusters, and developed an algorithm to identify these clusters. The percentage of vertices in a base cluster can give an estimate of the strength of the community of the network. This conjecture is supported by comparing the vertices in base clusters to the ones in large consensus communities. Additionally our algorithm returns zero base clusters for networks not known to have community structures.

Base clusters can also be used as a preprocessing step to improve the the value of modularity. We used this preprocessing in conjunction with two agglomerative methods, the CNM and the Louvain on 60 permutations per network. The improvement to CNM is higher than the improvement to the Louvain method, perhaps due to its self adjusting feature. We however note that the base clusters are identified orthogonal of any modularity values, and therefore the increase is perhaps due to the cluster representing a core kernel.

Our algorithm for identifying base clusters has room for improvement. First, we are only considering a vertex and its distance-1 neighbors as the base cluster. This kernel can be expanded to include vertices at longer distances to create a stronger base cluster. Additionally, we have observed that our method picks up some false positives if vertices of two nearby consensus communities are tightly connected. We plan to improve the conditions on the base cluster to reduce the false positives. Finally comparing base clusters over all vertices is still very expensive, particularly for large graphs and we are investigating better implementation practices to reduce the time.

References

[B1] A. L. Barabási, H. Jeong, Z. Néda, E. Ravasz, A. Schubert, and T. Vicsek, *Evolution of the social network of scientific collaborations*, Phys. A **311** (2002), no. 3-4, 590–614, DOI 10.1016/S0378-4371(02)00736-7. MR1943379

[B2] V.D. Blondel, J.-L. Guillaume, R. Lambiotte and E. Lefebvre. Fast unfolding of community hierarchies in large networks. *J. Stat. Mech.* (10) (2008)

[B3] *Statistical mechanics of complex networks*, Lecture Notes in Physics, vol. 625, Springer-Verlag, Berlin, 2003. Selected papers from the 18th Conference held in Sitges, June 10–14, 2002; Edited by R. Pastor-Satorras, M. Rubi and A. Diaz-Guilera. MR2179067 (2007e:82002)

[B4] M. Boguna, R. Pastor-Satorras, A. Diaz-Guilera and A. Arenas, *Physical Review E.*, vol. 70, 056122 (2004).

[B5] U. Brandes, D. Delling, M. Gaertler, R. Gorke, M. Hoefer, Z. Nikoloski, and D. Wagner. On modularity clustering. *IEEE Transactions on Knowledge and Data Engineering*, 20(2):172188, (2008)

[C] A. Clauset, M. E. J. Newman, and C. Moore, Finding community structure in very large networks. *Phys. Rev. E.* 70(6), 66111 (2004)

[D1] Dimacs Testbed http://www.cc.gatech.edu/dimacs10/downloads.shtml

[D2] J. Duch and A. Arenas. Community identification using Extremal Optimization , *Physical Review E.*, vol. 72, (2005)

[G1] M. Girvan and M. E. J. Newman, *Community structure in social and biological networks*, Proc. Natl. Acad. Sci. USA **99** (2002), no. 12, 7821–7826 (electronic), DOI 10.1073/pnas.122653799. MR1908073

[G1] Benjamin H. Good, Yves-Alexandre de Montjoye, and Aaron Clauset, *Performance of modularity maximization in practical contexts*, Phys. Rev. E (3) **81** (2010), no. 4, 046106, 19, DOI 10.1103/PhysRevE.81.046106. MR2736215 (2011g:05279)

[G2] P.Gleiser and L. Danon ,*Adv. Complex Syst.* 6, 565 (2003)

[G3] R. Guimera, L. Danon, A. Diaz-Guilera, F. Giralt and A. Arenas. *Physical Review E.* , vol. 68, 065103(R), (2003).

[K1] B. Karrer, E. Levina, and M. E. J. Newman. Robustness of community structure in networks *Physical Review E.* Vol. 77, No. 4. (2008)

[K2] V. Kreb. Books on US politics. http://www.orgnet.com/

[L1] A. Lancichinetti and S. Fortunato. Consensus clustering in complex networks. *Scientific Reports* Vol 2 (2012)

[L2] D. Lai, H. Lu and C. Nardini. Enhanced modularity-based community detection by random walk network preprocessing. *Phys. Rev. E.* 81, 066118 (2010)

[L3] D. Lusseau, K. Schneider, O. J. Boisseau, P. Haase, E. Slooten, and S. M. Dawson. *Behavioral Ecology and Sociobiology* 54, 396-405 (2003)

[N1] M. C. V. Nascimento and L. S. Pitsouli. Community Detection by Modularity Maximization using GRASP with Path Relinking. *10th DIMACS Implementation Challenge Graph Partitioning and Graph Clustering.* (2012)

[N2] M.E.J. Newman and M. Girvan. Finding and evaluating community structure in networks.*Phys. Rev. E.* 69(2), 026113 (2004)

[O] M. Ovelgonne and A. Geyer-Schulz. An Ensemble Learning Strategy for Graph Clustering. *10th DIMACS Implementation Challenge Graph Partitioning and Graph Clustering.* (2012)

[P] Mason A. Porter, Jukka-Pekka Onnela, and Peter J. Mucha, *Communities in networks*, Notices Amer. Math. Soc. **56** (2009), no. 9, 1082–1097. MR2568495

[R1] W. Rand, Objective criteria for the evaluation of clustering methods. *J. Am. Stat. Assoc.* 66 (336), 846850 (1971)

[R2] J. Reidy, D. A. Bader, K. Jiang, P. Pande and R. Sharma. Detecting Communities from Given Seeds in Social Networks. Technical Report. http://hdl.handle.net/1853/36980

[V] K. Voevodski, S. H. Teng, Y. Xia. Finding local communities in protein networks. BMC Bioinformatics 10(10), 297 (2009)

[W] D. J. Watts and S. H. Strogatz, *Nature* 393, 440-442 (1998).

[Z] W. W. Zachary, An information flow model for conflict and fission in small groups, *Journal of Anthropological Research* 33, 452-473 (1977)

Department of Computer Science, University of Nebraska, Omaha, Nebraska 68106
E-mail address: `sriramsrinivas@unomaha.edu`

Indian Institute of Technology, Kharagpur, West Bengal, India
E-mail address: `its_tanmoy@cse.iitkgp.ernet.in`

Department of Computer Science, University of Nebraska, Omaha, Nebraska 68106
E-mail address: `sbhowmick@unomaha.edu`

Complete hierarchical cut-clustering: A case study on expansion and modularity

Michael Hamann, Tanja Hartmann, and Dorothea Wagner

ABSTRACT. In this work we study the hierarchical cut-clustering approach introduced by Flake et al., which is based on minimum s-t-cuts. The resulting cut-clusterings stand out due to strong connections inside the clusters, which indicate a clear membership of the vertices to the clusters. The algorithm uses a parameter which controls the coarseness of the resulting partition and which can be used to construct a hierarchy of cut-clusterings. The parameter further provides a quality guarantee in terms of expansion, a quality measure for clusterings which in general is NP-hard to compute.

We conduct an experimental study on the expansion of cut-clusterings revealing that, compared to a trivial bound, the given guarantee allows for a deeper insight. Our experiment further documents that the true expansion even exceeds the guarantee. In a second experiment we investigate the quality of cut-clusterings with respect to the widely used measure modularity. In this study the cut-clustering algorithm competes surprisingly well with a greedy modularity-based heuristic, although it is not designed to optimize modularity. This attests a high trustability of the cut-clustering approach confirming that the cut-clustering algorithm returns nice clusterings if such clusterings are clearly indicated by the graph structure.

1. Introduction

The aim of graph clustering is to identify subgraphs of high internal connectivity that are only sparsely interconnected. This vague notion lead to countless attempts of formalizing properties that characterize a set of good clusters. The resulting variety of different quality measures still affects the design of algorithms, although for many measures the sufficiency of the underlying properties is not examined yet or has been even disproven. This is, a good clustering according to a non-sufficient quality measure might be still implausible with respect to the given graph structure. For example, Montgolfier et al. [4] showed that the asymptotic modularity of grids is 1, which is maximum since modularity ranges whithin $[-0.5, 1]$. However, by intuition the uniform structure of grids does not support any meaningful clustering, and thus, also a clustering of high modularity can not be plausible. Furthermore, common quality measures are most generally hard to optimize. Thus, heuristics are often used in practice.

2010 *Mathematics Subject Classification.* Primary 05C85, 05C75; Secondary 05C21, 05C40.

This work was partially supported by the DFG under grant WA 654/15 within the Priority Programme "Algorithm Engineering".

Motivated by these drawbacks of established clustering algorithms, we focus on a different approach postulated by Flake et al. [**5**]. Their algorithm exploits properties of minimum s-t-cuts in order to find clusterings where the membership of each set of vertices to a cluster is clearly indicated by the graph structure. More precisely, there are clusterings desired where each subset of a cluster is at least as strongly connected to the remaining vertices inside the cluster as to the vertices outside the cluster — a property that is not expressed by any of the common measures. The clusterings resulting from the cut-clustering approach are called *cut-clusterings.* This concept of cut-based clustering leads to a relatively strict behavior in the sense that vertices that can not be clearly assigned to any cluster remain *unclustered*, i.e., form singleton clusters. Such a behavior particularly prevents an arbitrary assignment of vertices to clusters, which is highly desirable for example in sociology applications where it is essential that ambiguous scenarios are interpreted by human experts instead of automated routines.

The algorithm of Flake et al. depends on a parameter, which controls the coarseness of the resulting clustering. Different parameter values result in at most $n-1$ different cut-clusterings. Those clusterings form a hierarchy where low parameter values create large clusters and high values result in fine clusterings. Having such a hierarchy at hand, it is then possible to choose the best clustering with respect to modularity, which is a feasible quality measure in this context, since the construction of the clusterings already guarantees their plausibility with respect to the graph structure. A high modularity value additionally implies nice properties, like for example decent and balanced cluster sizes.

The parameter finally also constitutes a guarantee on intra-cluster expansion and inter-cluster expansion[1], which are both cut-based quality indices. This is particularly remarkable since at least intra-cluster expansion is hard to compute.

Contribution. Flake et al. tested their algorithm on a citation network and a network of linked web pages with respect to the semantic meaning of the clusters. In this work we present an experimental analysis of the general behavior of cut-clusterings on benchmark instances proclaimed within the 10th DIMACS Implementation Challenge on Graph Partitioning and Graph Clustering [**2**]. After presenting a direct description of the cut-clustering algorithm in Section 2.2, we investigate the guaranteed expansion of the cut-clusterings in Section 3.1 as well as the modularity values that are reached by cut-clusterings in Section 3.2. Since intra-cluster expansion is hard to compute, we consider lower bounds in the analysis. Our study gives evidence that trivial bounds do not match up to the given guarantee. The analysis of a special non-trivial bound further indicates that the true intra-cluster expansion of the cut-clusterings even surpasses the guarantee. Also the inter-cluster expansion turns out to be better, i.e., lower than guaranteed.

Within the modularity analysis of the cut-clusterings we additionally consider reference clusterings obtained from a common modularity-based heuristic [**12**]. We took the implementation of this heuristical approach from Lisowski [**10**]. Our study documents that for many of the tested graphs the cut-clusterings reach modularity values quite close to the references. On the other hand the cut-clustering algorithm

[1]The inter-cluster expansion considered in this work is defined slightly different from the common inter-cluster expansion. The latter normalizes by the number of vertices on the smaller cut side while we count the vertices on the side that does not induce the cluster.

returns only fine clusterings with low modularity values if there are no other plausible clusterings supported by the graph structure. Based on this result we claim modularity applied to cut-clusterings as a good measure for how well a graph can be clustered.

2. Preliminaries

Throughout this work we consider a simple, undirected, weighted graph $G = (V, E, c)$ with vertex set V, edge set E and a non-negative edge cost function c. In unweighted graphs we assign cost 1 to each edge. We denote the number of vertices (edges) by $n := |V|$ ($m := |E|$) and the costs of a set $E' \subseteq E$ by $c(E') := \sum_{e \in E'} c(e)$. Whenever we consider the degree $\deg(v)$ of $v \in V$, we implicitly mean the sum of all edge costs incident to v. With $S, T \subset V$, $S \cap T = \emptyset$, we write $c(S,T)$ for the costs of all edges having one endpoint in S and one in T. If S, T induce a *cut* in G, i.e., $S \cup T = V$, $c(S,T)$ describes the costs of this cut.

Our understanding of a *clustering* $\Omega(G)$ is a partition of the vertex set V into subsets $C^1, \ldots, C^k$, which define vertex-induced subgraphs, called *clusters*. A cluster is called *trivial* if it corresponds to a connected component. A vertex that forms a non-trivial singleton cluster we consider as *unclustered*. A clustering is *trivial* if it consists of trivial clusters or if $k = n$, i.e., all vertices are unclustered. A *hierarchy of clusterings* is a sequence $\Omega_1(G) \leq \cdots \leq \Omega_r(G)$ such that $\Omega_i(G) \leq \Omega_j(G)$ implies that each cluster in $\Omega_i(G)$ is a subset of a cluster in $\Omega_j(G)$. We say $\Omega_i(G) \leq \Omega_j(G)$ are *hierarchically nested*.

2.1. Quality Measures. A *quality measure* for clusterings is a mapping to real numbers. Depending on the measure, either high or low values correspond to high quality. In this work we consider three quality measures, modularity, intra-cluster expansion and inter-cluster expansion. The former two indicate high quality by high values. Inter-cluster expansion indicates good quality by low values.

Modularity was first introduced by Newman and Girvan [11] and is based on the total edge costs covered by clusters. The values range between -0.5 and 1 and express the significance of a given clustering compared to a random clustering. Formally, the modularity $\mathcal{M}(\Omega)$ of a clustering Ω is defined as

$$\mathcal{M}(\Omega) := \sum_{C \in \Omega} c(E_C)/c(E) - \sum_{C \in \Omega} \Big(\sum_{v \in C} deg(v)\Big)^2 / 4c(E)^2,$$

where E_C denotes the set of edges with both endpoints in C.

The *inter-cluster expansion* of a cluster C is given by the costs for cutting off the cluster from the remaining graph divided by the number of vertices outside the cluster, i.e., $\Phi(C) := \frac{c(C, V \setminus C)}{|V \setminus C|}$. The inter-cluster expansion $\Phi(\Omega)$ of a clustering Ω is the maximum inter-cluster expansion of all clusters in Ω.

The *intra-cluster expansion* of a clustering derives from the expansion defined for cuts. The expansion $\Psi(S, C \setminus S)$ of a cut $(S, C \setminus S)$ in a cluster C evaluates the costs of the cut per vertex on the smaller cut side, i.e.,

$$\Psi(S, C \setminus S) := \frac{c(S, C \setminus S)}{\min\{|S|, |C \setminus S|\}}.$$

The intra-cluster expansion $\Psi(C)$ of a cluster equals the minimum expansion of all cuts in C. Note that intra-cluster expansion is not defined for singleton clusters. The intra-cluster expansion $\Psi(\Omega)$ of a clustering finally is the minimum intra-cluster

Algorithm 1: CUTC

Input: Graph $G_\alpha = (V_\alpha, E_\alpha, c_\alpha)$

1 $\Omega \leftarrow \emptyset$

2 **while** $\exists\, u \in V_\alpha \setminus \{t\}$ **do**

3 $\quad C^u \leftarrow$ community of u in G_α w.r.t. t

4 $\quad r(C^u) \leftarrow u$

5 $\quad$ **forall the** $C^i \in \Omega$ **do**

6 $\quad\quad$ **if** $r(C^i) \in C^u$ **then** $\Omega \leftarrow \Omega \setminus \{C^i\}$

7 $\quad \Omega \leftarrow \Omega \cup \{C^u\}$; $V_\alpha \leftarrow V_\alpha \setminus C^u$

8 **return** Ω

expansion of all non-singleton clusters in Ω. Unfortunately, computing $\Psi(C)$, and thus, also $\Psi(\Omega)$, is known to be NP-hard. Thus, in our analysis we consider bounds instead. These are introduced in Section 3.1.

2.2. The Hierarchical Cut-Clustering Algorithm. In this section we review the parametric cut-clustering approach of Flake et al., which returns hierarchically ordered clusterings for varying parameter values. In [**5**] Flake et al. develop their parametric cut-clustering algorithm step by step using an idea involving Gomory-Hu trees [**7**]. The final approach, however, just uses special minimum s-t-cuts, so called *community-cuts* in order to identify clearly indicated clusters in the graph structure. Let G denote a graph and (S, T) a minimum s-t-cut in G, with $s \in S$, $t \in T$. The cut (S, T) is the *community cut* of s with respect to t if $|S|$ is minimum for all minimum s-t-cuts in G. The set S is the unique *community* of s, and s is a *representative* of S, denoted by $r(S)$. Representatives are not necessarily unique. Communities of different vertices with respect to the same vertex t are either disjoint or nested. Otherwise either the intersection or the difference of the communities would be a smaller community for one of the considered vertices (for a detailed proof see [**7**] or [**5**]).

Based on this definition of communities we give a more direct description of the cut-clustering algorithm, to which we refer by CutC in the following. Given a graph $G = (V, E, c)$ and a parameter $\alpha > 0$, as a preprocessing step, augment G by inserting an artificial vertex t and connecting t to each vertex in G by an edge of costs α. Denote the resulting graph by $G_\alpha = (V_\alpha, E_\alpha, c_\alpha)$. Then apply CutC: iterate V and for each vertex u not yet contained in a previously computed community compute a community in G_α with respect to t. The vertex u becomes the representative of the newly computed community (cp. Algorithm 1, line 4). Since communities are either disjoint or nested we finally get a set Ω of (inclusion)maximal communities. Together these communities decompose V and thus induce a clustering for G.

Since the clusters in a cut-clustering are communities in G_α, each cluster C satisfies the *significance-property*, formally defined below, which says that any set S of vertices in C that does not contain the representative $r(C)$ is clearly assigned to C by connections into C that are at least as strong as those to the outside of C. Due to this property the membership of S to C is clearly indicated by the graph structure:

$$\exists\, r \in C: \;\; c(S, C \setminus S) \;\geq\; c(S, V \setminus C), \;\; \forall S \subseteq C \setminus \{r\} \quad \text{(significance-property)}$$

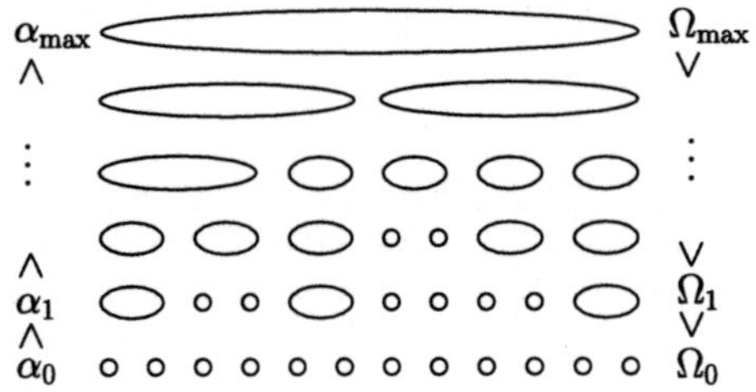

FIGURE 1. Clustering hierarchy after applying CutC iteratively. Note, that $\alpha_{\max} < \alpha_0$ whereas $\Omega_{\max} > \Omega_0$.

Otherwise, there would exist a set $S \subseteq C \backslash \{r(C)\}$ such that $c(S, C \backslash S) < c(S, V \backslash C)$. This implies that the cut $(C \setminus S, V \setminus (C \setminus S))$ is a cheaper $r(C)$-t-cut in G_α than the cut $(C, V \setminus C)$, which induces the cluster C. This contradicts the fact that C is the community of $r(C)$ in G_α. The costs of these cuts in G_α are

$$\begin{aligned} c(C \setminus S, V \setminus (C \setminus S)) + \alpha|C \setminus S| &= \\ c(C \setminus S, S) + c(C \setminus S, V \setminus C) + \alpha|C \setminus S| &< c(S, V \setminus C) + c(C \setminus S, V \setminus C) + \alpha|C| \\ &= c(C, V \setminus C) + \alpha|C|. \end{aligned}$$

With similar arguments Flake et al. have further proven that the parameter value α that was used to construct the augmented graph G_α constitutes a guarantee on intra-cluster expansion and inter-cluster expansion:

$$\Psi(\Omega) \geq \alpha \geq \Phi(\Omega).$$

Applying CutC iteratively with decreasing parameter values yields a hierarchy of at most n different clusterings (cp. Fig. 1). This is due to a further nesting property of communities, which is proven by Gallo et al. **[6]** as well as Flake et al. **[5]**: Let C_1 denote the community of a fixed vertex u in G_{α_1} and C_2 the community of u in G_{α_2}. Then it is $C_1 \subseteq C_2$ if $\alpha_1 \geq \alpha_2$. The hierarchy is bounded by two trivial clusterings, which we already know in advance. The clustering at the top consists of the connected components of G and is returned by CutC for $\alpha_{\max} = 0$, the clustering at the bottom consists of singletons and comes up if we choose α_0 equal to the maximum edge cost in G.

The crucial point with the construction of such a clustering hierarchy, however, is the choice of α. If we choose the next value too close to a previous one, we get a clustering we already know, which implies unnecessary effort. If we choose the next value too far from any previous value, we possibly miss a meaningful clustering. In our experiments we thus use a simple parametric search approach that returns a complete hierarchy without fail. For a detailed description of this approach see **[8]**. In order to find all different levels in the hierarchy, this approach constructs the breakpoints in the continuous parameter range between consecutive levels. This is, each clustering Ω_i is assigned to an interval $[\alpha_i, \alpha_{i-1})$ where CutC returns Ω_i. The breakpoint α_i marks the border to the next higher clustering Ω_{i+1}, whereas α_{i-1} is the breakpoint between Ω_i and the previous level Ω_{i-1}. Thus the guarantee on expansion given by the parameter can be extended to

$$\Psi(\Omega_i) \geq \alpha_{i-1} > \alpha_i \geq \Phi(\Omega_i)$$

for each cut-clustering Ω_i in the complete hierarchy. We call $[\alpha_i, \alpha_{i-1})$ the *guarantee interval* of Ω_i.

TABLE 1. Testbed encompassing real-world networks and randomly generated graphs.

graph	n	m	graph	n	m
karate	34	78	dolphins	62	159
lesmis	77	254	polbooks	105	441
adjnoun	112	425	football	115	613
jazz	198	2742	celegansneural	297	2148
celegans_metabolic	453	2025	delaunay_n10	1024	3056
email	1133	5451	polblogs	1490	16715
netscience	1589	2742	delaunay_n11	2048	6127
bo_cluster	2114	2203	data	2851	15093
delaunay_n12	4096	12264	dokuwiki_org	4416	12914
power	4941	6594	hep-th	8361	15751
PGPgiantcompo	10680	24316	astro-ph	16706	121251
cond-mat	16726	47594	as-22july06	22963	48436
cond-mat-2003	31163	120029	rgg_n_2_15_s0	32768	160240
cond-mat-2005	40421	175691	G_n_pin_pout	100000	501198

3. Experimental Study

The experiments in this work aim at two questions. The first question asks how much more information the given guarantee on expansion provides, compared to a trivial intra-cluster expansion bound that is easy to compute. Recall that computing the intra-cluster expansion of a clustering is NP-hard, and thus, bounds give at least an idea of the true values. Since we are nevertheless interested in the actual intra-cluster expansion of cut-clusterings, we consider a further, non-trivial lower bound, which is more costly to compute but also more precise than the trivial bound. Finally we also look at the inter-cluster expansion, which can be efficiently computed for a clustering. The second question focuses on the modularity values that can be reached by cut-clusterings, and the plausibility of these values with respect to the graph structure.

For our experiments we use real world instances as well as generated instances. Most instances are taken from the testbed of the 10th DIMACS Implementation Challenge [**1**], which provides benchmark instances for partitioning and clustering. Additionally, we consider the protein interaction network bo_cluster published by Jeong et al. [**9**], a snapshot of the linked wiki pages at www.dokuwiki.org, which we gathered ourselves, and 275 snapshots of the email-communication network of the Department of Informatics at KIT [**2**]. The latter have around 200 up to 400 vertices. The sizes of the remaining instances are listed in Table 1. Our analysis considers only one cut-clustering per instance, namely the cut-clustering with the best modularity value of all clusterings in the complete hierarchy. The results for the snapshots of the email network are depicted separately from the remaining instances in the following figures, for the sake of a better readability. Furthermore, the instances are decreasingly ordered by the amount of unclustered vertices in the cut-clusterings, which corresponds to an increasing order by coarseness. The instances, respectively their clusterings, are associated with points on the x-axis.

3.1. Expansion Analysis of Cut-Clusterings. We consider the true inter-cluster expansion, which is easy to compute, and two lower and one upper bound on intra-cluster expansion, since the true intra-cluster expansion is hard to compute. For a cluster C the first lower bound $B_\ell(C)$ and the upper bound $B_u(C)$ are trivially obtained from a global minimum cut $(M, C \setminus M)$ in C:

$$B_\ell(C) := \frac{c(M, C \setminus M)}{\lfloor |C|/2 \rfloor} \leq \Psi(C) \leq \frac{c(M, C \setminus M)}{\min\{|M|, |C \setminus M|\}} =: B_u(C).$$

Note that $B_u(C)$ is just the expansion of the global minimum cut. The corresponding bounds $B_\ell(\Omega)$ and $B_u(\Omega)$ for a whole clustering Ω are given by the respective minimum of all clusters. Figure 2 shows how these bounds behave compared to the guarantee interval; more precisely, to the upper interval boundary, which we normalized to 1 for a better comparability. All further values are displayed proportionally. The upper part of Figure 2 further shows the inter-cluster expansion $\Phi(\Omega)$ of the clusterings of the listed instances. Comparing these values to the lower boundary of the guarantee interval proves many clusterings to have a better inter-cluster quality than guaranteed. This particularly holds for the snapshots of the email network, but also for instances like "lesmis", "power" or "netscience". Due to a better readability, we omitted the presentation of the inter-cluster expansion for the snapshots of the email network.

Regarding the intra-cluster quality, we observe that for most instances the trivial lower bound $B_\ell(\Omega)$ stays below the upper boundary of the guarantee interval. This reveals a true advantage from knowing the guarantee besides the trivial bound. The few exceptions, see for example the "polbooks" instance, can be explained by the shape of the found cut-clusterings. If the clustering contains only small clusters, the value of the minimum cut in each cluster is only divided by a few vertices when computing the trivial lower bound. Particularly in unweighted graphs this often yields a value bigger than 1, i.e., bigger than the maximum edge costs. The upper boundary of the guarantee interval, however, can not exceed the maximum edge costs in the graph.

Whenever the upper bound $B_u(\Omega)$ meets the guarantee interval, the guarantee even equals the true intra-cluster expansion. These instances are marked by a star in the upper part of Figure 2. For the snapshots of the email network we counted 3.6% of the instances where the exact intra-cluster expansion is known. However, in most cases there is still a large gap between the guaranteed intra-cluster expansion and the upper bound.

In order to explore this gap, we further consider an alternative non-trivial lower bound $A_\ell(\Omega)$ on intra-cluster expansion. This bound results from individually applying the hierarchical cut-clustering algorithm to the subgraphs induced by the clusters in Ω. The algorithm returns a complete clustering-hierarchy of the subgraph, thereby finding the breakpoint between the most upper hierarchy level, which consists of connected components, and the next lower level. This breakpoint is the largest parameter value where CutC still returns connected components. Since Ω is a cut-clustering, the considered subgraph is connected. Otherwise it would not be induced by a cluster of Ω. Thus, there is only one cluster in the most upper hierarchy level corresponding to the whole subgraph. Hence, the found breakpoint constitutes a non-trivial lower bound $A_\ell(C)$ on the intra-cluster expansion of the considered cluster in Ω. This bound again expands to the whole clustering Ω by taking the minimum value of all clusters. Since this method considers the clusters as

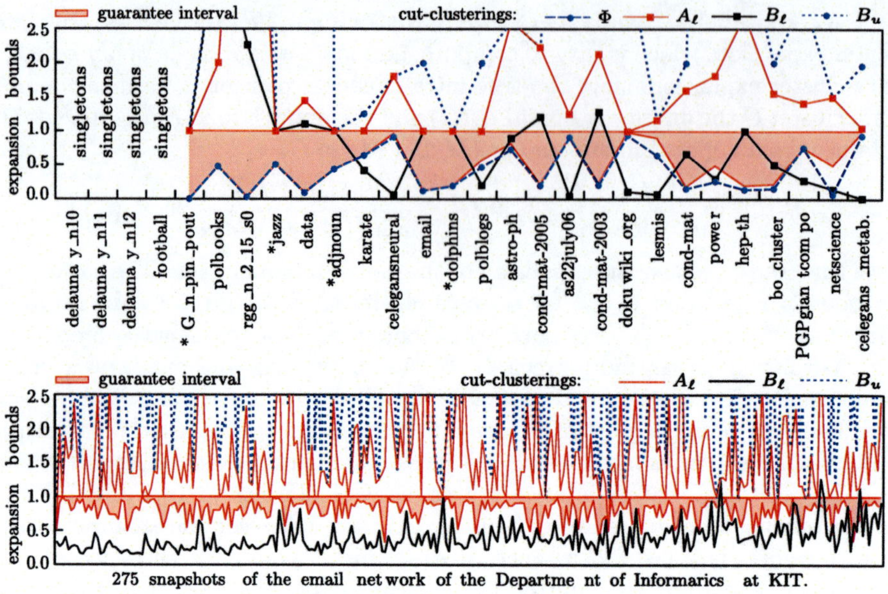

FIGURE 2. *Expansion Analysis of Cut-Clusterings:* Inter-cluster expansion (Φ) and bounds on intra-cluster expansion (B_ℓ, B_u trivial lower and upper bound based on minimum cut, A_ℓ alternative non-trivial lower bound). The upper boundary of the guarantee interval is normalized to 1, further values are displayed proportional. Instances where B_u meets the guarantee are marked by *. For the sake of readability Φ is omitted in the lower chart. Regarding the first four instances, the hierarchical cut-clustering algorithm returns only singletons, for which intra-cluster expansion is not defined.

independent instances ignoring the edges between the clusters, the resulting bound $A_\ell(\Omega)$ potentially lies above the guarantee interval, which is also confirmed by our experiment (cp. Figure 2). This is, most of the cut-clusterings are even better than guaranteed. Besides, by reaching the upper bound $B_u(\Omega)$ in some further cases, the bound $A_\ell(\Omega)$ increases the amount of instances for which we know the intra-cluster expansion for sure to 20%.

3.2. Modularity Analysis. In the following we examine the modularity values of the best cut-clusterings in the cut-clustering hierarchies. In order to justify whether a given modularity value is a good value in general, i.e., how far it is from a possibly better value of another clustering, we use a modularity-based greedy multilevel approach **[12]** to generate reference clusterings with generally good modularity values. This approach is widely used and has turned out to be reliable in former experiments. It starts with an initial clustering consisting of singleton clusters and moves vertices between clusters as long as this operation increases modularity. Then the found clusters are contracted and the algorithm continues on the next level. Finally the different levels are expanded top-down and the algorithm again allows single vertices to move in order to further increase modularity. Note, that computing a modularity-optimal clustering is NP-hard [**3**].

Since high modularity values are known to be misleading in some cases, we further establish a plausibility check by testing whether the clusters of the reference clusterings satisfy the significance-property, which guarantees that they are clearly indicated by the graph structure. Recall that the clusters of the cut-clusterings own this property due to their construction. Figure 3 shows the percentage amount of significant clusters, i.e., clusters with the significance-property, for the reference clusterings. To get also a better idea of the structure of the cut-clusterings, we present the percentage amount of unclustered vertices in these clusterings. Unclustered vertices may occur due to the strict behavior of the cut-clustering algorithm, which is necessary in oder to guarantee the significance-property. Note that in contrast none of the reference clusterings contains unclustered vertices. As a last structural information on the clusterings, Figure 3 depicts the cluster sizes in terms of whisker-bars.

With this bunch of information at hand we can say the following: In some cases the modularity of the cut-clusterings is quite low, however, it increases with the amount of clustered vertices and the size of the clusters. It also reaches very high values, in particular for the snapshots of the email network and the "netscience" instance. This is a rather unexpected behavior since the cut-clustering algorithm is not designed to optimize modularity. We further observe a gap between the modularity values of many cut-clusterings and those of the according reference clusterings. We conjecture that this is caused more by an implausibility of the modularity values of the reference clusterings than by an implausibility of the cut-clusterings. Our conjecture is based on the observation, that the more significant the clusters in the reference clustering are, the closer comes the references modularity to the modularity of the cut-clustering, suggesting that the cut-clusterings are more trustable.

Furthermore, the Delaunay triangulations and the snapshots of the email network are nice examples that also vividly reveal the meaningfulness and plausibility of the cut-clusterings. The latter consider emails that were sent at most 72 hours ago. In contrast to other email networks, which consider a longer period of time, this makes the snapshots very sparse and stresses recent communication links, which yields clear clusters of people that recently work together. Thus, we would expect any feasible clustering approach to return meaningful non-trivial clusters. This is exactly what the cut-clustering algorithm as well as the modularity-based greedy approach do. In contrast, the Delaunay triangulations generated from random points in the plane are quite uniform structures. By intuition significant clusters are rare therein. The cut-clustering algorithm confirms our intuition by leaving all vertices unclustered. This explains the low modularity values of these clusterings and indicates that the underlying graph can not be clustered well. The modularity-based reference clusterings, however, contradict the intuition, as they consist of large clusters containing at least 20 vertices.

3.3. Expansion Analysis of Modularity-Based Clusterings. For reasons of completeness and fairness we also examine whether the modularity-based greedy clusterings outperform the cut-clusterings in terms of intra-cluster expansion. To this end we study the same lower and upper bounds for these clusterings as considered in Section 3.1 for the cut-clusterings.

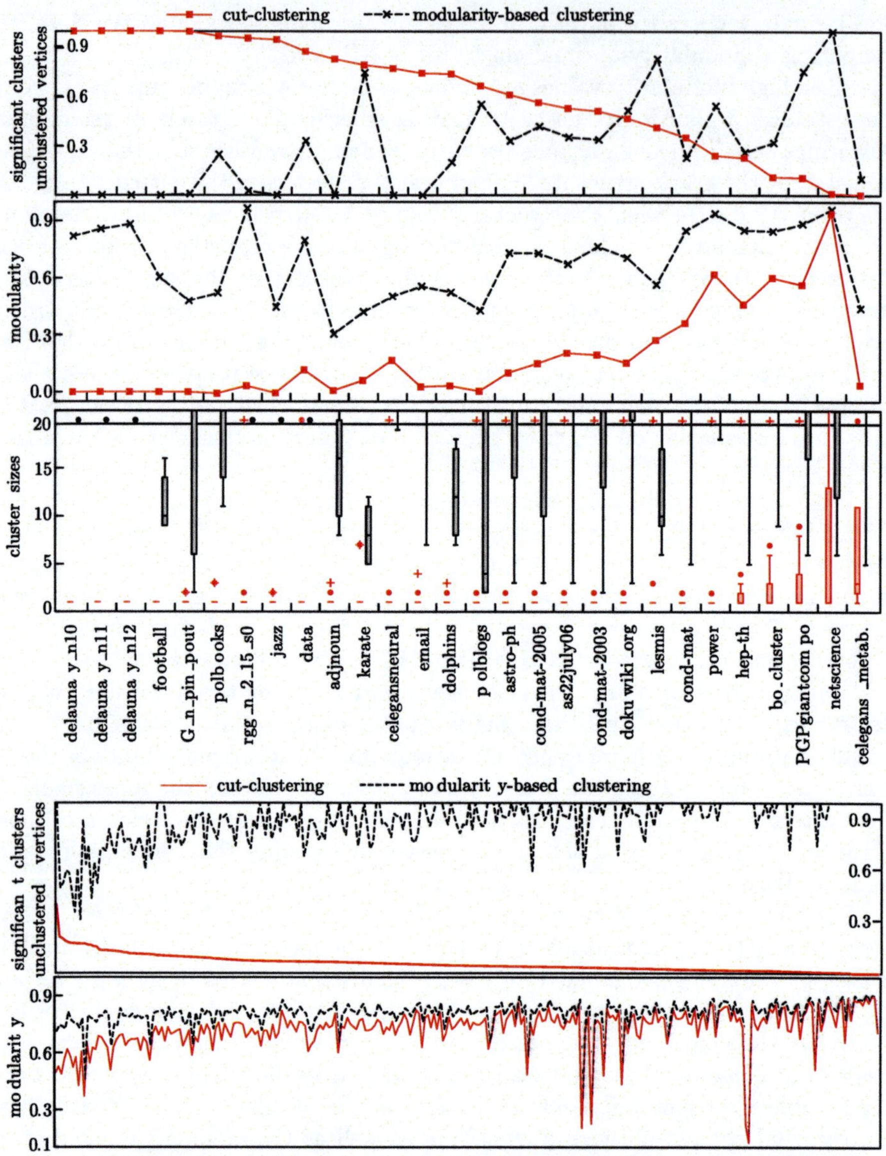

FIGURE 3. *Modularity Analysis:* Results for the best cut-clusterings and the modularity-based greedy clusterings. The upper charts in both parts of the figure show the ratio of unclustered vertices in the cut-clusterings and the ratio of nontrivial clusters that fulfill the significance-property in the modularity-based clusterings. In the upper part the cluster sizes for both types of clusterings (f.l.t.r. cut-clusterings and modularity-based clusterings) are shown by whisker-bars with maximum (+) and minimum (•) of the outliers. Values greater than 20 are placed at the edge of the displayed range. Due to the high number of email snapshots, we omitted whisker-bars there.

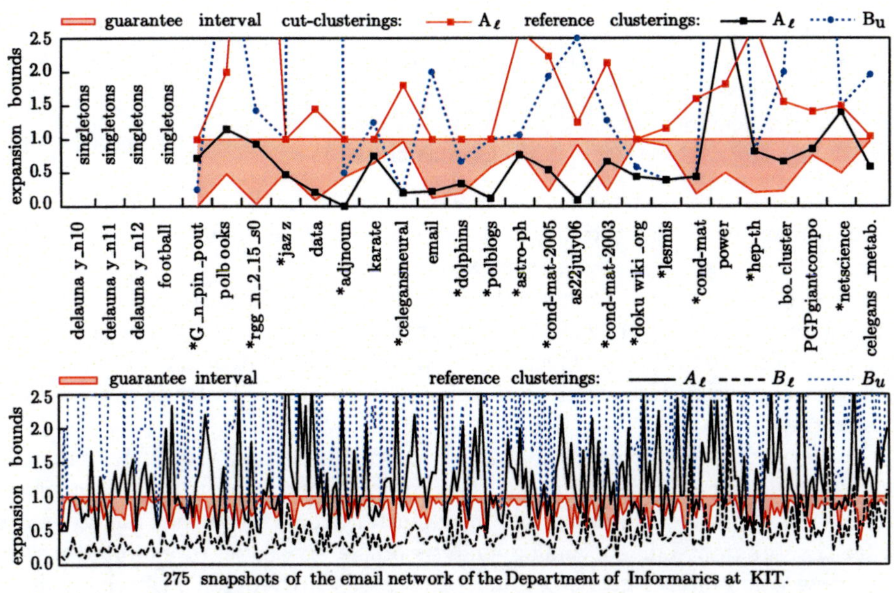

FIGURE 4. *Expansion Analysis of Modularity-Based Clusterings:* Guarantee interval and non-trivial bound (A_ℓ) for cut-clusterings, and bounds on intra-cluster expansion for the modularity-based references (B_ℓ, B_u trivial lower and upper bound based on minimum cut, A_ℓ alternative non-trivial lower bound); B_ℓ for the references and A_ℓ for the cut-clusterings are omitted in the upper and lower part, respectively. The upper boundary of the guarantee interval is normalized to 1, further values are displayed proportional. Instances where B_u drops below A_ℓ for the cut-clusterings in the upper part are marked by *. Regarding the first four instances, the hierarchical cut-clustering algorithm returns only singletons, for which intra-cluster expansion is not defined.

Figure 4 compares the guarantee interval and the alternative non-trivial lower bound $A_\ell(\Omega)$ for the cut-clusterings (already seen in Section 3.1) to the bounds for the modularity-based clusterings. Regarding the snapshots of the email network we omit depicting $A_\ell(\Omega)$ for the cut-clusterings.

We observe that the trivial lower bound $B_\ell(\Omega)$ stays clearly below the guarantee, and compared to the trivial bound for the cut-clusterings in Section 3.1 (cp. Figure 2) this behavior is even more evident. For the instances different from the snapshots of the email network the values of $B_\ell(\Omega)$ are so low so that we omit depicting them.

In contrast, the alternative non-trivial lower bound $A_\ell(\Omega)$ for the modularity-based clusterings often exceeds the guarantee interval, particularly for the snapshots. Nevertheless, it does rarely reach the corresponding bound for the cut-clusterings. For 85% of the instances it rather stays below the best lower bound for the cut-clustering. Thus, with respect to the lower bounds, there is no evidence that the intra-cluster expansion of the modularity-based clusterings surpasses that of the cut-clusterings. The upper bound $\Phi(\Omega)$, which drops below the best lower bound for the cut-clusterings in 23% of the cases, even *proves* a lower intra-cluster expansion for these clusterings. The according instances in the upper part of Figure 4 are marked by a star.

4. Conclusion

In this work we examined the behavior of the hierarchical cut-clustering algorithm of Flake et al. [5] in the light of expansion and modularity. Cut-clusterings are worth being studied since, in contrast to the results of other clustering approaches, they provide a guaranteed intra-cluster expansion and inter-cluster expansion and are clearly indicated by the graph structure. The latter materializes in the significance-property, which says that each set of vertices in a cluster is at least as strongly connected to the remaining vertices in the cluster as to the vertices outside the cluster.

Our experiments document that the given guarantee on intra-cluster expansion provides a deeper insight compared to a trivial bound that is easy to compute. The true intra-cluster expansion and inter-cluster expansion turned out to be even better than guaranteed. An analog analysis of the expansion of modularity-based clusterings could further give no evidence that modularity-based clusterings surpass cut-clusterings in terms of intra-cluster expansion. On the contrary, around one fourth of the considered modularity-based clusterings could be proven to be worse than the cut-clusterings.

Within the modularity analysis we could reveal that, although it is not designed to optimize modularity, the hierarchical cut-clustering algorithm fairly reliably finds clusterings of good modularity if those clusterings are structurally indicated. Otherwise, if no good clustering is clearly indicated, the cut-clustering algorithm returns only clusterings of low modularity. This confirms a high trustability of the cut-clustering algorithm and justifies the use of modularity applied to cut-clusterings as a feasible measure for how well a graph can be clustered.

Acknowledgements. We thank Markus Völker for technical support and Ignaz Rutter for proofreading and helpful suggestions on the structure of this paper. We further thank the anonymous reviewer for the thoughtful comments.

References

[1] *10th DIMACS Implementation Challenge – Graph Partitioning and Graph Clustering*, 2011, http://www.cc.gatech.edu/dimacs10/.

[2] *Dynamic network of email communication at the Department of Informatics at Karlsruhe Institute of Technology (KIT)*, 2011, Data collected, compiled and provided by Robert Görke and Martin Holzer of ITI Wagner and by Olaf Hopp, Johannes Theuerkorn and Klaus Scheibenberger of ATIS, all at KIT. i11www.iti.kit.edu/projects/spp1307/emaildata.

[3] Ulrik Brandes, Daniel Delling, Marco Gaertler, Robert Görke, Martin Höfer, Zoran Nikoloski, and Dorothea Wagner, *On Modularity Clustering*, IEEE Transactions on Knowledge and Data Engineering **20** (2008), no. 2, 172–188.

[4] Fabien de Montgolfier, Mauricio Soto, and Laurent Viennot, *Asymptotic Modularity of Some Graph Classes*, Proceedings of the 22nd International Symposium on Algorithms and Computation (ISAAC'11), 2011, pp. 435–444.

[5] Gary William Flake, Robert E. Tarjan, and Kostas Tsioutsiouliklis, *Graph clustering and minimum cut trees*, Internet Math. **1** (2004), no. 4, 385–408. MR2119992 (2005m:05210)

[6] Giorgio Gallo, Michael D. Grigoriadis, and Robert E. Tarjan, *A fast parametric maximum flow algorithm and applications*, SIAM J. Comput. **18** (1989), no. 1, 30–55, DOI 10.1137/0218003. MR978165 (90b:68038)

[7] R. E. Gomory and T. C. Hu, *Multi-terminal network flows*, J. Soc. Indust. Appl. Math. **9** (1961), 551–570. MR0135624 (24 #B1671)

[8] Michael Hamann, *Complete hierarchical Cut-Clustering: An Analysis of Guarantee and Quality*, Bachelor's thesis, Department of Informatics, Karlsruhe Institute of Technology (KIT), 2011, http://i11www.iti.uni-karlsruhe.de/teaching/theses/finished.

[9] Hawoong Jeong, Sean P. Mason, Albert-László Barabási, and Zoltan N. Oltvai, *Lethality and Centrality in Protein Networks*, Nature **411** (2001), 41–42.

[10] David Lisowski, *Modularity-basiertes Clustern von dynamischen Graphen im Offline-Fall*, Master's thesis, Department of Informatics, Karlsruhe Institute of Technology (KIT), 2011, http://i11www.iti.uni-karlsruhe.de/teaching/theses/finished.

[11] Mark E. J. Newman and Michelle Girvan, *Finding and evaluating community structure in networks*, Physical Review E **69** (2004), no. 026113, 1–16.

[12] Randolf Rotta and Andreas Noack, *Multilevel local search algorithms for modularity clustering*, ACM J. Exp. Algorithmics **16** (2011), Paper 2.3, 27, DOI 10.1145/1963190.1970376. MR2831090 (2012g:90232)

DEPARTMENT OF INFORMATICS, KARLSRUHE INSTITUTE OF TECHNOLOGY (KIT)
E-mail address: `mhamann@ira.uka.de`

DEPARTMENT OF INFORMATICS, KARLSRUHE INSTITUTE OF TECHNOLOGY (KIT)
E-mail address: `tanja.hartmann@kit.edu`

DEPARTMENT OF INFORMATICS, KARLSRUHE INSTITUTE OF TECHNOLOGY (KIT)
E-mail address: `dorothea.wagner@kit.edu`

A partitioning-based divisive clustering technique for maximizing the modularity

Ümit V. Çatalyürek, Kamer Kaya, Johannes Langguth, and Bora Uçar

ABSTRACT. We present a new graph clustering algorithm aimed at obtaining clusterings of high modularity. The algorithm pursues a divisive clustering approach and uses established graph partitioning algorithms and techniques to compute recursive bipartitions of the input as well as to refine clusters. Experimental evaluation shows that the modularity scores obtained compare favorably to many previous approaches. In the majority of test cases, the algorithm outperformed the best known alternatives. In particular, among 13 problem instances common in the literature, the proposed algorithm improves the best known modularity in 9 cases.

1. Introduction

Clustering graphs into disjoint vertex sets is a fundamental challenge in many areas of science [**3, 16, 22, 23**]. It has become a central tool in network analysis. With the recent rise in the availability of data on large scale real-world networks, the need for fast algorithms capable of clustering such instances accurately has increased significantly.

There is no generally accepted notion of what constitutes a good clustering, and in many cases the quality of a clustering is application specific. However, there are several widely accepted measurements for clustering quality called clustering indices. Among the most widespread clustering indices are *expansion*, *conductance*, and *modularity*. In the following, we will focus on modularity. See [**23**] for a discussion of the former two indices.

Modularity was proposed in [**32**] to analyze networks, and has recently grown in popularity as a clustering index [**15, 18–20, 27, 37**]. In addition, several heuristics based on greedy agglomeration [**11, 29**] and other approaches [**30, 34**] have been proposed for the problem. Although it was shown in [**7**] that these provide no approximation guarantee, for small real world instances the solutions produced by these heuristics are usually within a very small factor of the optimum.

In general there are two algorithmic approaches to community detection which are commonly known as agglomerative and divisive (see [**28**] for a short survey of general techniques). Agglomerative approaches start with every vertex in a separate cluster and successively merge clusters until the clustering can no longer

2010 *Mathematics Subject Classification.* Primary 91C20; Secondary 05C65, 05C70.
Key words and phrases. Clustering, modularity, graphs, hypergraphs.

be improved by merging pairs of clusters. The divisive approaches on the other hand consider removing edges to detect the communities. They start with the entire graph as a cluster and successively split clusters until further splitting is no longer worthwhile. We follow the divisive approach by making extensive use of graph partitioning algorithms and techniques. A similar approach which reduces the clustering problem to a variant of the well-known MinCut problem was recently proposed [**14**].

Finding a clustering that maximizes a clustering index is often NP-hard. In particular, finding a clustering of the maximum modularity in a graph was shown to be NP-hard [**7**]. It remains NP-hard even if the number of clusters is limited to 2. In addition, APX-hardness was established recently [**12**].

The remainder of this paper is organized as follows. We give some background in the next section. Section 3 contains the proposed divisive clustering method. The algorithm that we propose uses most of the standard ingredients of a graph (or hypergraph) partitioning tool: bisection, bisection refinement, and cluster refinement. We carefully put these together and explore the design space of a divisive clustering algorithm which makes use of those ingredients. In the same section, we discuss a contrived example which shows that the divisive approaches can be short-sighted. We evaluate the proposed divisive clustering algorithm in Section 4 with different parameter settings to explore the design space. We compare the resulting modularity scores with the best known ones from the literature. Section 5 concludes the paper. Some more results from the challenge data set are provided in the Appendix A.

2. Background

2.1. Preliminaries. In the following, $G = (V, E, \omega)$ is a weighted undirected graph with $\omega : E \to \mathbb{R}^+$ as the weight function. A *clustering* $\mathcal{C} = \{\mathcal{C}_1, \ldots, \mathcal{C}_K\}$ is a partition of the vertex set V. Each $\mathcal{C}_i$ is called a *cluster*. We use $G(\mathcal{C}_k)$ to denote the subgraph induced by the vertices in $\mathcal{C}_k$, that is $G(\mathcal{C}_k) = (\mathcal{C}_k, \mathcal{C}_k \times \mathcal{C}_k \cap E, \omega)$.

We define the weight of a vertex as the sum of the weights of its incident edges: $\psi(v) = \sum_{u \in V, \{u,v\} \in E} \omega(u, v)$, and we use $\psi(\mathcal{C}_\ell)$ to denote the sum of the weights of all vertices in a cluster $\mathcal{C}_\ell$. The sum of edge weights between two vertex sets U and T will be denoted by $\omega(U, T)$, that is $\omega(U, T) = \sum_{\{u,v\} \in U \times T \cap E} \omega(u, v)$. The sum of the weights of all edges is denoted by $\omega(E)$, and the sum of the weights of the edges whose both endpoints are in the same cluster $\mathcal{C}_\ell$ is denoted as $\omega(\mathcal{C}_\ell)$. Furthermore, by $cut(\mathcal{C})$ we denote the sum of the weights of all edges having vertices in two different clusters of $\mathcal{C}$.

2.2. Coverage and Modularity. We first define the *coverage* of a clustering, i.e., the fraction of edges that connect vertices in the same cluster:

$$cov(\mathcal{C}) = \frac{\sum_{\mathcal{C}_i \in \mathcal{C}} \omega(\mathcal{C}_i)}{\omega(E)}. \tag{2.1}$$

We can equivalently write that $cov(\mathcal{C}) = 1 - cut(\mathcal{C})/\omega(E)$. Obviously, a good clustering should have high coverage. However, since the number of clusters is not fixed, coverage can trivially be maximized by a clustering that consists of a single cluster. It is therefore not a suitable clustering index. By adding a penalty term for larger clusters, we obtain the *modularity* score of a clustering:

$$p(\mathcal{C}) = cov(\mathcal{C}) - \frac{\sum_{\mathcal{C}_i \in \mathcal{C}} \psi(\mathcal{C}_i)^2}{4 \times \omega(E)^2} \tag{2.2}$$

The penalty term is such that the trivial clustering, i.e., $\mathcal{C} = \{\mathcal{C}_1\}$, $\mathcal{C}_1 = V$, has a modularity of 0. Like other clustering indices, modularity captures the inherent trade-off between increasing the number of clusters and keeping the size of the cuts between clusters small. Almost all clustering indices require algorithms to face such a trade-off.

3. Algorithms

We follow the divisive approach to devise an algorithm for obtaining a clustering with high modularity. The main motivation for choosing this approach is that for a clustering $\mathcal{C}$ with two clusters, the coverage is just $1 - \text{cut}(\mathcal{C})/\omega(E)$ and the second term in (2.2) is minimized when clusters have equal weights. In other words, in splitting a graph into two clusters so as to maximize the modularity, heuristics for the NP-complete minimum bisection problem should be helpful (a more formal discussion is given by Brandes et al. [**7**, Section 4.1]). We can therefore harness the power and efficiency of the existing graph and hypergraph (bi-)partitioning routines such as MeTiS [**25**], PaToH [**10**], and Scotch [**33**] in a divisive approach to clustering for modularity.

Algorithm 1 Divisive clustering approach using graph/hypergraph bisection heuristics

Input: An edge weighted graph $G = (V, E, \omega)$
Output: $K^\star$: the number of clusters; $\mathcal{C}^\star = \{\mathcal{C}_1^\star, \mathcal{C}_2^\star, \ldots, \mathcal{C}_{K^\star}^\star\}$: the clusters;
$p^\star$: the modularity score

$K \leftarrow 1; \quad p \leftarrow 0$
$\mathcal{C} \leftarrow \{\mathcal{C}_1 = \{v_1, v_2, \ldots, v_n\}\}$ ▶ a single cluster
while there is an eligible cluster to consider **do**
  Let $\mathcal{C}_k$ be an eligible cluster with the largest vertex weight
  $\langle \mathcal{C}_{k_1}, \mathcal{C}_{k_2} \rangle \leftarrow$ BISECT$(\mathcal{C}_k, G)$
  if $\frac{\omega(\mathcal{C}_{k_1}, \mathcal{C}_{k_2})}{\omega(E)} < \frac{\psi(\mathcal{C}_k)^2 - \psi(\mathcal{C}_{k_1})^2 - \psi(\mathcal{C}_{k_2})^2}{4 \times \omega(E)^2}$ **then**
    $K \leftarrow K + 1$
    $p \leftarrow p - \frac{\omega(\mathcal{C}_{k_1}, \mathcal{C}_{k_2})}{\omega(E)} + \frac{\psi(\mathcal{C}_k)^2 - \psi(\mathcal{C}_{k_1})^2 - \psi(\mathcal{C}_{k_2})^2}{4 \times \omega(E)^2}$ ▶ update the modularity
    $\mathcal{C} \leftarrow \mathcal{C} \setminus \{\mathcal{C}_k\} \cup \{\mathcal{C}_{k_1}, \mathcal{C}_{k_2}\}$ ▶ replace $\mathcal{C}_k$ with two clusters
  else
    Mark $\mathcal{C}_k$ as ineligible for BISECT
$\langle K^\star, \mathcal{C}^\star, p^\star \rangle \leftarrow$ REFINECLUSTERS$(G, K, \mathcal{C}, p)$
return $\langle K^\star, \mathcal{C}^\star, p^\star \rangle$

Algorithm 1 displays the proposed approach. The algorithm accepts a weighted graph $G = (V, E, \omega)$, and returns the number of clusters $K^\star$ and the clustering $\mathcal{C}^\star = \{\mathcal{C}_1^\star, \mathcal{C}_2^\star, \ldots, \mathcal{C}_{K^\star}^\star\}$. It uses a bisection heuristic to compute a clustering of the given graph. Initially, all the vertices are in a single cluster. At every step, the heaviest cluster, say $\mathcal{C}_k$, is selected and split into two (by the subroutine BISECT), if $|\mathcal{C}_k| > 2$. If the bisection is acceptable, that is if the bisection improves the

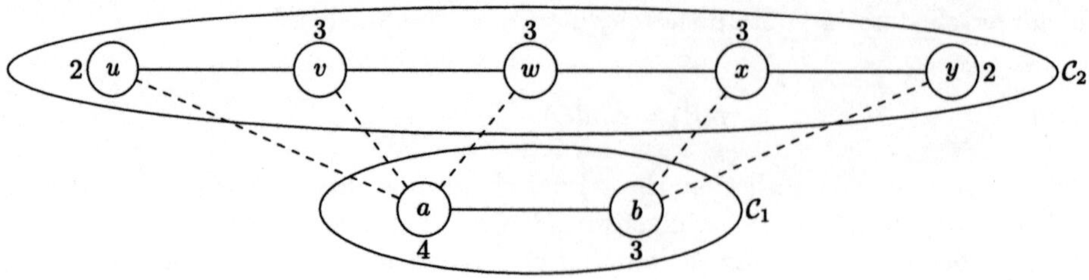

FIGURE 1. *Clustering $\mathcal{C}$. All edges have weight* 1. *Vertex weights are given.*

modularity (see the line 6), the cluster $\mathcal{C}_k$ is replaced by the two clusters resulting from the bisection. If not, the cluster $\mathcal{C}_k$ remains as is. The algorithm then proceeds to another step to pick the heaviest cluster. The clustering $\mathcal{C}$ found during the bisections is then refined in the subroutine REFINECLUSTERS that starts just after the bisections.

The computational core of the algorithm is the BISECT routine. This routine accepts a graph and splits that graph into two clusters using existing tools that are used for the graph/hypergraph bisection problem. We have instrumented the code in such a way that one can use MeTiS, PaToH, or Scotch quite effectively at this point.

Unfortunately, there is no guarantee that it is sufficient to stop bisecting a cluster as soon as a split on it reduced the modularity score. As finding a bipartition of maximum modularity is NP-hard [**7**], it is possible that a BISECT step which reduces modularity can be followed by a second BISECT step that increases it beyond its original value. As an example, consider the graph in Fig. 1 which shows a clustering, albeit a suboptimal one, that we will call $\mathcal{C}$ where $\mathcal{C} = \{\mathcal{C}_1, \mathcal{C}_2\}$. This clustering has the following modularity score

$$p(\mathcal{C}) = \frac{5}{10} - \frac{(3+4)^2 + (2+3+3+3+2)^2}{4 \times 10^2} = -\frac{18}{400}\,.$$

Since a trivial clustering $\{V\}$ has modularity $p(\{V\}) = 0$, we can easily see that the clustering $\mathcal{C}$ reduces the modularity to negative. Now, consider the clustering $\mathcal{C}' = \{\mathcal{C}_1, \mathcal{C}_{2_1}, \mathcal{C}_{2_2}\}$ which is obtained via a bipartition of $\mathcal{C}_2$ as shown in Fig. 2. Clustering $\mathcal{C}'$ has the following modularity:

$$p(\mathcal{C}') = \frac{4}{10} - \frac{(3+4)^2 + (2+3+3)^2 + (3+2)^2}{4 \times 10^2} = \frac{22}{400}\,.$$

Thus, clustering $\mathcal{C}_2$ has higher modularity than the initial trivial clustering $\{V\}$. Of course, this effect is due to the suboptimal clustering $\mathcal{C}$. However, since the bipartitioning algorithm provides no approximation guarantee, we cannot preclude this. Therefore, not bisecting a cluster anymore when a BISECT operation on it reduces the modularity score has its drawbacks.

3.1. The bisection heuristic. Our bisection heuristic is of the form shown in Algorithm 2 whose behavior is determined by a set of four parameters: a, imb, b, and e. The first one, a, chooses which algorithm to use as a bisector. We have integrated MeTiS, PaToH, and Scotch as bisectors. The bisection heuristics in PaToH and Scotch accept a parameter imb that defines the allowable imbalance

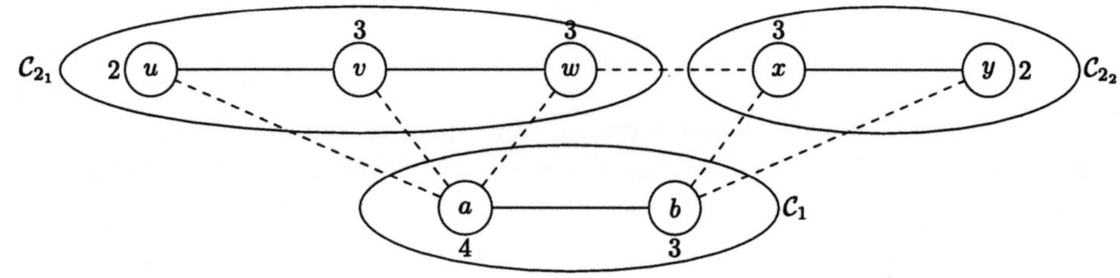

FIGURE 2. *Clustering* $\mathcal{C}'$. *All edges have weight* 1. *Vertex weights are given.*

between the part weights. We modified a few functions in the MeTiS 4.0 library to make the bisection heuristics accept the parameter imb. The other parameters are straightforwardly used as follows: the bisection heuristic (Algorithm 2) applies the bisector b times, refines each bisection e times and chooses the one that has the best modularity.

Algorithm 2 The bisection heuristics BISECT(U, G)

Input: A vertex set U, an edge weighted graph $G = (V, E, \omega)$
Output: $\langle L^\star, R^\star \rangle$ a bisection of the vertices U into two parts $L^\star$ and $R^\star$

1: mostIncrease $\leftarrow -\infty$
2: **for** imb $\in \{0.05, 0.10, 0.20, 0.40\}$ **do**
3:   **for** $i = 1$ to b **do**
4:     $\langle L, R \rangle \leftarrow$ apply BISECTOR a to G with imbalance tolerance imb
5:     **for** $j = 1$ to e **do**
6:       $\langle L, R \rangle \leftarrow$ REFINEBISECTION$(L, R, G(U))$
7:       **if** $\frac{\psi(U)^2 - \psi(L)^2 - \psi(R)^2}{4 \times \omega(E)^2} - \frac{\omega(L,R)}{\omega(E)} >$ mostIncrease **then**
8:         mostIncrease $\leftarrow \frac{\psi(U)^2 - \psi(L)^2 - \psi(R)^2}{4 \times \omega(E)^2} - \frac{\omega(L,R)}{\omega(E)}$
9:         $\langle L^\star, R^\star \rangle \leftarrow \langle L, R \rangle$

As shown in Algorithm 1, the bisection heuristic is called for a cluster $\mathcal{C}_k$ of a clustering $\mathcal{C}$ with the modularity score p. Note that $\mathcal{C}_k$ contributes $\frac{\omega(\mathcal{C}_k)}{\omega(E)} - \frac{\psi(\mathcal{C}_k)^2}{4 \times \omega(E)^2}$ to the modularity score. When we bisect $\mathcal{C}_k$ into $\mathcal{C}_{k_1}$ and $\mathcal{C}_{k_2}$, the coverage of the clustering $\mathcal{C} \leftarrow \mathcal{C} \setminus \{\mathcal{C}_k\} \cup \{\mathcal{C}_{k_1}, \mathcal{C}_{k_2}\}$ becomes $\omega(\mathcal{C}_{k_1}, \mathcal{C}_{k_2})$ less than the coverage of $\mathcal{C}$, and the new clusters $\mathcal{C}_{k_1}$ and $\mathcal{C}_{k_2}$ contribute $\frac{\omega(\mathcal{C}_{k_1}) + \omega(\mathcal{C}_{k_2})}{\omega(E)} - \frac{\psi(\mathcal{C}_{k_1})^2 + \psi(\mathcal{C}_{k_2})^2}{4 \times \omega(E)^2}$ to the modularity score. The difference between the modularity scores is therefore the formula used at line 7 of Algorithm 2.

The vertex weights that are passed to the bisectors are simply the weights $\psi(\cdot)$ defined on the original graph. Balancing the sums of weights of the vertices in the two parts will likely reduce the squared part weights and will likely yield better modularity. This however, is not guaranteed, as the increase in the modularity score is also affected by the cut of the bisection. That is why trying a few imbalance parameters (controlled by imb), running the bisector multiple times (controlled by b) with the same imbalance parameter, and refining those bisections (controlled by e) is a reasonable approach.

The algorithm REFINEBISECTION$(L, R, G = (L \cup R, E))$ is a variant of Fiduccia-Mattheyses [**17**] (FM) heuristic. Given two clusters, FM computes a gain associated with moving a vertex from one cluster to the other one. The efficiency of the method is achieved by keeping these gains up-to-date after every move. In the standard application of this refinement heuristic (for the graph and hypergraph bipartitioning problems, see e.g., [**9**, **24**]), moving a vertex changes the gains associated with the adjacent vertices only. This is not true during the refinement process for improving the the modularity score. Consider a given weighted graph $G = (V, E, \omega)$ and a bipartition L, R of V. The contribution of the clusters L and R to the modularity score is

$$\frac{\omega(L) + \omega(R)}{\omega(E)} - \frac{\psi(L)^2 + \psi(R)^2}{4 \times \omega(E)^2}\,. \tag{3.1}$$

When we move a vertex v from L to R, the new modularity score becomes

$$\frac{\omega(L \setminus \{v\}) + \omega(R \cup \{v\})}{\omega(E)} - \frac{\psi(L \setminus \{v\})^2 + \psi(R \cup \{v\})^2}{4 \times \omega(E)^2}\,. \tag{3.2}$$

Subtracting (3.1) from (3.2) we obtain the gain of moving v from L to R

$$\text{gain}(v, L \mapsto R) = \frac{\sum_{u \in R} \omega(v,u) - \sum_{u \in L} \omega(v,u)}{\omega(E)} + 2 \times \psi(v) \frac{\psi(L) + \psi(R) - \psi(v)}{4 \times \omega(E)^2}\,. \tag{3.3}$$

As the gain of a move includes the cluster weights, a single move necessitates gain updates for all vertices. Thus, it is not very practical to choose the move with the highest gain in modularity at every step. We therefore designed the following alternative. We keep two priority queues, one for each side of the partition, where the key values of the moves are the reduction in the cut size (that is, the key values are the standard FM gains). Assuming uniform vertex weights, among the moves of the form $L \mapsto R$, the one with the maximum gain in the modularity will be the vertex move with the maximum gain in the cut size. This is due to the fact that the second term in (3.3) will be the same for all vertices in L. Similarly, among the moves of the form $R \mapsto L$, the one with the maximum gain in the cut size will be the one with the maximum gain in the modularity. Since vertex weights are not uniform, we need to be a little careful about choosing which vertex to move. Every time we look for a vertex move, we check the first move of both priority queues and compute the actual gain (3.3) and perform the better move tentatively. Realizing the maximum gain sequence of these tentative moves is done in the same way as in the standard FM heuristic.

The bisectors (MeTiS, PaToH, or Scotch) are generally accepted to be of linear time complexity. The time complexity of the REFINEBISECTION is, due to the use of priority queues and gain update operations, $\mathcal{O}(|E| \log |V|)$ for a graph $G = (V, E)$. Therefore, the running time of bisection step is $\mathcal{O}(|E| \log |V|)$. However, we should note that depending on the parameter settings the constant hidden in the formula can be large.

3.2. Refining the clustering. The last ingredient of the proposed clustering algorithm is REFINECLUSTERS$(G, K, \mathcal{C}, p)$. It aims to improve the clustering found during the bisections. Unlike the REFINEBISECTION algorithm, this algorithm visits the vertices in random order. At each vertex v, the gain values associated with moving v from its current cluster to all others are computed. Among all those moves, the most beneficial one is performed if doing so increases the modularity score. If not, the vertex v remains in its own cluster. We repeat this process several times (we use m as a parameter to control the number of such passes). The time complexity of a pass is $\mathcal{O}(|V|K + |E|)$ for a K-way clustering of a graph with $|V|$ vertices and $|E|$ edges.

4. Experiments

We perform a series of experiments to measure the effect of the various parameters on the modularity scores of the solutions produced by the algorithm, and to evaluate overall performance of the approach.

To this end, we use a set of 29 popular test instances which have been used in the past to study the modularity scores achieved by clustering algorithms. The instances are from various resources **[1, 2, 4–6, 21, 31, 35, 36]** and are available at http://www.cc.gatech.edu/dimacs10/.

We first test the algorithm using the standard parameter combination. It consists of m=5 refinement rounds at the end and a bipartition parameter of b=1. No refinements are performed during the algorithm (e=0). The results using PaToH, MeTiS, and Scotch partitioning are shown in Table 1 below.

As expected, the results for each instance are very close, with a maximum difference of less than 0.04. All partitioners provide good results, with PaToH delivering somewhat higher modularity scores. However, using MeTiS consistently yielded slightly inferior results. The same was true in preliminary versions of the experiments described below. Thus, MeTiS was not considered in the following experiments.

The slightly higher scores of PaToH can be explained by the fact that unlike SCOTCH, it uses randomization. Even though this is not intended by the algorithm design, when performing multiple partitioning runs during the BISECT routine, the randomized nature of the PaToH gives it a slightly higher chance to find a superior solution, which is generally kept by the algorithm.

In the next experiment, we investigate the effect of the refinement algorithm REFINECLUSTERS on the final result. Table 2 shows the refined modularity scores using a maximum of $\mathsf{m} = 5$ refinement steps at the end of Algorithm 1 for PaToH and Scotch partitioning, as opposed to the unrefined results ($\mathsf{m} = 0$). On average, the effect of the clustering refinement step (REFINECLUSTERS) amounts to an improvement of about 0.01 for Scotch and 0.0042 for PaToH. Our preliminary experiments showed that increasing the number of refinement steps beyond $\mathsf{m} = 5$ improves the end result only marginally in both cases. Although the improvement for Scotch is slightly larger, the difference is not sufficient to completely equalize the gap between the unrefined results for PaToH and Scotch. Since the computational cost of the refinement heuristic is low, we will continue to use it in the following experiments.

Furthermore, we investigate the influence of the number of repetitions of the BISECTOR step on the modularity score by increasing the parameter b from 1 to 5. Results are shown in Table 3 where we observe a slight positive effect for $\mathsf{b} = 5$ as compared to $\mathsf{b} = 1$. It is interesting to note that even though PaToH is randomized, selecting the best out of 5 bisections has almost no effect. This is partially because the REFINECLUSTERS operation finds the same improvements. Due to the refinement, the total effect can even be negative since a different clustering might

TABLE 1. Modularity scores obtained by the basic algorithm before the refinement. The difference between PaToH, MeTiS, and Scotch is visible. The best modularity for each row is marked as bold.

Instance	Vertices	Edges	Modularity score		
			PaToH	Scotch	MeTiS
adjnoun	112	425	**0.2977**	0.2972	0.2876
as-22july06	22963	48436	**0.6711**	0.6578	0.6486
astro-ph	16706	121251	**0.7340**	0.7238	0.7169
caidaRouterLevel	192244	609066	**0.8659**	0.8540	0.8495
celegans_metabolic	453	2025	0.4436	0.4407	**0.4446**
celegansneural	297	2148	0.4871	**0.4939**	0.4754
chesapeake	39	170	0.2595	**0.2624**	0.2595
citationCiteseer	268495	1156647	**0.8175**	0.8119	0.8039
cnr-2000	325557	2738969	**0.9116**	0.9026	0.8819
coAuthorsCiteseer	227320	814134	**0.8982**	0.8838	0.8853
coAuthorsDBLP	299067	977676	**0.8294**	0.8140	0.8117
cond-mat	16726	47594	**0.8456**	0.8343	0.8309
cond-mat-2003	31163	120029	**0.7674**	0.7556	0.7504
cond-mat-2005	40421	175691	**0.7331**	0.7170	0.7152
dolphins	62	159	**0.5276**	0.5265	0.5246
email	1133	5451	**0.5776**	0.5748	0.5627
football	115	613	**0.6046**	**0.6046**	0.6019
G_n_pin_pout	100000	501198	**0.4913**	0.4740	0.4825
hep-th	8361	15751	**0.8504**	0.8409	0.8342
jazz	198	2742	0.4450	**0.4451**	0.4447
karate	34	78	**0.4198**	**0.4198**	0.3843
lesmis	77	254	**0.5658**	0.5649	0.5656
netscience	1589	2742	**0.9593**	0.9559	0.9533
PGPgiantcompo	10680	24316	**0.8831**	0.8734	0.8687
polblogs	1490	16715	**0.4257**	**0.4257**	**0.4257**
polbooks	105	441	**0.5269**	**0.5269**	0.4895
power	4941	6594	**0.9398**	0.9386	0.9343
preferentialAttachment	100000	499985	**0.3066**	0.2815	0.2995
smallworld	100000	499998	**0.7846**	0.7451	0.7489
Average			**0.6507**	0.6430	0.6373

preclude a particularly effective refinement. Overall, we conclude that $\mathsf{b} = 5$ is worthwhile for Scotch, but not for PaToH.

We also study the influence of applying the refinement process during the algorithm, as opposed to the standard refinement after termination of the main algorithm. This is done by setting the parameter $\mathsf{e} = 5$, i.e., we use 5 passes of REFINEBISECTION after each call of BISECTOR in Algorithm 2. Results that are displayed in Table 4 show that this approach does not help to improve modularity. Most likely, improvements that can be found in this manner can also be found by calling REFINECLUSTERS at the end of the algorithm. In addition, this technique is computationally expensive, and therefore it should not be used.

Finally, we compare modularity scores obtained by our algorithm with previous results found in literature. We compare the best score found by our algorithms with the best score found in the literature. These results are shown in Table 5. Compared to previous work, our algorithms perform quite well. For the small instances *dolphins, karate, polbooks*, and *football* the previous values are optimal,

TABLE 2. Modularity scores and improvement after the application of the REFINECLUSTERS algorithm at $m = 5$. Improvements for Scotch partitioning are larger than those for PaToH. The improvements are given in the column "Improv.". The best modularity for each row is marked as bold.

	PaToH			Scotch		
Instance	Unrefined	Refined	Improv.	Unrefined	Refined	Improv.
adjnoun	0.2945	**0.2977**	0.0033	0.2946	0.2972	0.0026
as-22july06	0.6683	**0.6711**	0.0028	0.6524	0.6578	0.0054
astro-ph	0.7295	**0.7340**	0.0046	0.7183	0.7238	0.0055
caidaRouterLevel	0.8641	**0.8659**	0.0019	0.8506	0.8540	0.0035
celegans_metabolic	0.4318	**0.4436**	0.0118	0.4343	0.4407	0.0064
celegansneural	0.4855	0.4871	0.0016	0.4905	**0.4939**	0.0034
chesapeake	0.2495	0.2595	0.0100	0.2624	**0.2624**	0.0000
citationCiteseer	0.8160	**0.8175**	0.0015	0.8094	0.8119	0.0025
cnr-2000	0.9116	**0.9116**	0.0000	0.8981	0.9026	0.0045
coAuthorsCiteseer	0.8976	**0.8982**	0.0005	0.8826	0.8838	0.0012
coAuthorsDBLP	0.8281	**0.8294**	0.0013	0.8115	0.8140	0.0025
cond-mat-2003	0.8443	**0.8456**	0.0013	0.8329	0.8343	0.0013
cond-mat-2005	0.7651	**0.7674**	0.0023	0.7507	0.7556	0.0049
cond-mat	0.7293	**0.7331**	0.0038	0.7084	0.7170	0.0086
dolphins	0.5155	**0.5276**	0.0121	0.5265	0.5265	0.0000
email	0.5733	**0.5776**	0.0043	0.5629	0.5748	0.0120
football	0.6009	**0.6046**	0.0037	0.6009	**0.6046**	0.0037
G_n_pin_pout	0.4565	**0.4913**	0.0347	0.3571	0.4740	0.1169
hep-th	0.8494	**0.8504**	0.0010	0.8392	0.8409	0.0016
jazz	0.4330	0.4450	0.0120	0.4289	**0.4451**	0.0162
karate	0.4188	**0.4198**	0.0010	0.4188	**0.4198**	0.0010
lesmis	0.5658	**0.5658**	0.0000	0.5540	0.5649	0.0108
netscience	0.9593	**0.9593**	0.0000	0.9559	0.9559	0.0000
PGPgiantcompo	0.8830	**0.8831**	0.0001	0.8726	0.8734	0.0008
polblogs	0.4257	**0.4257**	0.0000	0.4247	**0.4257**	0.0010
polbooks	0.5266	**0.5269**	0.0004	0.5242	**0.5269**	0.0027
power	0.9394	**0.9398**	0.0003	0.9384	0.9386	0.0002
preferentialAttachment	0.3013	**0.3066**	0.0053	0.2461	0.2815	0.0353
smallworld	0.7838	**0.7846**	0.0008	0.7061	0.7451	0.0390
Average	0.6465	**0.6507**	0.0042	0.6329	0.6430	0.0101

and the algorithms come quite close, deviating by only 0.00047 from the optimum values on average. The instance *lesmis* is a weighted graph and was treated as such here. Therefore the modularity score obtained is higher than the unweighted optimum computed in [8]. It is included here for the sake of completeness, but it is not considered for the aggregated results.

For larger instances, obtaining optimum values is computationally infeasible. Thus, the scores given here represent the best value found by other clustering algorithms. Our algorithm surpasses those in 9 out of 13 instances, and its average modularity score surpasses the best reported values by 0.01. Naturally, most clustering algorithms will be quite close in such a comparison, which renders the difference quite significant.

Summing up, we conclude that the optimum configuration for our algorithm uses PaToH for partitioning with REFINECLUSTERS at $m = 5$. For the bipartition

TABLE 3. Comparison between bipartition parameter setting of b = 1 and b = 5. Using 5 steps improves the end result slightly. The best modularity for each tool with b = 1 and b = 5 is marked as bold.

	PaToH			Scotch		
Instance	b=1	b=5	Difference	b=1	b=5	Difference
adjnoun	0.2977	**0.2990**	0.0012	0.2972	**0.2999**	0.0027
as-22july06	0.6711	**0.6722**	0.0011	**0.6578**	0.6503	-0.0075
astro-ph	0.7340	**0.7353**	0.0012	0.7238	**0.7261**	0.0023
caidaRouterLevel	0.8659	**0.8677**	0.0018	0.8540	**0.8576**	0.0035
celegans_metabolic	0.4436	**0.4454**	0.0017	0.4407	**0.4467**	0.0060
celegansneural	0.4871	**0.4945**	0.0074	0.4939	**0.4942**	0.0004
chesapeake	0.2595	**0.2624**	0.0029	**0.2624**	**0.2624**	0.0000
citationCiteseer	**0.8175**	0.8166	-0.0009	0.8119	**0.8141**	0.0022
cnr-2000	0.9116	**0.9119**	0.0003	0.9026	**0.9052**	0.0026
coAuthorsCiteseer	0.8982	**0.8994**	0.0012	0.8838	**0.8872**	0.0033
coAuthorsDBLP	0.8294	**0.8306**	0.0011	0.8140	**0.8180**	0.0040
cond-mat	0.8456	**0.8469**	0.0013	0.8343	**0.8378**	0.0035
cond-mat-2003	0.7674	**0.7692**	0.0018	0.7556	**0.7593**	0.0037
cond-mat-2005	0.7331	**0.7338**	0.0007	0.7170	**0.7248**	0.0078
dolphins	**0.5276**	0.5265	-0.0011	**0.5265**	**0.5265**	0.0000
email	**0.5776**	0.5768	-0.0008	0.5748	**0.5770**	0.0022
football	**0.6046**	**0.6046**	0.0000	**0.6046**	**0.6046**	0.0000
G_n_pin_pout	0.4913	**0.4915**	0.0002	0.4740	**0.4844**	0.0104
hep-th	0.8504	**0.8506**	0.0002	0.8409	**0.8425**	0.0017
jazz	**0.4450**	**0.4450**	0.0000	**0.4451**	**0.4451**	0.0000
karate	**0.4198**	**0.4198**	0.0000	**0.4198**	**0.4198**	0.0000
lesmis	**0.5658**	**0.5658**	0.0000	**0.5649**	**0.5649**	0.0000
netscience	**0.9593**	**0.9593**	0.0000	0.9559	**0.9591**	0.0032
PGPgiantcompo	0.8831	**0.8834**	0.0004	0.8734	**0.8797**	0.0063
polblogs	**0.4257**	**0.4257**	0.0000	**0.4257**	**0.4257**	0.0000
polbooks	**0.5269**	**0.5269**	0.0000	**0.5269**	**0.5269**	0.0000
power	**0.9398**	0.9397	-0.0001	0.9386	**0.9398**	0.0012
preferentialAttachment	**0.3066**	0.3065	-0.0001	0.2815	**0.2887**	0.0073
smallworld	0.7846	**0.7850**	0.0004	0.7451	**0.7504**	0.0053
Average	0.6507	**0.6514**	0.0008	0.6430	**0.6455**	0.0025

parameter, exceeding b = 1 is hardly worthwhile. In this configuration, REFINEBISECTION should not be used, i.e., e = 0 should be selected.

5. Conclusion

We have presented a new algorithm for finding graph clusterings of high modularity. It follows a divisive approach by applying recursive bipartition to clusters. In addition, it makes use of a standard refinement heuristic. It can be implemented efficiently by making use of established partitioning software.

We experimentally established that the best modularity scores can be obtained by choosing the best out of multiple partitionings during the bipartitioning step and applying the refinement heuristic at the end of the algorithm. The modularity scores obtained in this manner surpass those of previously known clustering algorithms.

A possible variant of the proposed algorithm that can be further studied would accept bipartitions of inferior modularity for a limited number of recursion steps, thereby alleviating the problem described in Section 3.

TABLE 4. Modularity scores for refinement steps during the algorithm. After every bisection, up to 5 refinement steps are performed. The best modularity for each tool with e = 0 and e = 5 is marked as bold.

	PaToH			Scotch		
Instance	e=0	e=5	Diff.	e=0	e=5	Diff.
adjnoun	0.2977	**0.3014**	0.0037	**0.2972**	0.2941	-0.0031
as-22july06	**0.6711**	0.6653	-0.0058	0.6578	**0.6581**	0.0003
astro-ph	**0.7340**	0.7283	-0.0058	**0.7238**	0.7204	-0.0034
caidaRouterLevel	**0.8659**	0.8627	-0.0033	**0.8540**	0.8483	-0.0058
celegans_metabolic	**0.4436**	0.4430	-0.0007	0.4407	**0.4433**	0.0026
celegansneural	0.4871	**0.4945**	0.0074	0.4939	**0.4944**	0.0005
chesapeake	0.2595	**0.2658**	0.0063	0.2624	**0.2658**	0.0034
citationCiteseer	**0.8175**	0.8145	-0.0030	**0.8119**	0.8088	-0.0031
cnr-2000	**0.9116**	0.9050	-0.0066	**0.9026**	0.9019	-0.0007
coAuthorsCiteseer	**0.8982**	0.8971	-0.0011	**0.8838**	0.8829	-0.0009
coAuthorsDBLP	**0.8294**	0.8276	-0.0018	**0.8140**	0.8106	-0.0033
cond-mat	**0.8456**	0.8424	-0.0031	**0.8343**	0.8333	-0.0010
cond-mat-2003	**0.7674**	0.7643	-0.0031	**0.7556**	0.7532	-0.0023
cond-mat-2005	**0.7331**	0.7309	-0.0022	**0.7170**	0.7142	-0.0028
dolphins	**0.5276**	0.5265	-0.0011	**0.5265**	**0.5265**	0.0000
email	**0.5776**	0.5748	-0.0028	**0.5748**	0.5647	-0.0101
football	**0.6046**	0.6032	-0.0013	**0.6046**	0.6032	-0.0013
G_n_pin_pout	0.4913	**0.4921**	0.0009	0.4740	**0.4872**	0.0132
hep-th	**0.8504**	0.8472	-0.0031	0.8409	**0.8412**	0.0003
jazz	0.4450	**0.4451**	0.0001	**0.4451**	0.4271	-0.0181
karate	**0.4198**	**0.4198**	0.0000	**0.4198**	**0.4198**	0.0000
lesmis	**0.5658**	**0.5658**	0.0000	0.5649	**0.5652**	0.0003
netscience	**0.9593**	0.9551	-0.0042	**0.9559**	0.9558	-0.0001
PGPgiantcompo	**0.8831**	0.8791	-0.0040	**0.8734**	0.8732	-0.0002
polblogs	**0.4257**	**0.4257**	0.0000	**0.4257**	**0.4257**	0.0000
polbooks	**0.5269**	0.5108	-0.0161	**0.5269**	0.5108	-0.0161
power	**0.9398**	0.9373	-0.0024	**0.9386**	0.9346	-0.0040
preferentialAttachment	**0.3066**	0.3058	-0.0008	0.2815	**0.2952**	0.0137
smallworld	0.7846	**0.7851**	0.0005	0.7451	**0.7857**	0.0406
Average	**0.6507**	0.6488	-0.0018	**0.6430**	0.6429	0.0001

TABLE 5. Comparison between modularity score obtained by our algorithm and and scores reported in previous work. An asterisk indicates that this instances has been solved optimally. The best modularity for each row is marked as bold.

Instance	Best found Modularity	Best known Modularity	Source	Difference
adjnoun	0.3014	**0.3080**	[26]	-0.0066
caidaRouterLevel	**0.8677**	0.8440	[13]	0.0237
celegans_metabolic	**0.4467**	0.4350	[8]	0.0117
celegansneural	**0.4945**	0.4010	[26]	0.0935
citationCiteseer	**0.8175**	0.8037	[13]	0.0138
cnr-2000	0.9119	**0.9130**	[13]	-0.0011
coAuthorsDBLP	**0.8306**	0.8269	[13]	0.0037
dolphins*	0.5276	**0.5290**	[8]	-0.0014
email	**0.5776**	0.5738	[15]	0.0038
football*	0.6046	**0.6050**	[8]	-0.0004
jazz	0.4451	**0.4452**	[15]	-0.0001
karate*	**0.4198**	**0.4198**	[8]	0.0000
lesmis*	**0.5658**	0.5600	[8]	0.0058
netscience	**0.9593**	0.9540	[26]	0.0053
PGPgiantcompo	**0.8834**	0.8550	[8]	0.0284
polblogs	0.4257	**0.4260**	[26]	-0.0003
polbooks*	0.5269	**0.5270**	[8]	-0.0001
power	**0.9398**	0.9390	[8]	0.0008
Average	**0.6414**	0.6314		0.0100

Acknowledgment

This work was supported in parts by the DOE grant DE-FC02-06ER2775 and by the NSF grants CNS-0643969, OCI-0904809, and OCI-0904802.

References

[1] A. Arenas, *Network data sets*, available at http://deim.urv.cat/ aarenas/data/welcome.htm, October 2011.
[2] Albert-László Barabási and Réka Albert, *Emergence of scaling in random networks*, Science **286** (1999), no. 5439, 509–512, DOI 10.1126/science.286.5439.509. MR2091634
[3] M. Bern and D. Eppstein, *Approximation algorithms for geometric problems*, Approximation Algorithms for NP-Hard Problems (D. S. Hochbaum, ed.), PWS Publishing Co., Boston, MA, USA, 1997, pp. 296–345.
[4] P. Boldi, B. Codenotti, M. Santini, and S. Vigna, *Ubicrawler: A scalable fully distributed web crawler*, Software: Practice & Experience **34** (2004), no. 8, 711–726.
[5] P. Boldi, M. Rosa, M. Santini, and S. Vigna, *Layered label propagation: A multiresolution coordinate-free ordering for compressing social networks*, Proceedings of the 20th international conference on World Wide Web, ACM Press, 2011.
[6] P. Boldi and S. Vigna, *The WebGraph framework I: Compression techniques*, Proc. of the Thirteenth International World Wide Web Conference (WWW 2004) (Manhattan, USA), ACM Press, 2004, pp. 595–601.
[7] U. Brandes, D. Delling, M. Gaertler, R. Görke, M. Hoefer, Z. Nikoloski, and D. Wagner, *On finding graph clusterings with maximum modularity*, Graph-theoretic concepts in computer science, Lecture Notes in Comput. Sci., vol. 4769, Springer, Berlin, 2007, pp. 121–132, DOI 10.1007/978-3-540-74839-7_12. MR2428570 (2009j:05215)
[8] S. Cafieri, P. Hansen, and L. Liberti, *Locally optimal heuristic for modularity maximization of networks*, Phys. Rev. E **83** (2011), 056105.
[9] Ü. V. Çatalyürek and C. Aykanat, *Hypergraph-partitioning-based decomposition for parallel sparse-matrix vector multiplication*, IEEE Transactions on Parallel and Distributed Systems **10** (1999), no. 7, 673–693.
[10] ———, *PaToH: A multilevel hypergraph partitioning tool, version 3.0*, Bilkent University, Department of Computer Engineering, Ankara, 06533 Turkey. PaToH is available at http://bmi.osu.edu/ umit/software.htm, 1999.
[11] A. Clauset, M. E. J. Newman, and C. Moore, *Finding community structure in very large networks*, Phys. Rev. E **70** (2004), 066111.
[12] B. Dasgupta and D. Desai, *On the complexity of Newman's finding approach for biological and social networks*, arXiv:1102.0969v1, 2011.
[13] D. Delling, R. Görke, C. Schulz, and D. Wagner, *Orca reduction and contraction graph clustering*, Proceedings of the 5th International Conference on Algorithmic Aspects in Information and Management (Berlin, Heidelberg), AAIM '09, Springer-Verlag, 2009, pp. 152–165.
[14] H. N. Djidjev and M. Onuş, *Scalable and accurate graph clustering and community structure detection*, IEEE Transactions on Parallel and Distributed Systems, **99**, Preprints, (2012).
[15] J. Duch and A. Arenas, *Community detection in complex networks using extremal optimization*, Phys. Rev. E **72** (2005), 027104.
[16] B. Everitt, *Cluster analysis*, 2nd ed., Social Science Research Council Reviews of Current Research, vol. 11, Heinemann Educational Books, London, 1980. MR592781 (82a:62082)
[17] C. M. Fiduccia and R. M. Mattheyses, *A linear-time heuristic for improving network partitions*, DAC '82: Proceedings of the 19th Conference on Design Automation (Piscataway, NJ, USA), IEEE Press, 1982, pp. 175–181.
[18] P. F. Fine, E. Di Paolo, and A. Philippides, *Spatially constrained networks and the evolution of modular control systems*, From Animals to Animats 9: Proceedings of the Ninth International Conference on Simulation of Adaptive Behavior (S. Nolfi, G. Baldassarre, R. Calabretta, J. Hallam, D. Marocco, O. Miglino, J. A. Meyer, and D. Parisi, eds.), Springer Verlag, 2006, pp. 546–557.
[19] S. Fortunato and M. Barthélemy, *Resolution limit in community detection*, Proceedings of the National Academy of Sciences **104** (2007), no. 1, 36–41.

[20] M. Gaertler, R. Görke, and D. Wagner, *Significance-driven graph clustering*, Proceedings of the 3rd International Conference on Algorithmic Aspects in Information and Management (AAIM'07), Lecture Notes in Computer Science, Springer, June 2007, pp. 11–26. MR2393900 (2008i:68005)

[21] R. Geisberger, P. Sanders, and D. Schultes, *Better approximation of betweenness centrality*, Proceedings of the Ninth Workshop on Algorithm Engineering and Experiments (ALENEX), 2008.

[22] A. K. Jain and R. C. Dubes, *Algorithms for clustering data*, Prentice Hall Advanced Reference Series, Prentice Hall Inc., Englewood Cliffs, NJ, 1988. MR999135 (91h:68155)

[23] R. Kannan, S. Vempala, and A. Vetta, *On clusterings: good, bad and spectral*, J. ACM **51** (2004), no. 3, 497–515, DOI 10.1145/990308.990313. MR2145863 (2006m:05153)

[24] G. Karypis and V. Kumar, *A fast and high quality multilevel scheme for partitioning irregular graphs*, SIAM J. Sci. Comput. **20** (1998), no. 1, 359–392 (electronic), DOI 10.1137/S1064827595287997. MR1639073 (99f:68158)

[25] G. Karypis and V. Kumar, *MeTiS: A software package for partitioning unstructured graphs, partitioning meshes, and computing fill-reducing orderings of sparse matrices version 4.0*, University of Minnesota, Department of Comp. Sci. and Eng., Army HPC Research Center, Minneapolis, 1998.

[26] W. Li. and D. Schuurmans, *Modular community detection in networks*, IJCAI (T. Walsh, ed.), IJCAI/AAAI, 2011, pp. 1366–1371.

[27] S. Muff, F. Rao, and A. Caflisch, *Local modularity measure for network clusterizations*, Phys. Rev. E **72** (2005), 056107.

[28] M. E. J. Newman, *Detecting community structure in networks*, The European Physical Journal B - Condensed Matter and Complex Systems **38** (2004), 321–330.

[29] ———, *Fast algorithm for detecting community structure in networks*, Phys. Rev. E **69** (2004), 066133.

[30] ———, *Modularity and community structure in networks*, Proc. Natl. Acad. Sci, USA **103** (2006), 8577.

[31] ———, *Network data*, http://www-personal.umich.edu/ mejn/netdata/, October 2011.

[32] M. E. J. Newman and M. Girvan, *Finding and evaluating community structure in networks*, Phys. Rev. E **69** (2004), 026113.

[33] F. Pellegrini, *SCOTCH 5.1 User's Guide*, Laboratoire Bordelais de Recherche en Informatique (LaBRI), 2008.

[34] Jörg Reichardt and Stefan Bornholdt, *Statistical mechanics of community detection*, Phys. Rev. E (3) **74** (2006), no. 1, 016110, 14, DOI 10.1103/PhysRevE.74.016110. MR2276596 (2007h:82089)

[35] C. Staudt and R. Görke, *A generator for dynamic clustered random graphs*, http://i11www.iti.uni-karlsruhe.de/en/projects/spp1307/dyngen, 2009.

[36] D. J. Watts and S. H. Strogatz, *Collective dynamics of 'small-world' networks.*, Nature **393** (1998), no. 6684, 440–442.

[37] E. Ziv, M. Middendorf, and C. H. Wiggins, *Information-theoretic approach to network modularity*, Phys. Rev. E (3) **71** (2005), no. 4, 046117, 9, DOI 10.1103/PhysRevE.71.046117. MR2139990 (2005m:94013)

Appendix A. DIMACS Challenge results

After DIMACS Challenge was completed, we run the proposed divisive clustering algorithm on the challenge instances (we skipped a few of the largest graphs). In these runs, we did not try to explicitly optimize the other challenge metrics. Doing so would require some trivial changes to the proposed framework for each metric: the evaluation functions measuring the modularity, the functions that update the modularity score, and the refinement functions should now measure the desired metric(s). Without doing these changes, we measured the other clustering scores (of the partitions that took the modularity score as the criterion). The results are shown in Table 6. The proposed algorithm is referred to as ParMod in this appendix.

TABLE 6. The clustering scores of the partitions.

Instance	mod	mid	aixc	aixe	perf	cov	$\overline{\text{cov}}$
333SP	0.989095	0.000218	0.006843	0.040992	1.991958	0.993118	1.991960
as-22july06	0.673605	0.000712	0.205813	0.740325	1.846038	0.768127	1.846236
astro-ph	0.736729	0.007728	0.022127	0.294091	1.934897	0.788472	1.935894
audikw1	0.917323	0.002133	0.050979	4.163516	1.936944	0.948903	1.937029
belgium.osm	0.994887	0.000411	0.003260	0.007006	1.996223	0.996795	1.996224
cage15	0.898534	0.000072	0.080895	1.479740	1.950329	0.924107	1.950333
caidaRouterLevel	0.868406	0.000655	0.039705	0.204044	1.956779	0.896955	1.956814
celegans_metabolic	0.446655	0.062234	0.402824	3.488044	1.729581	0.590123	1.752573
citationCiteseer	0.818692	0.000417	0.107262	0.793361	1.911682	0.872232	1.911715
coAuthorsCiteseer	0.899906	0.000602	0.074358	0.483233	1.977108	0.911773	1.977142
cond-mat-2005	0.738004	0.001516	0.017326	0.132351	1.935902	0.787838	1.936149
coPapersDBLP	0.858850	0.002412	0.125238	4.922138	1.968125	0.881037	1.968238
email	0.579957	0.045179	0.344470	3.140532	1.800371	0.692350	1.809870
eu-2005	0.940386	0.000286	0.029615	1.500914	1.927900	0.971262	1.927942
G_n_pin_pout	0.493583	0.001623	0.492139	4.914818	1.968009	0.509639	1.968155
in-2004	0.980272	0.000232	0.005798	0.082436	1.987802	0.993129	1.987816
kron_g500-simple-logn16	0.064586	0.000284	0.734346	59.815302	1.556555	0.287037	1.558008
kron_g500-simple-logn20	0.048710	0.000059	0.807405	68.818142	1.706426	0.200393	1.706548
ldoor	0.969370	0.002566	0.019954	0.954162	1.976477	0.981167	1.976527
luxembourg.osm	0.989312	0.002636	0.006909	0.014389	1.992082	0.993331	1.992100
memplus	0.697240	0.003627	0.282562	1.806004	1.939520	0.742195	1.939931
PGPgiantcompo	0.884130	0.007515	0.059229	0.218017	1.947627	0.924864	1.948063
polblogs	0.425691	0.020995	0.077119	1.843465	0.998813	0.927430	0.999905
power	0.940119	0.011169	0.035545	0.092199	1.944968	0.968759	1.945495
preferentialAttachment	0.308605	0.000310	0.566181	5.665209	1.749771	0.433877	1.749903
rgg_n_2_17_s0	0.977658	0.006684	0.014969	0.166193	1.984094	0.985676	1.984179
smallworld	0.787234	0.010091	0.210331	2.103284	1.992484	0.791019	1.992605
uk-2002	0.976712	0.000003	0.042345	0.663675	1.973205	0.978745	1.973207

A comparison of the modularity scores obtained in the implementation challenge by the three best algorithms is shown in Table 7. Even though our *ParMod* algorithm provides the best score only for two instances, the differences in comparison to the scores of the best ranked algorithm *CGGCi_RG* are very small. In fact they are smaller than the impact of several of the parameters studied in Section 4. The second ranked algorithm, *VNS_quality* provides many top ranked results. However, due to an extreme outlier for the instance *cage15*, the average value is lower than that of the other algorithms. When disregarding the outlier, the average lies between those of *ParMod* and *CGGCi_RG*. We also give the execution time of *ParMod* for some small challenge instances in Table 8.

TABLE 7. The modularity scores of our algorithm *ParMod* in comparison to the two best ranked submissions *CGGCi_RG* and *VNS_quality*.

Instance	*ParMod*	*CGGCi_RG*	*VNS_quality*
333SP	**0.9891**	0.9887	0.9884
as-22july06	0.6736	**0.6783**	0.6776
astro-ph	0.7367	0.7438	**0.7446**
audikw1	0.9173	0.9174	**0.9180**
belgium.osm	0.9949	**0.9949**	0.9948
cage15	0.8985	**0.9032**	0.3438
caidaRouterLevel	0.8684	**0.8720**	0.8709
celegans_metabolic	0.4467	0.4521	**0.4532**
citationCiteseer	0.8187	**0.8239**	0.8217
coAuthorsCiteseer	0.8999	**0.9053**	0.9039
cond-mat-2005	0.7380	**0.7463**	0.7451
coPapersDBLP	0.8589	**0.8668**	0.8650
email	0.5800	0.5819	**0.5828**
eu-2005	0.9404	**0.9416**	0.9413
G_n_pin_pout	0.4936	**0.5001**	0.4993
in-2004	0.9803	**0.9806**	0.9805
kron_g500-simple-logn16	0.0646	0.0637	**0.0651**
kron_g500-simple-logn20	0.0487	**0.0504**	0.0494
ldoor	**0.9694**	0.9689	0.9691
luxembourg.osm	0.9893	0.9895	**0.9896**
memplus	0.6972	**0.7005**	0.6953
PGPgiantcompo	0.8841	**0.8866**	0.8861
polblogs	0.4257	0.4271	**0.4271**
power	0.9401	0.9403	**0.9409**
preferentialAttachment	0.3086	0.3023	**0.3160**
rgg_n_2_17_s0	0.9777	0.9781	**0.9783**
smallworld	0.7872	0.7930	**0.7930**
uk-2002	0.9767	**0.9903**	0.9901
Average	0.7466	0.7496	0.7297

TABLE 8. The execution time of *ParMod* for a few challenge instances. The times are the averages of 5 executions and given for reference purposes.

Instance	Time(sec)	Instance	Time(sec)
celegans_metabolic	1.02	email	4.02
power	11.18	polblogs	6.40
PGPgiantcompo	27.44	as-22july06	58.73
astro-ph	112.45	cond-mat-2005	232.82
preferentialAttachement	524.85	smallworld	397.49

DEPARTMENT OF BIOMEDICAL INFORMATICS AND DEPARTMENT OF ELECTRICAL & COMPUTER ENGINEERING, THE OHIO STATE UNIVERSITY
E-mail address: `umit@bmi.osu.edu`

DEPARTMENT OF BIOMEDICAL INFORMATICS, THE OHIO STATE UNIVERSITY
E-mail address: `kamer@bmi.osu.edu`

SIMULA RESEARCH LABORATORY, FORNEBU, NORWAY
E-mail address: `langguth@simula.no`

CNRS AND LIP, ENS LYON, LYON 69364, FRANCE
E-mail address: `bora.ucar@ens-lyon.fr`

An ensemble learning strategy for graph clustering

Michael Ovelgönne and Andreas Geyer-Schulz

ABSTRACT. This paper is on a graph clustering scheme inspired by ensemble learning. In short, the idea of ensemble learning is to learn several weak classifiers and use these weak classifiers to form a strong classifier. In this contribution, we use the generic procedure of ensemble learning and determine several weak graph clusterings (with respect to the objective function). From the partition given by the maximal overlap of these clusterings (the cluster cores), we continue the search for a strong clustering. We demonstrate the performance of this scheme by using it to maximize the modularity of a graph clustering. We show, that the quality of the initial weak clusterings is of minor importance for the quality of the final result of the scheme if we iterate the process of restarting from maximal overlaps. In addition to the empirical evaluation of the clustering scheme, we will link its search behavior to global analysis. With help of Morse theory and a discussion of the path space of the search heuristics we explain the superior search performance of this clustering scheme.

1. Introduction

Graph clustering, i.e. the identification of cohesive submodules or 'natural' groups in graphs, is an important technique in several domains. The identification of functional groups in metabolic networks [**GA05**] and the identification of social groups in friendship networks are two popular application areas of graph clustering.

Here we define graph clustering as the task of simultaneously detecting the number of submodules in a graph and detecting the submodules themselves. In contrast, we use the term graph partitioning for the problem of identifying a parametrized number of partitions where usually additional restrictions apply (usually, that all submodules are of roughly equal size). Two recent review articles on graph clustering by Schaeffer [**Sch07**] and Fortunato [**For10**] provide a good overview on graph clustering techniques as well as on related topics like evaluating and benchmarking clustering methods.

Graph clustering by optimizing an explicit objective function became popular with the introduction of the modularity measure [**NG04**]. Subsequently, a number of variations of modularity [**MRC05, LZW$^+$08**] have been proposed to address shortcomings of modularity such as its resolution limit [**FB07**]. The identification of a graph clustering by finding a graph partition with maximal modularity is NP-hard [**BDG$^+$08**]. Therefore, finding clusterings of a problem instance with more

2000 *Mathematics Subject Classification.* Primary 05C85; Secondary 05C70, 68R10, 90C27.

than a few hundred vertices has to be based on good heuristics. A large number of modularity optimization heuristics has been proposed in recent years, but most of them have a poor optimization quality.

The objective of this contribution is to present a new graph clustering scheme, called the Core Groups Graph Clustering (CGGC) scheme, which is able to find high quality clustering by using an ensemble learning approach. In [**OGS10**] we presented an algorithm called RG+ for maximizing the modularity of a graph partition via an intermediate step of first identifying core groups of vertices. The RG+ algorithm was able to outperform all previously published heuristics in terms of optimization quality. This paper deals with a generalization of this optimization approach.

The paper has been organized in the following way. First, we briefly discuss ensemble learning in Section 2. Then, we introduce the CGGC scheme in Section 3 and modularity maximization algorithms in Section 4. In Section 5, we evaluate the performance of the CGGC scheme using modularity maximization algorithms within the scheme. We discuss the scheme from the viewpoint of global analysis in Section 6. Finally, a short conclusion follows in Section 7.

2. Ensemble Learning

Ensemble based systems have been used in decision making for quite some time. Ensemble learning is a paradigm in machine learning, where several intermediate classifiers (called weak or base classifiers) are generated and combined to finally get a single classifier. The algorithms used to compute the weak classifiers are called weak learners. An important notion is, that even if a weak learner has only a slightly better accuracy than random choice, by combining several classifiers created by this weak learner, a strong classifier can be created [**Sch90**]. For a good introduction to this topic, see the review article by Polikar [**Pol06**].

Two examples of ensemble learning strategies are bagging and boosting. A bagging algorithm for supervised classification trains several classifiers from bootstraps of the training data. The combined classifier is computed by simple majority voting of the ensemble of base classifiers, i.e. a data item gets the label the majority of base classifiers assigns to that data item. A simple boosting algorithm (following [**Pol06**]) works with classifiers trained from three subsets of the training data. The first dataset is a random subset of the training data of arbitrary size. The second dataset is created so that the classifier trained with the first dataset classifies half of the data items correctly and the other half wrong. The third dataset consists of the data items the classifiers trained by the first and the second dataset disagree on. The strong classifier is the majority vote of the three classifiers.

Another ensemble learning strategy called Stacked Generalization has been proposed by Wolpert [**Wol92**]. This strategy is based on the assumption that some data points are more likely to be misclassified than others, because they are near to the boundary that separates different classes of data points. First, an ensemble of classifiers is trained. Then, using the output of the classifiers a second level of classifiers is trained with the outputs of the ensemble of classifiers. In other words, the second level of classifiers learns for which input a first level classifier is correct or how to combine the "guesses" of the first level classifiers.

An ensemble learning strategy for clustering has been used by Fred and Jain [**FJ05**], first. They called this approach evidence accumulation. They worked on clustering data points in an Euclidean space. Initially, the data points are clustered several times based on their distance and by means of an algorithm like k-means. The ensemble of generated clusterings is used to create a new distance matrix called the co-association matrix. The new similarity between two data points is the fraction of partitions that assign both data points to the same cluster. Then, the data points are clustered on basis of the co-association matrix.

3. Core Groups Graph Clustering Scheme

Let us restrict our considerations to the problem of whether a pair of vertices should belong to the same cluster or to different clusters. Making this decision is complicated. Many algorithms get misled during their search so that sometimes bad decisions are made. But what if we have one or more algorithms that find several clusterings with fair quality but still a lot of non-optimal decisions on whether a pair of vertices belongs to the same cluster? If all clusterings agree on whether a pair of vertices belongs to the same cluster, we can be pretty sure that this decision is correct. However, if the clusterings disagree, we should have a second look at this pair.

Based on these considerations, we propose the CGGC scheme. We use the agreements of several clusterings with fair quality to decide whether a pair of vertices should belong to the same cluster. The groups of vertices which are assigned to the same cluster in every clustering (i.e. the maximal overlaps of the clusterings) are denoted as core groups. To abstract from any specific quality measure, we use the term *good* partition for a partition that has a good quality according to an arbitrary quality measure. The CGGC scheme consists of the following steps:

(1) Create a set S of k *good* partitions of G with base algorithm $A_{initial}$
(2) Identify the partition $\hat{P}$ of the maximal overlaps in S
(3) Create a graph $\hat{G}$ induced by the partition $\hat{P}$
(4) Use base algorithm A_{final} to search for a *good* partition of $\hat{G}$
(5) Project partition of $\hat{G}$ back to G

Initially, a set S of k partitions of G is created. That means, one non-deterministic clustering algorithm is started k times to create the graph partitions, k deterministic but different algorithms are used or a combination of both is used. In terms of ensemble learning, the used algorithms are the base algorithms or weak learners and the computed clusterings are the weak classifiers.

Next, we combine the information of the weak classifiers: We calculate the maximal overlap of the clusterings in S. Let $c_P(v)$ denote the cluster that vertex v belongs to in partition P. We create from a set S of partitions $\{P_1, \ldots, P_k\}$ of V a new partition $\hat{P}$ of V so that

$$\forall v, w \in V : (\bigwedge_{i=1}^{k} c_{P_i}(v) = c_{P_i}(w)) \Leftrightarrow c_{\hat{P}}(v) = c_{\hat{P}}(w)$$

Extracting the maximum overlap of an ensemble of partitions creates an intermediate solution which is used as the starting point for the base algorithm A_{final} to calculate the final clustering. The base algorithm used in this phase could be an algorithm used in step 1 or any other algorithm appropriate to optimize the objective function. For example, algorithms that are not able to cluster the original network in reasonable time could be used to cluster the smaller graph $\hat{G} = (\hat{V}, \hat{E})$

induced by $\hat{P}$. To create the induced graph, all vertices in a cluster in $\hat{P}$ are merged to one vertex in $\hat{G}$. Accordingly, $\hat{G}$ has as many vertices as there are clusters in $\hat{P}$. An edge $(v, w) \in \hat{E}$ has the weight of the combined weights of all edges in G that connect vertices in the clusters represented by v and w. Then, the clustering of $\hat{G}$ would have to be projected back to G to get a clustering of the original graph.

Agglomerative hierarchical optimization schemes often show the best scalability for clustering algorithms as they usually make local decisions. A partial explanation is that the number of partitions of n nodes in k classes grows as a Stirling number of the second kind $S(n,k) = \frac{1}{k!}\sum_{j=0}^{k}(-1)^j\binom{k}{r}(k-r)^n$ and that this implies that growth of the search space is smaller in the bottom-up direction than in the top-down direction [**Boc74**, p. 110]. For the example shown in Figure 1, we have 10 partitions (5 objects in 4 clusters) for the first bottom-up decision versus 15 partitions (5 objects in 2 clusters) for the first top-down decision.

While using only local information increases the scalability, it is a source of globally poor decisions, too. Extracting the overlap of an ensemble of clusterings provides a more global view. Figure 1 shows the complete merge lattice of an example graph of 5 vertices. An agglomerative hierarchical algorithm always starts with the partition into singletons (shown at the bottom) and merges in some way the clusters until only one cluster containing all vertices remains (shown at the top). Every merge decision means going one level up in the lattice. Restarting the search at the maximal overlap of several partitions in an ensemble means to go back to a point in the lattice from which all of the partitions in this ensemble can be reached. If we restart the search for a good partition from this point, we will most probably be able to reach other good partitions than those in the ensemble, too. In fact, reaching other good or even better partitions than those in the ensemble will be easier than starting from singletons as poor cluster assignments in the ensemble have been leveled out.

3.1. The Iterated Approach. Wolpert [**Wol92**] discussed the problem that some data points are harder to assign to the correct cluster than others. Data points at the natural border of two clusters are harder to assign than those inside. For the specific case of modularity maximization with agglomerative hierarchical algorithms, we discussed the influence of prior merge decision on all later merges in [**OGS12**]. Often, the order of the merge operations influences which side of the border a vertex is assigned to. Node 3 in Figure 1 is an example for this effect.

With the help of the maximal overlaps of the CGGC scheme we try to separate the cores of the cluster from the boundaries. The harder decisions on which clusters contain the vertices at the boundaries are made, when the knowledge of the cores provides additional information. This idea of separating cores and boundaries can be iterated in the following way (subsequently denoted as the CGGCi scheme):

(1) Set P^{best} to the partition into singletons and set $\hat{G}$ to G
(2) Create a set S of k *(fairly) good* partitions of $\hat{G}$ with base algorithm $A_{initial}$
(3) Identify the partition $\hat{P}$ of the maximal overlaps in S
(4) If $\hat{P}$ is a better partition than P^{best}, set $P^{best} = \hat{P}$, create the graph $\hat{G}$ induced by $\hat{P}$ and go back to step 2
(5) Use base algorithm A_{final} to search for a *good* partition of $\hat{G}$
(6) Project partition of $\hat{G}$ back to G

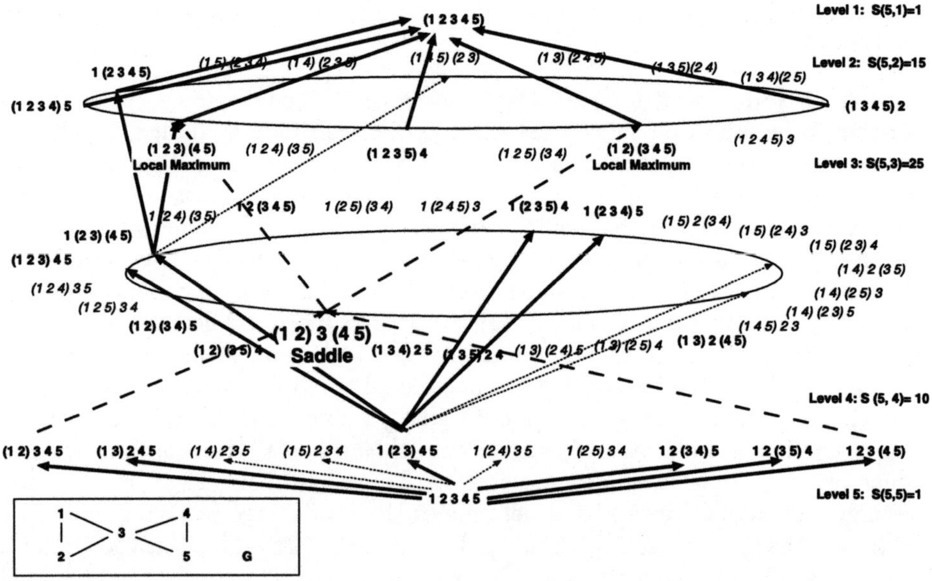

FIGURE 1. Graph G with 5 vertices and its merge lattice. The edges indicate the possible merge paths of hierarchical clustering algorithms (not all edges drawn). The stroked edges indicate paths leading through the saddle point (1 2) 3 (4 5) to the two local maxima. The dotted edges and italic partitions can be neglected as they correspond to merging clusters that are not adjacent. Merging non-adjacent clusters will always decrease the modularity.

In every new clustering P^{best} some more vertices or groups of vertices have been merged or rearranged. So, every new clustering is likely to provide more accurate information on the structure of the graph for the succeeding iterations.

4. Modularity and its Optimization

Modularity is a popular objective function for graph clustering that measures the non-randomness of a graph partition. Let $G = (V, E)$ be an undirected, unweighted graph, $n := |V|$ the number of vertices, $m := |E|$ the number of edges and $P = \{C_1, \ldots, C_k\}$ a partition of V, i.e. $\cup_{i=1}^{k} C_i = V$ and $\forall_{i \neq j \in \{1,\ldots,k\}} C_i \cap C_j = \emptyset$. The modularity Q of the partition P of graph G is defined as

$$Q(G,P) = \frac{1}{2m} \sum_{v_x, v_y} \left(w_{xy} - \frac{s_x s_y}{2m} \right) \delta(c_P(v_x), c_P(v_y)) \tag{1}$$

where w_{xy} is an element in the adjacency matrix of G, s_x is the degree of vertex v_x, $c_P(v_x)$ is the cluster of v_x in partition P and the Kronecker symbol $\delta(c(v_x), c(v_y)) = 1$ when v_x and v_y belong to the same cluster and $\delta(c(v_x), c(v_y)) = 0$ otherwise.

Research on modularity maximization algorithms has been very popular in the last years and a lot of heuristic algorithms have been proposed. In the following, we discuss a randomized greedy and a label propagation algorithm in detail, as we will use them exemplarily to evaluate the CGGC scheme. We will give a brief summary of other algorithms which could be used as base algorithms for the CGGC scheme as well. For an extensive overview on modularity maximization algorithms, see [**For10**].

4.1. Randomized Greedy (RG). Newman [**New04**] proposed the first algorithm to be used to identify clusterings by maximizing modularity. The hierarchical agglomerative algorithm starts with a partition into singletons and merges in each step one pair of clusters that causes the maximal increase in modularity. The result is the cut of the dendrogram with the maximal modularity. This algorithm is slow, as it considers to merge every pair of adjacent clusters in every step. The complete search over all adjacent pairs also leads to an unbalanced merge process. Some clusters grow faster than others and the size difference is a bias for later merge decisions. Large clusters are merged with many small clusters in their neighborhood, whether this is good from a global perspective or not [**OGS12**].

The randomized greedy algorithm [**OGS10**] is a fast agglomerative hierarchical algorithm that has a very similar structure to Newman's algorithm but does not suffer from an unbalanced merge process. This algorithm selects in every step a small sample of k vertices and determines the best merge involving one of the vertices in the sample (see Algorithm 1). Because of the sampling, the algorithm can be implemented quite efficiently and has a complexity of $O(m \ln n)$ (see [**OGS10**]).

Algorithm 1: Randomized Greedy (RG) algorithm

```
Input: undirected, connected graph G = (V, E), sample size k
Output: clustering
▼ Initialize
    forall the v ∈ V do
        forall the neighbors n of v do
            e[v, n] ← 1/(2 * edgecount);
        a[v] ← rowsum(e[v])
▼ Build Dendrogram (Randomized Greedy)
    for i = 1 to rank(e)-1 do
        maxDeltaQ ← −∞;
        for j = 1 to k do //search among k communities for best join
            c1 ← random community;
            for all communities c2 connected to c1 do
                deltaQ ← 2(e[c1, c2] − (a[c1] * a[c2]));
                if deltaQ > maxDeltaQ then
                    maxDeltaQ ← deltaQ;
                    nextjoin ← (c1, c2);
        join(nextjoin);
        joinList ← joinList + nextjoin;
    clusters ← extractClustersFromJoins(joinList) ;
```

In [**OGS10**] we also introduced the RG+ (improved randomized greedy) algorithm, which we generalized to the CGGC scheme in this contribution. The RG+ algorithm uses the RG algorithm as its base clustering algorithm to create the weak

Algorithm 2: Label Propagation (LP) algorithm

```
Input: undirected, connected graph G = (V, E) set of labels LP
Output: clustering
▼ Initialize
    forall the v ∈ V do
        label[v] ← getUniqueID(LP);
▼ Propagate Labels
    majorityLabelCount ← 0;
    while majorityLabelCount ≠ |V| do
        majorityLabelCount ← 0;
        forall the v ∈ V at random do
            label[v] ← argmax_{l∈LP} Σ_{n∈neighbors(v)} δ(l, label[n]) ;
            if Σ_{n∈N(v)} δ(label[v], label[n]) ≥ |V|/2 then
                majorityLabelCount ← majorityLabelCount + 1;
```

classifiers and for the final clustering starting from the maximal overlap of these partitions. To obtain a standardized naming of all other CGGC scheme algorithms in this article we will denote this algorithms as $CGGC_{RG}$ in the following.

4.2. Label Propagation (LP). Raghavan et al. [**RAK07**] proposed a label propagation algorithm for graph clustering. This algorithm initializes every vertex of a graph with a unique label. Then, in iterative sweeps over the set of vertices the vertex labels are updated. A vertex gets the label that the maximum number of its neighbors have. Ties are broken arbitrarily. The procedure is stopped when every vertex has the label that at least half of its neighbors have. The pseudocode of the LP algorithm is shown in Algorithm 2.

This procedure does not explicitly or implicitly maximize modularity. It is especially interesting, because it has a near linear time complexity. Every sweep has a complexity of $O(m)$ and Raghavan et al. report that 95% of the vertices have a label the majority of its neighbors have in only about 5 iterations.

As we will show in Section 5, the CGGC scheme is able to find good final clusterings from weak results of intermediate runs of base algorithms. It does not matter if the algorithm is stopped prior to its originally defined stopping criterion.

4.3. Other Modularity Maximization Algorithms. A very fast agglomerative hierarchical algorithm has been developed by Blondel et al. [**BGLL08**]. The algorithm starts with singleton clusters. Every step of the algorithm consists of two phases. At first, all vertices are sequentially and iteratively moved between their current and a neighboring cluster, if this increases the modularity. In the case that several moves have a positive influence on the modularity, the one with the highest modularity increase is chosen. To speed up this process, a threshold is introduced to determine, when to stop the first phase based on the relative increase in modularity. In the second phase of each step, the result of the first phase is used to create a new graph, where all vertices that have been assigned to the same cluster in the first phase are represented by one vertex. The edge weights between the original vertices are summed up and give the new edge weights between the new vertices. Then, the algorithm returns to the first phase and moves the new vertices between clusters.

Noack and Rotta [**NR09**] experimentally investigated a framework of hierarchical agglomerative modularity optimization algorithms. While most algorithms only use the modularity increase as the priority criterion, they analyzed several other priority criteria that weight modularity increase in some way. Furthermore, they considered merging more than one pair of vertices in every step and locally refining the intermediate partitions regularly during the merging process (multi-level refinement). With the best configuration of their framework Noack and Rotta achieve significantly better results than Blondel et al. [**BGLL08**] at the price of a much higher runtime.

Another well performing algorithm is the MOME algorithm by Zhu et al. [**ZWM$^+$08**]. In a first phase, the coarsening phase, the algorithm recursively creates a set of graphs. Starting with the input graph, each vertex of the graph will be merged with the neighbor that yields the maximal increase in modularity. If the modularity delta is negative for all neighbors, the vertex will be left as it is. The resulting graph will be recursively processed until the graph can not be contracted any more. Subsequently, in the uncoarsening phase, the set of successively collapsed graphs will be expanded while the clustering gets refined by moving vertices between neighboring clusters.

Many other algorithms have been proposed. For practical usage and to be used within the CGGC scheme most of them are of no interest due to their inferior performance in terms of modularity maximization or runtime efficiency. Among these algorithms are several spectral algorithms ([**WS05**], [**New06**], [**RZ07**], [**RZ08**]) and algorithms based on generic meta heuristics like iterated tabu search [**MLR06**], simulated annealing [**MAnD05**], or mean field annealing [**LH07**]. Formulations of modularity maximization as an integer linear program (e.g. [**AK08**], [**BDG$^+$07**]) allow finding an optimal solution without enumerating all possible partitions. However, processing networks with as few as 100 vertices is already a major problem for current computers.

4.3.1. *Refinement.* The results of most modularity maximization algorithms can be improved by a local vertex mover strategy. Noack and Rotta [**NR09**] surveyed the performance of several strategies inspired by the famous Kernighan-Lin algorithm [**KL70**]. We employ the fast greedy vertex movement strategy to the results of all evaluated algorithms, because all other strategies scale much worse without providing significant improvements in quality. The fast greedy vertex mover strategy sweeps iteratively over the set of vertices as long as moving a vertex to one of its neighboring clusters improves modularity.

5. Evaluation

The clustering scheme is evaluated by means of real-world and artificial networks from the testbed of the 10th DIMACS implementation challenge on graph partitioning and graph clustering. Memory complexity is a bigger issue than time complexity for our algorithms and we had to omit the two largest datasets from the category *Clustering Instances* because of insufficient main memory. We also omitted the small networks with less than 400 vertices where many algorithms are able to find the optimal partitions [**OGS10**].

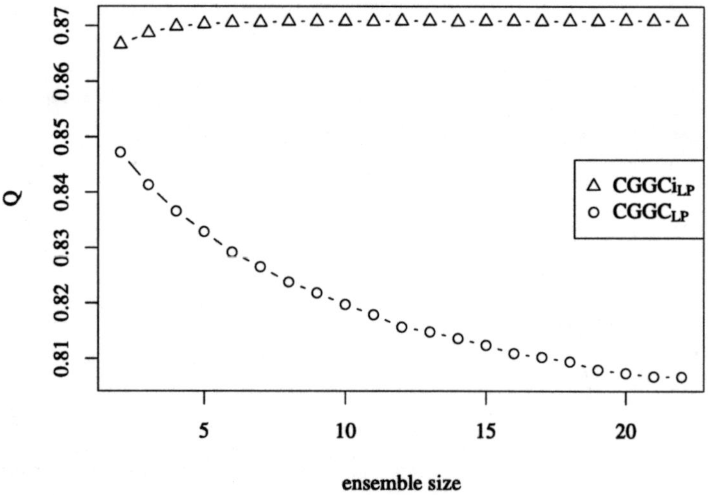

FIGURE 2. Average modularity of 30 test runs of the CGGC- and CGGCi-scheme using LP as the base algorithm subject to the ensemble size k for the dataset *caidaRouterLevel*.

Before we conducted the evaluation, we first determined the best choice for the number of partitions in the ensembles. The results of our tests (see Figure 3) show that the ensemble size should be roughly $\ln n$ for all algorithms but $CGGC_{LP}$. When using LP as the base algorithm, the quality improves with increasing ensemble size for the iterated scheme but heavily decreases for the non-iterated scheme (see Figure 2). This seems to be a result of the weak learning quality of LP. A larger ensemble size results in more and smaller core groups in the maximal overlap partition. LP is not able to find a good clustering from finer decompositions when not iteratively applied as in the CGGCi scheme.

The results in Table 2 show the average optimization quality and therefore the quality we can expect when using the algorithm in a practical context. In Table 1 we show the boundary of the scheme, i.e. the best optimization quality we were able to achieve using the scheme given much time.

While the iterated CGGCi scheme does not provide much improvement compared to the non-iterated scheme when used with the RG algorithm ($CGGCi_{RG}$ vs. $CGGC_{RG}$), its improvement for the LP algorithm is significant ($CGGCi_{LP}$ vs. $CGGC_{LP}$). There is still a difference between the $CGGCi_{RG}$ and $CGGCi_{LP}$. But for most networks, $CGGCi_{LP}$ achieves better results than the standalone *RG* algorithm which showed to be a quite competitive algorithm [**OGS10**] among non-CGGC scheme algorithms.

A notable result is that the LP algorithm performs extremely bad on the *preferentialAttachment* network (*pref.Attach.*). This network is the result of a random network generation process where iteratively edges are added to the network and the probability that an edge is attached to one vertex depends on the current degree of the vertex. The average modularity for the standalone LP on the *preferentialAttachment* network is extremely low as the algorithm identified only in 1 of 100 test runs a community structure. In all other cases the identified clusterings were partitions into singletons. Therefore, using LP within the CGGC scheme failed as well.

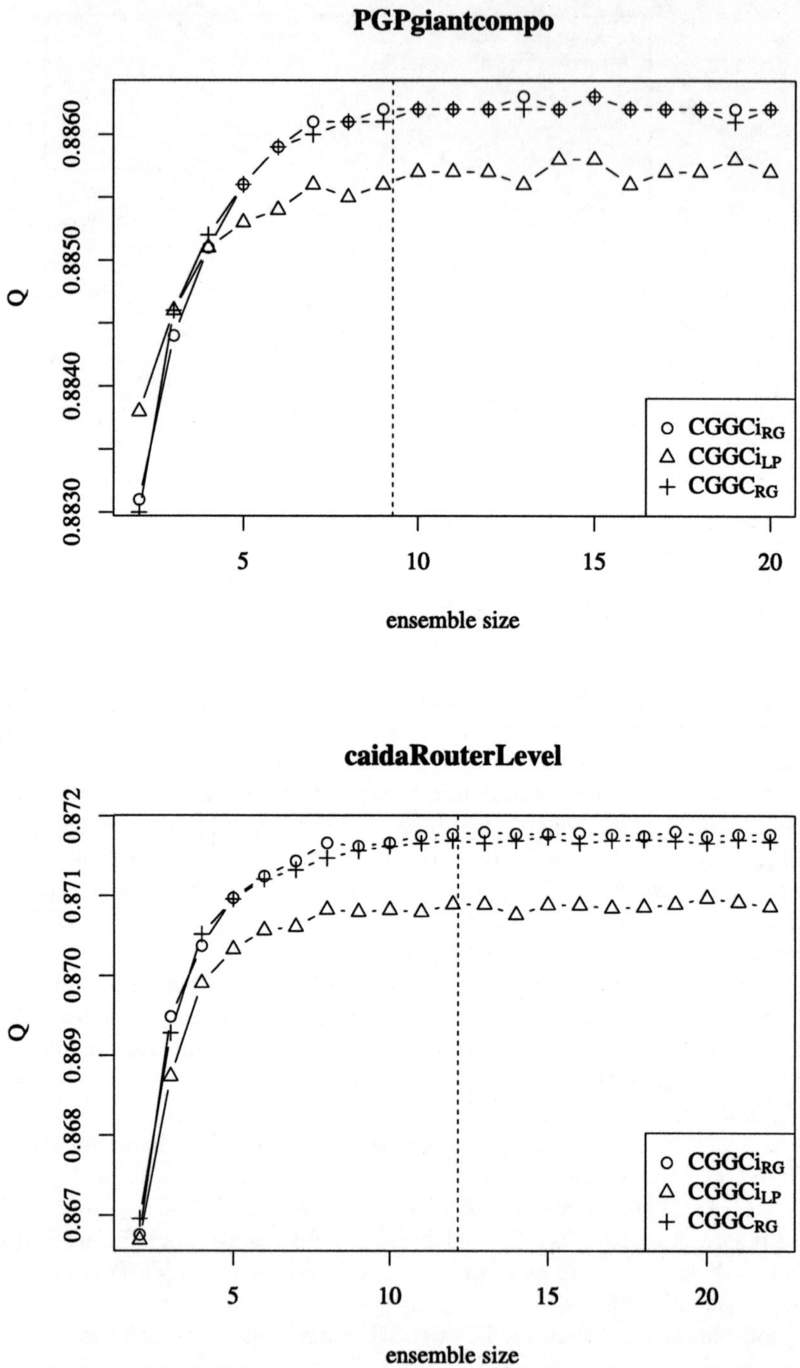

FIGURE 3. Average modularity of 30 test runs of the CGGC/CGGCi-scheme algorithms subject to the ensemble size k for the two datasets *PGPgiantcompo* and *caidaRouterLevel*. The dotted vertical line shows the value of ln n (where n is the number of vertices)

TABLE 1. Best modularity of a clustering computed for networks from the DIMACS testbed categories *Clustering Instances* and *Coauthors*. All partitions have been identified with help of the CGGCi scheme and the denoted base algorithm.

Network	Max Modularity	Alg.	Network	Max Modularity	Alg.
celegans_metabolic	0.4526664838	RG	eu-2005	0.9415630621	RG
Email	0.5828085954	RG	in-2004	0.9806076266	RG
PGPgiantcompo	0.8865501696	RG	road_central	0.9976280448	RG
as-22july06	0.6783599573	RG	road_usa	0.9982186002	RG
astro-ph	0.7444262906	RG	caidaRouterLevel	0.8720295371	RG
cond-mat	0.8530972563	RG	pref.Attach.	0.3048516381	RG
cond-mat-2003	0.7786823470	RG	smallworld	0.7930994465	LP
cond-mat-2005	0.7464446826	RG	G_n_pin_pout	0.5002934104	LP
hep-th	0.8565536355	RG	citationCiteseer	0.8241257861	RG
netscience	0.9598999889	RG	coAuthorsCiteseer	0.9053559700	RG
polblogs	0.4270879141	RG	coAuthorsDBLP	0.8415177919	RG
power	0.9404810777	RG	coPapersCiteseer	0.9226201646	RG
cnr-2000	0.9131075546	RG	coPapersDBLP	0.8667751453	RG

However, we can argue that trying to find a significant community structure in a random network should fail.

The clustering process of the iterated CGGC scheme is shown by example in Figure 4. The LP algorithm is a much weaker learner than the RG algorithm and initially finds clusterings with very low modularity. But after a few iterations the modularity of the core groups partitions of both base algorithms are about the same. But although the quality of the final core groups for both base algorithms is similar, the core groups are different. The final core groups identified from the ensemble generated with the LP algorithm are a weaker restart point than those identified with RG. If we use RG as the base algorithm for the final clustering (A_{final}) to start from the LP core groups, the identified partitions have about the same modularity than those identified with LP. Because of page limitations we omit detailed results.

In Table 3 runtime results for the base algorithm RG are shown. Due to ensemble learning approach the CGGC/CGGCi scheme has a runtime that is a multiple of the runtime of the base algorithm. However, our implementation does not make use of parallelization. Because all partitions for the ensembles can be computed independently, parallelization is straightforward.

6. A Global Analysis View on the CGGC Scheme

We already gave an intuitive explanation of the way the CGGC scheme works in Section 3. Now, we want to provide a link to global analysis and Morse theory.

The merge lattice shown in Figure 1 shows the space of all paths an agglomerative hierarchical algorithm can follow. The level number k corresponds to the number of clusters of the partition(s) at level k in the merge lattice: from the singleton partition (the inf of the lattice) at level $k = n$ to the partition with a single

TABLE 2. Average modularity of the results of 100 test runs (10 test runs for very large networks marked with *) on networks from the DIMACS testbed categories *Clustering Instances* and *Coauthors*. $CGGC_X$ and $CGGCi_X$ denote the usage of an base algorithm X within the CGGC and the iterated CGGC scheme, respectively.

	RG	$CGGC_{RG}$	$CGGCi_{RG}$	LP	$CGGC_{LP}$	$CGGCi_{LP}$
celegans_metabolic	0.43674	0.45021	0.45019	0.37572	0.43856	0.44343
Email	0.57116	0.57986	0.58012	0.41595	0.55750	0.55668
PGPgiantcompo	0.86436	0.88616	0.88617	0.76512	0.85431	0.88565
as-22july06	0.66676	0.67742	0.67747	0.54930	0.61205	0.67316
astro-ph	0.69699	0.74275	0.74277	0.67511	0.70272	0.74143
cond-mat	0.82975	0.85240	0.85242	0.75661	0.79379	0.85116
cond-mat-2003	0.75715	0.77754	0.77755	0.67852	0.70551	0.77524
cond-mat-2005	0.72203	0.74543	0.74550	0.64184	0.67453	0.74199
hep-th	0.83403	0.85577	0.85575	0.76102	0.80614	0.85463
netscience	0.94037	0.95975	0.95974	0.92477	0.95375	0.95933
polblogs	0.42585	0.42678	0.42680	0.42610	0.42635	0.42633
power	0.92818	0.93962	0.93966	0.72124	0.79601	0.93794
cnr-2000	0.91266	0.91302	0.91309	0.86887	0.90603	0.91284
eu-2005	0.93903	0.94114	0.94115	0.85291	0.90610	0.93982
in-2004	0.97763	0.97832	0.98057	0.92236	0.97086	0.97791
road_central*	0.99716	0.99761	0.99767	0.70863	0.94351	0.99749
road_usa*	0.99786	0.99821	0.99825	0.72234	0.94682	0.99812
caidaRouterLevel	0.86136	0.86762	0.87172	0.76353	0.81487	0.87081
pref.Attach.	0.27984	0.29389	0.30099	0.00202	0.00000	0.00000
smallworld	0.78334	0.79289	0.79300	0.66687	0.69181	0.79307
G_n_pin_pout	0.47779	0.49991	0.50006	0.30609	0.34639	0.50023
citationCiteseer	0.80863	0.82333	0.82336	0.66184	0.72256	0.82064
coAuthorsCiteseer	0.89506	0.90507	0.90509	0.79549	0.83862	0.90360
coAuthorsDBLP	0.82081	0.83728	0.84055	0.71502	0.75108	0.83661
coPapersCiteseer	0.91626	0.92168	0.92221	0.85653	0.89921	0.92162
coPapersDBLP	0.85383	0.86471	0.86655	0.77918	0.82674	0.86540

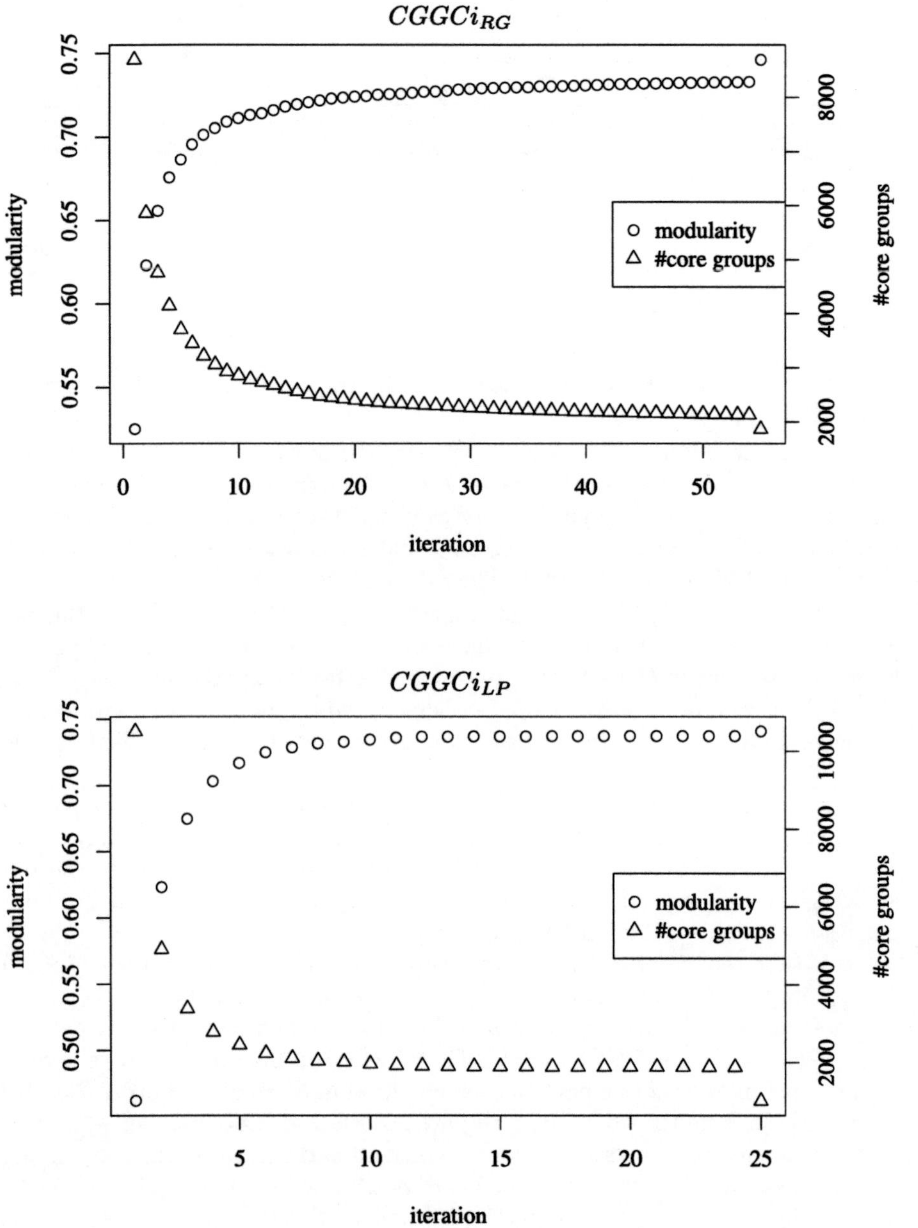

FIGURE 4. The clustering process of the iterated CGGCi scheme on the *cond-mat-2005* dataset for the base algorithms RG (top) and LP (bottom). All points but the last one are core groups, i.e. maximal overlaps of the k partitions in the ensemble. The last points are the results for the final clustering run and after applying the refinement procedure.

TABLE 3. Average runtime results (in sec.) for selected networks.

	#Vertices	#Edges	RG	$CGGC_{RG}$	$CGGCi_{RG}$
polblogs	1490	16715	0.02	0.11	0.16
power	4941	6594	0.02	0.12	0.49
cond-mat-2005	31163	120029	0.53	6.18	24.0
caidaRouterLevel	192244	609066	2.23	30.1	83.0
eu-2005	862664	16138468	32.1	466	505

cluster (the sup of the lattice) at level $k = 1$. For a partition with k clusters we have $\frac{k(k-1)}{2}$ merge choices.

In each iteration of the CGGCi scheme, the search starts at some partition P at level k_P and goes up in the lattice to identify several new local maxima (the partitions in the ensemble S). For example, the algorithm starts twice at the partition 1 2 3 4 5 at level 5 in Figure 1 and reaches the two local maxima (1 2 3)(4 5) and (1 2)(3 4 5) at level 2. Then the algorithm goes down in the lattice to the maximal overlap partition $\hat{P}$ at a level $k_{\hat{P}} \leq k_P$. In the example, this is the partition (1 2) 3 (4 5) at level 3. In the worst case, when the ensemble of partitions S created starting at P does not agree on any vertex, the maximal overlap is again P and the core groups search stops. Otherwise, when the ensemble agrees on how to merge at least one vertex, a new core groups partition is identified at a level $k_{\hat{P}} < k_P$.

If a local optimum P has been reached by a hill-climbing method, all partitions that have been visited on the way through the merge lattice to P have a lower objective function value than the local optima. As can be seen from the merge lattice given in Figure 1, there are usually many paths to get from the bottom partition on level n to any other partition.

A path in the merge lattice can be identified by an ordered set of partitions. Let F_{P_i} denote the set of all paths that connect the singleton partition to the partition P_i, let Ω denote all partitions of a set of vertices V, and S be a set of partitions. Then, $\mathcal{P}(S) = \{P \in \Omega \mid \forall P_i \in S \, \exists D \in F_{P_i} : P \in D\}$ is the set of all partitions that are included in at least one path to each partition in S. In other words, $\mathcal{P}(S)$ is the set of all branch points from which all partitions in S can be reached. $\mathcal{P}(S)$ always contains at least the singleton partition which all paths share as the starting point. The maximal overlap $\hat{P}$ of the ensemble of partitions in S is the partition in $\mathcal{P}(S)$ with the minimal number of clusters. That means, $\hat{P}$ is the latest point from where a hierarchical agglomerative algorithm can reach all partitions in the ensemble. We see that the core groups partition of the maximal overlap is a special partition as it is a branching point in the merge path of the ensemble S.

For a moment, we put the merge path discussion aside and discuss Morse theory which originates from the work of Morse on the topology of manifolds [**Mor34**]. Although the theory originally has been developed for continuous function spaces, and we are dealing with discrete optimization, Morse theory provides a suitable means to understand the topology of high-dimensional non-linear functions. In the following, we assume that the discrete points (the partitions of a graph) are embedded in a continuous space in such a way that the critical points (maxima,

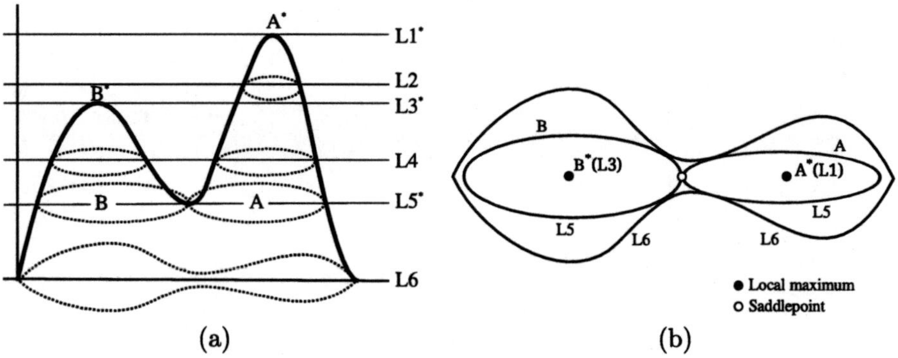

FIGURE 5. Graph (a) and respective level line (b). The levels marked with * are critical levels with Karush-Kuhn-Tucker points.

minima, saddle-points) of the discrete modularity maximization problem are also critical points in the continuous version of the problem.

Following the discussion in [**JJT00**], let us assume we have a continuous function $f : \mathbb{R}^2 \to \mathbb{R}$ as in Figure 5a. The inverse of f, f^{-1}, then gives the level line of all points having the same value of f. We denote $f^{-1}(y)$ as the level line at level y. While the level line is continuously deformed while going along y, whenever a level passes a stationary or Karush-Kuhn-Tucker point (local minimum, local maximum or saddle point), its topology changes. Figure 5b shows the level lines at critical levels of the function in Figure 5a. At level 5 the level line connected at lower levels separates into the lines A and B, i.e. at level 5 the two lines are glued together by the saddle point and above level 5 they are unconnected.

This analysis of the level lines is important for optimization, as a greedy algorithm starting at level 6 can potentially reach any point, while for a hill-climbing algorithm starting from a point at level 5 the starting point determines the reachable parts of the search space. The separated level lines (A and B) create basins of attraction for the respective local optima (A* and B*). At level 5, the only point with a gradient path to both local maxima is the saddle point. Let us assume we have a deterministic, hill-climbing algorithm. Then, the result of the algorithm is determined by the starting point. Each local optimum has a basin of attraction, i.e. a non-empty set of points from which the algorithm goes to the respective local optimum.

In Figure 6 we show the basins of attraction for two functions in $\mathbb{R}^1$ and $\mathbb{R}^2$. Consider the bounded non-linear function in $\mathbb{R}^1$ shown in Figure 6a. Maxima and minima alternate when going from one end of the interval to the other. The minima are critical points for gradient algorithms, because they separate the basins of attraction (labeled A-D). Starting at a minimum, a gradient algorithm can reach either the maximum to its left or the one to its right. In addition, the intermediate value theorem tells us that in between two maxima, there must be at least one minimum.

In Figure 6b a similar situation in $\mathbb{R}^2$ is shown. In contrast to the situation in $\mathbb{R}^1$, for the higher dimensional space the borders of the basins of attraction are glued together at saddle points. Again, these saddle points are important starting points for randomized gradient algorithms, because when starting from these points

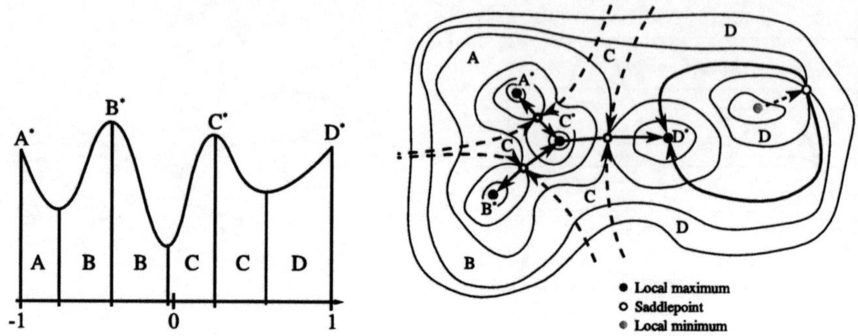

(a) (b)

FIGURE 6. Basins of attraction for a non-linear function f in $\mathbb{R}^1$ (a) and for a non-linear function f in $\mathbb{R}^2$ (b). In (b) the broken arrows indicate the trajectories to the saddle points, the full arrows the trajectories to the local maxima A^*, B^*, C^*, and D^* (rough sketch). They separate the (open) basins of attraction A, B, C, and D. We call this graph a Morse Graph. The dotted arrow in D goes from the local minimum to a saddle point. (Subfigure (b) is a variation of [**JJT00**, Fig. 1.4.2])

TABLE 4. Properties of strict critical points. Let $C^k(M,\mathbb{R})$ be the space of k-times continuously differentiable functions on M with M open and $N(\cdot)$ be the join neighborhood of a partition. $Df(x)$ is the row vector of the gradient $\left(\frac{\delta}{\delta x_1} f(x), \dots \frac{\delta}{\delta x_n} f(x),\right)$ and $D^2 f(x)$ is the Hessian matrix $\left(\frac{\delta^2}{\delta x_i \delta x_j} f(x)\right)_{i,j=1,\dots,n}$.

	$C^k(M,\mathbb{R}), k > 2$	Merge lattice of partitions
Local maximum	$Df(x_c) = 0$ $D^2 f(x_c)$ negative definite	$\forall x \in N(x_c) : f(x_c) > f(x)$
Local minimum	$Df(x_c) = 0$ $D^2 f(x_c)$ positive definite	$\forall x \in N(x_c) : f(x_c) < f(x)$
Saddle point	$Df(x_c) = 0$ $D^2 f(x_c)$ non degenerate and not positive definite	x_c is a point at which the merge paths to more than one critical point split

different local optima can be reached depending on the direction the randomized gradient algorithm follows. In contrast, gradient algorithms starting at points in the interior of basins of attraction lead to one local maximum - even if they are randomized.

Table 4 compares the properties of strict critical points for at least 2-times continuously differentiable spaces with the properties of critical points in the merge lattice of agglomerative hierarchical modularity clustering algorithms. Note, that

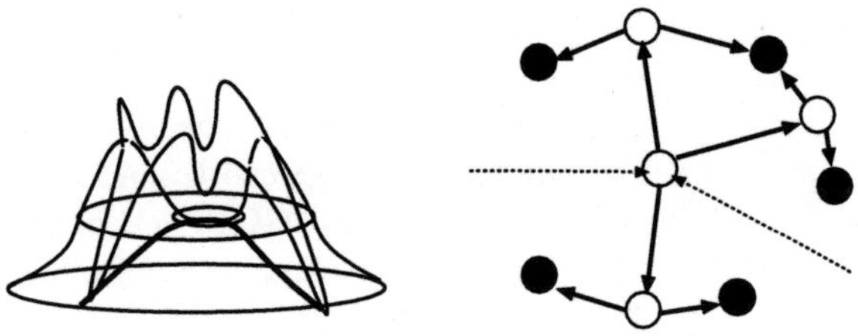

FIGURE 7. Rugged mountain saddle and its Morse graph

saddle points are characterized as split points of algorithm paths to critical points. In Figure 1 such a path split occurs at the partition (1 2) 3 (4 5) with two paths leading to the two local maxima (1 2 3)(4 5) and (1 2)(3 4 5).

Thus, the core groups partitions correspond to saddle points as in the path space of a graph the core groups are branch points where the join-paths to local maxima separate. As the core groups partitions correspond to saddle points, they are good start points for randomized greedy algorithms. For other classifiers, e.g. the label propagation algorithm, core groups partitions work well as long as the classifiers reach points in different basins of attraction which is a weaker condition than the requirement of reaching a local maximum. Obviously, in order to be a good restart point in the CGGC scheme, other local optima need to be reachable from a core group than those used to create the core groups, too. The rugged mountain saddle shown in Figure 7 is a familiar example for such a branch point in $\mathbb{R}^3$. By iteratively identifying core groups of increasing modularity, we identify saddle points that lead to higher and higher local maxima.

In summary, through the theoretical considerations of this section (and supported by the evaluation in Section 5) our explanation for the high optimization quality of the CGGC scheme is:

- The operation of forming core groups partitions from sets of locally (almost) maximal partitions identifies (some) critical points on the merge lattice of partitions.
- Core group partitions are good points for restarting randomized greedy algorithms, because a core groups partition is a branch point (saddle point) in the search space where different basins of attraction meet.

7. Conclusion

In this paper we have shown that learning several weak classifiers has a number of advantages for graph clustering. The maximal overlap of several weak classifiers is a good restart point for further search. Depending on the viewpoint, this approach can be regarded as a way to make first the 'easy' decisions on which pairs of vertices belong together and make 'harder' decisions not before the unambiguous ones have been made. When looking at the search space, maximal overlaps seem to be capable of identifying those critical points from which especially randomized gradient algorithms can find good local maxima.

As it turned out, when using the CGGCi scheme, the choice of base algorithm has no major impact on the clustering quality. This is an important notion. Using the core groups scheme, the base algorithm(s) can be selected because of other considerations. For example, for most so far developed algorithms for modularity maximization an efficient implementation for distributed computer environments (e.g. a Hadoop cluster) would be very hard. However, the label propagation algorithm seems to be very suitable for this kind of environment. Propagating labels requires only to pass the label information between the nodes of a computer cluster. Thus, this algorithm can be used in the CGGCi scheme and in a distributed computing environment to find high quality clusterings of billion-edge networks.

For greedy base algorithms, we showed that the CGGC scheme explores the Morse graph of critical points. That explains why the scheme is able to achieve high optimization performance even in huge graphs with modest effort.

However, an open question is the theoretical justification of the size of the ensemble which is used for the determination of the maximal overlap partition.

References

[AK08] G. Agarwal and D. Kempe, *Modularity-maximizing graph communities via mathematical programming*, Eur. Phys. J. B **66** (2008), no. 3, 409–418, DOI 10.1140/epjb/e2008-00425-1. MR2465245 (2009k:91130)

[BDG+07] Ulrik Brandes, Daniel Delling, Marco Gaertler, Robert Görke, Martin Hoefer, Zoran Nikoloski, and Dorothea Wagner, *On finding graph clusterings with maximum modularity*, Graph-theoretic concepts in computer science, Lecture Notes in Comput. Sci., vol. 4769, Springer, Berlin, 2007, pp. 121–132, DOI 10.1007/978-3-540-74839-7_12. MR2428570 (2009j:05215)

[BDG+08] Ulrik Brandes, Daniel Delling, Marco Gaertler, Robert Gorke, Martin Hoefer, Zoran Nikoloski, and Dorothea Wagner, *On modularity clustering*, IEEE Transactions on Knowledge and Data Engineering **20** (2008), no. 2, 172–188.

[BGLL08] Vincent D Blondel, Jean-Loup Guillaume, Renaud Lambiotte, and Etienne Lefebvre, *Fast unfolding of communities in large networks*, Journal of Statistical Mechanics: Theory and Experiment **2008** (2008), no. 10, P10008.

[Boc74] Hans Hermann Bock, *Automatische Klassifikation*, Vandenhoeck & Ruprecht, Göttingen, 1974. Theoretische und praktische Methoden zur Gruppierung und Strukturierung von Daten (Cluster-Analyse); Studia Mathematica/Mathematische Lehrbücher, Band XXIV. MR0405723 (53 #9515)

[FB07] Santo Fortunato and Marc Barthélemy, *Resolution limit in community detection*, Proceedings of the National Academy of Sciences of the United States of America **104** (2007), no. 1, 36–41.

[FJ05] Ana L. N. Fred and Anil K. Jain, *Combining multiple clusterings using evidence accumulation*, IEEE Trans. Pattern Anal. Mach. Intell. **27** (2005), 835–850.

[For10] Santo Fortunato, *Community detection in graphs*, Phys. Rep. **486** (2010), no. 3-5, 75–174, DOI 10.1016/j.physrep.2009.11.002. MR2580414 (2011d:05337)

[GA05] R. Guimera and LAN Amaral, *Functional cartography of complex metabolic networks*, Nature **433** (2005), 895–900.

[JJT00] Hubertus Th. Jongen, Peter Jonker, and Frank Twilt, *Nonlinear optimization in finite dimensions*, Nonconvex Optimization and its Applications, vol. 47, Kluwer Academic Publishers, Dordrecht, 2000. Morse theory, Chebyshev approximation, transversality, flows, parametric aspects. MR1794354 (2001i:90002)

[KL70] B.W. Kernighan and S. Lin, *An efficient heuristic procedure for partitioning graphs*, The Bell System Technical Journal **49** (1970), no. 1, 291–307.

[LH07] S. Lehmann and L.K. Hansen, *Deterministic modularity optimization*, The European Physical Journal B - Condensed Matter and Complex Systems **60** (2007), no. 1, 83–88.

[LZW+08] Zhenping Li, Shihua Zhang, Rui-Sheng Wang, Xiang-Sun Zhang, and Luonan Chen, *Quantitative function for community detection*, Physical Review E **77** (2008), no. 3, 036109.

[MAnD05] A. Medus, G. Acuña, and C.O. Dorso, *Detection of community structures in networks via global optimization*, Physica A: Statistical Mechanics and its Applications **358** (2005), no. 2-4, 593–604.

[MLR06] Alfonsas Misevicius, Antanas Lenkevicius, and Dalius Rubliauskas, *Iterated tabu search: an improvement to standard tabu search*, Information Technology and Control **35** (2006), 187–197.

[Mor34] Marston Morse, *The calculus of variations in the large*, Colloquium Publications of the American Mathematical Society, vol. 18, American Mathematical Society, New York, 1934.

[MRC05] Stefanie Muff, Francesco Rao, and Amedeo Caflisch, *Local modularity measure for network clusterizations*, Physical Review E **72** (2005), no. 5, 056107.

[New04] Mark E. J. Newman, *Fast algorithm for detecting community structure in networks*, Physical Review E **69** (2004), no. 6, 066133.

[New06] ———, *Modularity and community structure in networks*, Proceedings of the National Academy of Sciences of the United States of America **103** (2006), no. 23, 8577–8582.

[NG04] Mark E. J. Newman and Michelle Girvan, *Finding and evaluating community structure in networks*, Physical Review E **69** (2004), no. 2, 026113.

[NR09] Andreas Noack and Randolf Rotta, *Multi-level algorithms for modularity clustering*, Proceedings of the 8th International Symposium on Experimental Algorithms, Lecture Notes in Computer Science, vol. 5526, Springer Berlin / Heidelberg, 2009, pp. 257–268.

[OGS10] Michael Ovelgönne and Andreas Geyer-Schulz, *Cluster cores and modularity maximization*, ICDMW '10. IEEE International Conference on Data Mining Workshops, 2010, pp. 1204–1213.

[OGS12] Michael Ovelgönne and Andreas Geyer-Schulz, *A comparison of agglomerative hierarchical algorithms for modularity clustering*, Challenges at the Interface of Data Analysis, Computer Science, and Optimization, Studies in Classification, Data Analysis, and Knowledge Organization, Springer Berlin Heidelberg, 2012, pp. 225–232.

[Pol06] R. Polikar, *Ensemble based systems in decision making*, IEEE Circuits and Systems Magazine **6** (2006), no. 3, 21–45.

[RAK07] Usha Nandini Raghavan, Réka Albert, and Soundar Kumara, *Near linear time algorithm to detect community structures in large-scale networks*, Physical Review E **76** (2007), no. 3, 036106.

[RZ07] Jianhua Ruan and Weixiong Zhang, *An efficient spectral algorithm for network community discovery and its applications to biological and social networks*, ICDM 2007, Seventh IEEE International Conference on Data Mining, 2007, pp. 643–648.

[RZ08] Jianhua Ruan and Weixiong Zhang, *Identifying network communities with a high resolution*, Physical Review E **77** (2008), 016104.

[Sch90] Robert E. Schapire, *The strength of weak learnability*, Machine Learning **5** (1990), 197–227.

[Sch07] Satu Elisa Schaeffer, *Graph clustering*, Computer Science Review **1** (2007), no. 1, 27–64.

[Wol92] David H. Wolpert, *Stacked generalization*, Neural Networks **5** (1992), no. 2, 241–259.

[WS05] S. White and P. Smyth, *A spectral clustering approach to finding communities in graphs*, Proceedings of the Fifth SIAM International Conference on Data Mining, SIAM, 2005, pp. 274–285.

[ZWM+08] Zhemin Zhu, Chen Wang, Li Ma, Yue Pan, and Zhiming Ding, *Scalable community discovery of large networks*, WAIM '08: Proceedings of the 2008 International Conference on Web-Age Information Management, 2008, pp. 381–388.

UMIACS, University of Maryland, College Park, Maryland
E-mail address: `mov@umiacs.umd.edu`

Institute of Information Systems and Management, Karlsruhe Institute of Technology, Karlsruhe, Germany
E-mail address: `andreas.geyer-schulz@kit.edu`

Parallel community detection for massive graphs

E. Jason Riedy, Henning Meyerhenke, David Ediger, and David A. Bader

ABSTRACT. Tackling the current volume of graph-structured data requires parallel tools. We extend our work on analyzing such massive graph data with a massively parallel algorithm for community detection that scales to current data sizes, clustering a real-world graph of over 100 million vertices and over 3 billion edges in under 500 seconds on a four-processor Intel E7-8870-based server. Our algorithm achieves moderate parallel scalability without sacrificing sequential operational complexity. Community detection partitions a graph into subgraphs more densely connected within the subgraph than to the rest of the graph. We take an agglomerative approach similar to Clauset, Newman, and Moore's sequential algorithm, merging pairs of connected intermediate subgraphs to optimize different graph properties. Working in parallel opens new approaches to high performance. We improve performance of our parallel community detection algorithm on both the Cray XMT2 and OpenMP platforms and adapt our algorithm to the DIMACS Implementation Challenge data set.

1. Communities in Graphs

Graph-structured data inundates daily electronic life. Its volume outstrips the capabilities of nearly all analysis tools. The Facebook friendship network has over 845 million users [**9**]. Twitter boasts over 140 million new messages each day [**34**], and the NYSE processes over 300 million trades each month [**25**]. Applications of analysis range from database optimization to marketing to regulatory monitoring. Global graph analysis kernels at this scale tax current hardware and software architectures due to the size *and* structure of typical inputs.

One such useful analysis kernel finds smaller communities, subgraphs that locally optimize some connectivity criterion, within these massive graphs. We extend the boundary of current complex graph analysis by presenting the first algorithm for detecting communities that scales to graphs of practical size, over 100 million vertices and over three billion edges in less than 500 seconds on a shared-memory parallel architecture with 256 GiB of memory.

2010 *Mathematics Subject Classification.* Primary 68R10, 05C85; Secondary 68W10, 68M20.

This work was supported in part by the Pacific Northwest National Lab (PNNL) Center for Adaptive Supercomputing Software for MultiThreaded Architectures (CASS-MT), NSF Grant CNS-0708307, and the Intel Labs Academic Research Office for the Parallel Algorithms for Non-Numeric Computing Program.

Community detection is a graph clustering problem. There is no single, universally accepted definition of a community within a social network. One popular definition is that a community is a collection of vertices more strongly connected than would occur from random chance, leading to methods based on modularity [**22**]. Another definition [**28**] requires vertices to be more connected to others within the community than those outside, either individually or in aggregate. This aggregate measure leads to minimizing the communities' conductance. We consider disjoint partitioning of a graph into connected communities guided by a local optimization criterion. Beyond obvious visualization applications, a disjoint partitioning applies usefully to classifying related genes by primary use [**36**] and also to simplifying large organizational structures [**18**] and metabolic pathways [**29**]. We report results for maximizing modularity, although our implementation also supports minimizing conductance.

Contributions. We present our previously published parallel agglomerative community detection algorithm, adapt the algorithm for the DIMACS Implementation Challenge, and evaluate its performance on two multi-threaded platforms. Our algorithm scales to practical graph sizes on available multithreaded hardware while keeping the same sequential operation complexity as current state-of-the-art algorithms. Our approach is both natively parallel and simpler than most current sequential community detection algorithms. Also, our algorithm is agnostic towards the specific criterion; any criterion expressible as individual edge scores can be optimized locally with respect to edge contractions. Our implementation supports both maximizing modularity and minimizing conductance.

Capability and performance. On an Intel-based server platform with four 10-core processors and 256 GiB of memory, our algorithm extracts modular communities from the 105 million vertex, 3.3 billion edge `uk-2007-05` graph in under 500 seconds. A 2 TiB Cray XMT2 requires around 2 400 seconds on the same graph. Our edge-list implementation scales in execution time up to 80 OpenMP threads and 64 XMT2 processors on sufficiently large graphs.

Outline. Section 2 presents our high-level algorithm and describes our current optimization criteria. Section 3 discusses implementation and data structure details for our two target threaded platforms. Section 4 considers parallel performance and performance on different graph metrics for two of the DIMACS Implementation Challenge graphs; full results are in the workshop report [**32**]. Section 5 discusses related work, and Section 6 considers future directions.

2. Parallel Agglomerative Community Detection

Agglomerative clustering algorithms begin by placing every input graph vertex within its own unique community. Then neighboring communities are merged to optimize an objective function like maximizing modularity [**2, 21, 22**] (internal connectedness) or minimizing conductance (normalized edge cut) [**1**]. Here we summarize the algorithm and break it into primitive operations. Section 3 then maps each primitive onto our target threaded platforms.

We consider maximizing metrics (without loss of generality) and target a local maximum rather than a global, possibly non-approximable, maximum. There are a wide variety of metrics for community detection [**12**]. We discuss two, modularity and conductance, in Section 2.1.

Our algorithm maintains a *community graph* where every vertex represents a community, edges connect communities when they are neighbors in the input graph, and weights count the number of input graph edges either collapsed into a single community graph edge or contained within a community graph vertex. We currently do not require counting the vertices in each community, but such an extension is straight-forward.

From a high level, our algorithm repeats the following steps until reaching some termination criterion:

(1) associate a score with each edge in the community graph, exiting if no edge has a positive score,
(2) greedily compute a weighted maximal matching using those scores, and
(3) contract matched communities into a new community graph.

Each step serves as our primitive parallel operations.

The first step scores edges by how much the optimization metric would change if the two adjacent communities merge. Computing the change in modularity and conductance requires only the weight of the edge and the weight of the edge's adjacent communities. The change in conductance is negated to convert minimization into maximization.

The second step, a greedy approximately maximum weight maximal matching, selects pairs of neighboring communities where merging them will improve the community metric. The pairs are independent; a community appears at most once in the matching. Properties of the greedy algorithm guarantee that the matching's weight is within a factor of two of the maximum possible value [**27**]. Any positive-weight matching suffices for optimizing community metrics. Some community metrics, including modularity [**6**], form NP-complete optimization problems. Additional work computing our heuristic by improving the matching may not produce better results. Our approach follows existing parallel algorithms [**15, 20**]. Differences appear in mapping the matching algorithm to our data structures and platforms.

The final step contracts the community graph according to the matching. This contraction primitive requires the bulk of the time even though there is little computation. The impact of contraction's intermediate data structure on improving multithreaded performance is explained in Section 3.

Termination occurs either when the algorithm finds a local maximum or according to external constraints. If no edge score is positive, no contraction increases the objective, and the algorithm terminates at a local maximum. In our experiments with modularity, our algorithm frequently assigns a single community per connected component, a useless local maximum. Real applications will impose additional constraints like a minimum number of communities or maximum community size. Following the DIMACS Implementation Challenge rules [**3**], Section 4's performance experiments terminate once at least half the initial graph's edges are contained within the communities, a coverage ≥ 0.5.

Assuming all edges are scored in a total of $O(|E_c|)$ operations and some heavy weight maximal matching is computed in $O(|E_c|)$ [**27**] where E_c is the edge set of the current community graph, each iteration of our algorithm's loop requires $O(|E|)$ operations. As with other algorithms, the total operation count depends on the community growth rates. If our algorithm halts after K contraction phases, our algorithm runs in $O(|E| \cdot K)$ operations where the number of edges in the original graph, $|E|$, bounds the number of edges in any community graph. If the community graph is halved with each iteration, our algorithm requires $O(|E| \cdot \log |V|)$ operations, where $|V|$ is the number of vertices in the input graph. If the graph is a star, only two vertices are contracted per step and our algorithm requires $O(|E| \cdot |V|)$ operations. This matches experience with the sequential CNM algorithm [35].

2.1. Local optimization metrics. We score edges for contraction by modularity, an estimate of a community's deviation from random chance [2, 22], or conductance, a normalized edge cut [1]. We maximize modularity by choosing the largest independent changes from the current graph to the new graph by one of two heuristics explained below. Minimization measures like conductance involve maximizing changes' negations.

Modularity. Newman [21]'s modularity metric compares the connectivity within a collection of vertices to the expected connectivity of a random graph with the same degree distribution. Let m be the number of edges in an undirected graph $G = G(V, E)$ with vertex set V and edge set E. Let $S \subset V$ induce a graph $G_S = G(S, E_S)$ with $E_S \subset E$ containing only edges where both endpoints are in S. Let m_S be the number of edges $|E_S|$, and let $\overline{m_S}$ be an expected number of edges in S given some statistical background model. Define the modularity of the community induced by S as $Q_S = \frac{1}{m}(m_S - \overline{m_S})$. Modularity represents the deviation of connectivity in the community induced by S from an expected background model. Given a partition $V = S_1 \cup S_2 \cup \cdots \cup S_k$, the modularity of that partitioning is $Q = \sum_{i=1}^{k} Q_{S_i}$.

Newman [21] considers the specific background model of a random graph with the same degree distribution as G where edges are independently and identically distributed. If x_S is the total number of edges in G where either endpoint is in S, then we have $Q_S = (m_S - x_S^2/4m)/m$ as in [2]. A subset S is considered a module when there are more internal edges than expected, $Q_S > 0$. The m_S term encourages forming large modules, while the x_S term penalizes modules with excess external edges. Maximizing Q_S finds communities with more internal connections than external ones. Expressed in matrix terms, optimizing modularity is a quadratic integer program and is an NP-complete optimization problem [6]. We compute a local maximum and not a global maximum. Different operation orders produce different locally optimal points.

Section 3's implementation scores edges by the change in modularity produced by contracting that one edge, analogous to the sequential CNM algorithm. Merging the vertex U into a disjoint set of vertices $W \in C$, requires that the change $\Delta Q(W, U) = Q_{W \cup U} - (Q_W + Q_U) > 0$. Expanding the expression for modularity,

$$\begin{aligned} m \cdot \Delta Q(W,U) &= m\,(Q_{W\cup U} - (Q_W + Q_U)) \\ &= (m_{W\cup U} - (m_W + m_U) - \\ &\qquad (\overline{m_{W\cup U}} - (\overline{m_W} + \overline{m_U})) \\ &= m_{W\leftrightarrow U} - \overline{(m_{W\cup U} - (m_W + m_U))}, \end{aligned}$$

where $m_{W\leftrightarrow U}$ is the number of edges between vertices in sets W and U. Assuming the edges are independent and identically distributed across vertices respecting their degrees [8],

$$\overline{(m_{W\cup U} - (m_W + m_U))} = m \cdot \frac{x_W}{2m} \cdot \frac{x_U}{2m}, \text{ and}$$

$$\Delta Q(W, U) = \frac{m_{W\leftrightarrow U}}{m} - \frac{x_W}{2m} \cdot \frac{x_U}{2m}. \tag{1}$$

We track $m_{W\leftrightarrow U}$ and x_W in the contracted graph's edge and vertex weights, respectively. The quantity x_W equals the sum of W's degrees or the volume of W. If we represent the graph G by an adjacency matrix A, then ΔQ is the rank-one update $A/m - (v/2m) \cdot (v/2m)^T$ restricted to non-zero, off-diagonal entries of A. The data necessary for computing the score of edge $\{i, j\}$ are $A(i, j)$, $v(i)$, and $v(j)$, similar in spirit to a rank-one sparse matrix-vector update.

Modularity can be defined slightly differently depending on whether you double-count edges within a community by treating an undirected graph as a directed graph with edges in both directions. The DIMACS Implementation Challenge uses this variation, and we have included an option to double-count edges.

Modularity has known limitations. Fortunato and Barthélemy [11] demonstrate that global modularity optimization cannot distinguish between a single community and a group of smaller communities. Berry *et al.* [4] provide a weighting mechanism that overcomes this resolution limit. Instead of this weighting, we compare CNM with the modularity-normalizing method of McCloskey and Bader [2]. Lancichinetti and Fortunato [17] show that multi-resolution modularity can still have problems, *e.g.* merging small clusters and splitting large ones.

McCloskey and Bader's algorithm (MB) only merges vertices into the community when the change is deemed statistically significant against a simple statistical model assuming independence between edges. The sequential MB algorithm computes the mean $\overline{\Delta Q(W, :)}$ and standard deviation $\sigma(\Delta Q(W, :))$ of all changes adjacent to community W. Rather than requiring only $\Delta Q(W, U) > 0$, MB requires a tunable level of statistical significance with $\Delta Q(W, U) > \overline{\Delta Q(W, :)} + k \cdot \sigma(\Delta Q(W, :))$. Section 4 sets $k = -1.5$. Sequentially, MB considers only edges adjacent to the vertex under consideration and tracks a history for wider perspective. Because we evaluate merges adjacent to all communities at once by matching, we instead filter against the threshold computed across all current potential merges.

Conductance. Another metric, graph conductance, measures a normalized cut between a graph induced by vertex set S and the graph induced by the remaining vertices $V \setminus S$. Denote the cut induced by a vertex set $S \subset V$ by

$$\partial(S) = \{\{u, v\} | \{u, v\} \in E, u \in S, v \notin S\},$$

and the size of the cut by $|\partial(S)|$. Then the conductance of S is defined [1] as

$$\phi(S) = \frac{|\partial(S)|}{\min\{\text{Vol}(S), \text{Vol}(V \setminus S)\}}. \tag{2}$$

If $S = V$ or $S = \emptyset$, let $\phi(S) = 1$, the largest obtainable value.

The minimum conductance over all vertex sets $S \subset V$ is the graph's conductance. Finding a subset with small conductance implies a bottleneck between the subset's induced subgraph and the remainder. Random walks will tend to stay in the induced subgraph and converge rapidly to their stationary distribution [5]. Given a partition $V = S_1 \cup S_2 \cup \cdots \cup S_k$, we evaluate the conductance of that partitioning as $\sum_{i=1}^{k} \phi(S_i)$.

We score an edge $\{i, j\}$ by the negation of the change from old to new, or $\phi(S_i) + \phi(S_j) - \phi(S_i \cup S_j)$. We again track the edge multiplicity in the edge weight and the volume of the subgraph in the vertex weight.

3. Mapping the Agglomerative Algorithm to Threaded Platforms

Our implementation targets two multithreaded programming environments, the Cray XMT [**16**] and OpenMP [**26**], both based on the C language. Both provide a flat, shared-memory view of data but differ in how they manage parallelism. However, in our use, both environments intend that ignoring the parallel directives produces correct although sequential C code. The Cray XMT environment focuses on implicit, automatic parallelism, while OpenMP requires explicit management.

The Cray XMT architecture tolerates high memory latencies from physically distributed memory using massive multithreading. There is no cache in the processors; all latency is handled by threading. Programmers do not directly control the threading but work through the compiler's automatic parallelization with occasional pragmas providing hints to the compiler. There are no explicit parallel regions. Threads are assumed to be plentiful and fast to create. Current XMT and XMT2 hardware supports over 100 hardware thread contexts per processor. Unique to the Cray XMT are full/empty bits on every 64-bit word of memory. A thread reading from a location marked empty blocks until the location is marked full, permitting very fine-grained synchronization amortized over the cost of memory access. The full/empty bits permit automatic parallelization of a wider variety of data dependent loops. The Cray XMT provides one additional form of parallel structure, futures, but we do not use them here.

The widely-supported OpenMP industry standard provides more traditional, programmer-managed threading. Parallel regions are annotated explicitly through compiler pragmas. Every loop within a parallel region must be annotated as a work-sharing loop or else every thread will run the entire loop. OpenMP supplies a lock data type which must be allocated and managed separately from reading or writing the potentially locked memory. OpenMP also supports tasks and methods for interaction, but our algorithm does not require them.

3.1. Graph representation. We use the same core data structure as our earlier work [**30, 31**] and represent a weighted, undirected graph with an array of triples (i, j, w) for edges between vertices i and j with $i \neq j$. We accumulate repeated edges by adding their weights. The sum of weights for self-loops, $i = j$, are stored in a $|V|$-long array. To save space, we store each edge *only once*, similar to storing only one triangle of a symmetric matrix.

Unlike our initial work, however, the array of triples is kept in buckets defined by the first index i, and we hash the order of i and j rather than storing the strictly lower triangle. If i and j both are even or odd, then the indices are stored such that $i < j$, otherwise $i > j$. This scatters the edges associated with high-degree vertices across different source vertex buckets.

The buckets need not be sequential. We store both beginning and ending indices into the edge array for each vertex. In a traditional sparse matrix compressed format, the entries adjacent to vertex $i + 1$ would follow those adjacent to i. Permitting non-sequential buckets reduces synchronization within graph contraction.

Storing both i and j enables direct parallelization across the entire edge array. Because edges are stored only once, edge $\{i, j\}$ can appear in the bucket for either i or j but not both.

A graph with $|V|$ vertices and $|E|$ non-self, unique edges requires space for $3|V| + 3|E|$ 64-bit integers plus a few additional scalars to store $|V|$, $|E|$, and other book-keeping data. Section 3.4 describes cutting some space by using 32-bit integers for some vertex information.

3.2. Scoring and matching. Each edge's score is an independent calculation for our metrics. An edge $\{i, j\}$ requires its weight, the self-loop weight for i and j, and the graph's total weight. Parallel computation of the scores is straight-forward, and we store the edge scores in an $|E|$-long array of 64-bit floating point data.

Computing the heavy maximal matching is less straight-forward. We repeatedly sweep across the vertices and find the best adjacent match until all vertices are either matched or have no potential matches. The algorithm is non-deterministic when run in parallel. Different executions on the same data may produce different matchings. This does not affect correctness but may lead to different communities.

Our earlier implementation iterated in parallel across all of the graph's edges on each sweep and relied heavily on the Cray XMT's full/empty bits for synchronization of the best match for each vertex. This produced frequent hot spots, memory locations of high contention, but worked sufficiently well with nearly no programming effort. The hot spots crippled an explicitly locking OpenMP implementation of the same algorithm on Intel-based platforms.

We have updated the matching to maintain an array of currently unmatched vertices. We parallelize across that array, searching each unmatched vertex u's bucket of adjacent edges for the highest-scored unmatched neighbor, v. Once each unmatched vertex u finds its best current match, the vertex checks if the other side v (also unmatched) has a better match. If the current vertex u's choice is better, it claims both sides using locks or full/empty bits to maintain consistency. Another pass across the unmatched vertex list checks if the claims succeeded. If not and there was some unmatched neighbor, the vertex u remains on the list for another pass. At the end of all passes, the matching will be maximal. Strictly this is not an $O(|E|)$ algorithm, but the number of passes is small enough in social network graphs that it runs in effectively $O(|E|)$ time.

If edge $\{i, j\}$ dominates the scores adjacent to i and j, that edge will be found by one of the two vertices. The algorithm is equivalent to a different ordering of existing parallel algorithms [**15, 20**] and also produces a maximal matching with weight (total score) within a factor of 0.5 of the maximum. Our non-deterministic algorithm matches our shared-memory execution platform and does not introduce synchronization or static data partitioning to duplicate deterministic message-passing implementations.

Social networks often follow a power-law distribution of vertex degrees. The few high-degree vertices may have large adjacent edge buckets, and not iterating across the bucket in parallel may decrease performance. However, neither the Cray XMT nor OpenMP implementations currently support efficiently composing general, nested, light-weight parallel loops. Rather than trying to separate out the high-degree lists, we scatter the edges according to the graph representation's hashing. This appears sufficient for high performance in our experiments.

Our improved matching's performance gains over our original method are marginal on the Cray XMT but drastic on Intel-based platforms using OpenMP. The original method followed potentially long chains of pointers, an expensive operation on Intel-based platforms. Scoring and matching together require $|E| + 4|V|$ 64-bit integers plus an additional $|V|$ locks on OpenMP platforms.

3.3. Graph contraction. Contracting the agglomerated community graph requires from 40% to 80% of the execution time. Our previous implementation was relatively efficient on the Cray XMT but infeasible on OpenMP platforms. We use the bucketing method to avoid locking and improve performance for both platforms.

Our current implementation relabels the vertex endpoints and re-orders their storage according to the hashing. We then roughly bucket sort by the first stored vertex in each edge. If a stored edge is $(i, j; w)$, we place $(j; w)$ into a bucket associated with vertex i but leave i implicitly defined by the bucket. Within each bucket, we sort by j and accumulate identical edges, shortening the bucket. The buckets then are copied back out into the original graph's storage, filling in the i values. This requires $|V|+1+2|E|$ storage, more than our original implementation, but permits much faster operation on both the XMT2 and Intel-based platforms.

Because the buckets need not be stored contiguously in increasing vertex order, the bucketing and copying do not need to synchronize beyond an atomic fetch-and-add. Storing the buckets contiguously requires synchronizing on a prefix sum to compute bucket offsets. We have not timed the difference, but the technique is interesting.

3.4. DIMACS adjustments. Our original implementation uses 64-bit integers to store vertex labels. All of the graphs in the DIMACS Implementation Challenge, however, require only 32-bit integer labels. Halving the space required for vertex labels fits the total size necessary for the largest challenge graph, `uk-2007-05`, in less than 200 GiB of RAM. Note that indices into the edge list must remain 64-bit integers. We also keep the edge scores in 64-bit binary floating-point, although only 32-bit floating-point may suffice.

Surprisingly, we found no significant performance difference between 32-bit and 64-bit integers on smaller graphs. The smaller integers should decrease the bandwidth requirement but not the number of memory operations. We conjecture our performance is limited by the latter.

The Cray XMT's full/empty memory operations work on 64-bit quantities, so our Cray XMT2 implementation uses 64-bit integers throughout. This is not a significant concern with 2 TiB of memory.

4. Parallel Performance

We evaluate parallel performance on two different threaded hardware architectures, the Cray XMT2 and an Intel-based server. We highlight two graphs, one real and one artificial, from the Implementation Challenge to demonstrate scaling and investigate performance properties. Each experiment is run three times to capture some of the variability in platforms and in our non-deterministic algorithm. Our current implementation achieves speed-ups of up to 13× on a four processor, 40-physical-core Intel-based platform. The Cray XMT2 single-processor times are too slow to evaluate speed-ups on that platform.

TABLE 1. Sizes of graphs used for performance evaluation.

Graph	$|V|$	$|E|$
uk-2002	18 520 486	261 787 258
kron_g500-simple-logn20	1 048 576	44 619 402

4.1. Evaluation platforms. The next generation Cray XMT2 is located at the Swiss National Supercomputing Centre (CSCS). Its 64 processors run at 500 MHz and support four times the memory density of the Cray XMT for a total of 2 TiB. These 64 processors support over 6 400 hardware thread contexts. The improvements over the XMT also include additional memory bandwidth within a node, but exact specifications are not yet officially available.

The Intel-based server platform is located at Georgia Tech. It has four ten-core Intel Xeon E7-8870 processors running at 2.40 GHz with 30 MiB of L3 cache per processor. The processors support HyperThreading, so the 40 physical cores appear as 80 logical cores. This server, mirasol, is ranked #17 in the November 2011 Graph 500 list and is equipped with 256 GiB of 1 067 MHz DDR3 RAM.

Note that the Cray XMT allocates entire processors, each with at least 100 threads, while the OpenMP platforms allocate individual threads which are mapped to cores. Results are shown per-Cray-XMT processor and per-OpenMP-thread. We run up to the number of physical Cray XMT processors or logical Intel cores. Intel cores are allocated in a round-robin fashion across sockets, then across physical cores, and finally logical cores.

4.2. Test graphs. We evaluate on two DIMACS Implementation Challenge graphs. Excessive single-processor runs on highly utilized resources are discouraged, rendering scaling studies using large graphs difficult. We cannot run the larger graph on a single XMT2 processor within a reasonable time. Table 1 shows the graphs' names and number of vertices and edges. The workshop report [**32**] contains maximum-thread and -processor timings for the full DIMACS Implementation Challenge. Additionally, we consider execution time on the largest Challenge graph, uk-2007-05. This graph has 105 896 555 vertices and 3 301 876 564 edges.

4.3. Time and parallel speed-up. Figure 1 shows the execution time as a function of allocated OpenMP thread or Cray XMT processor separated by platform and graph. Figure 2 translates the time into speed-up against the best single-thread execution time on the Intel-based platform. The execution times on a single XMT2 processor are too large to permit speed-up studies on these graphs. The results are the best of three runs maximizing modularity with our parallel variant of the Clauset, Newman, and Moore heuristic until the communities contain at least half the edges in the graph. Because fewer possible contractions decrease the conductance, minimizing conductance requires three to five times as many contraction steps and a proportionally longer time.

Maximizing modularity on the 105 million vertex, 3.3 billion edge uk-2007-05 requires from 496 seconds to 592 seconds using all 80 hardware threads of the Intel E7-8870 platform. The same task on the Cray XMT2 requires from 2 388 seconds to 2 466 seconds.

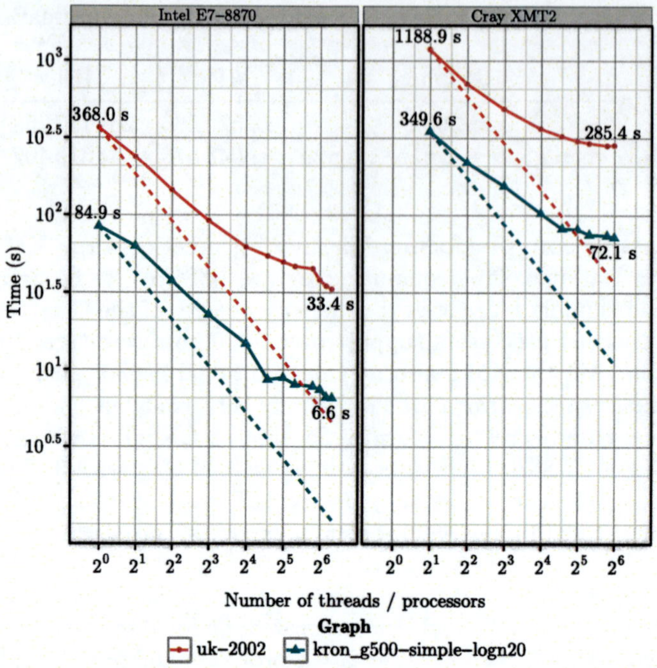

FIGURE 1. Execution time against allocated OpenMP threads or Cray XMT processors per platform and graph. The best single-processor and overall times are noted in the plot. The dashed lines extrapolate perfect speed-up from the time on the least number of processors.

4.4. Community quality. Computing communities quickly is only good if the communities themselves are useful. Full details are in the workshop report [**32**]. Figure 3 shows the results from two different modularity-maximizing heuristics and one conductance-minimizing heuristic. The real-world `uk-2002` graph shows non-trivial community structure, but the artificial `kron_g500-simple-logn20` lacks such structure [**33**]. There appears to be a significant trade-off between modularity and conductance which should be investigated further. Subsequent work has improved modularity results through better convergence criteria than coverage.

5. Related Work

Graph partitioning, graph clustering, and community detection are tightly related topics. A recent survey by Fortunato [**12**] covers many aspects of community detection with an emphasis on modularity maximization. Nearly all existing work of which we know is sequential and targets specific contraction edge scoring mechanisms. Many algorithms target specific contraction edge scoring or vertex move mechanisms [**14**]. Our previous work [**30, 31**] established and extended the first parallel agglomerative algorithm for community detection and provided results on the Cray XMT. Prior modularity-maximizing algorithms sequentially maintain and update priority queues [**8**], and we replace the queue with a weighted graph matching. Separately from this work, Fagginger Auer and Bisseling developed a similar

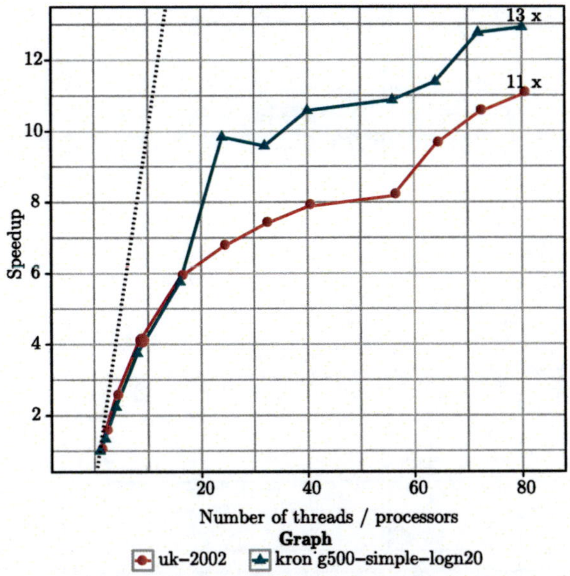

FIGURE 2. Parallel speed-up relative to the best single-threaded execution. The best achieved speed-up is noted on the plot. The dotted line denotes perfect speed-up matching the number of processors.

modularity-optimizing clustering algorithm [**10**]. Their algorithm uses more memory, is more synchronous, and targets execution on GPUs. Fagginger Auer and Bisseling's algorithm performs similarly to ours and includes an interesting star detection technique.

Zhang *et al.* [**37**] recently proposed a parallel algorithm that identifies communities based on a custom metric rather than modularity. Gehweiler and Meyerhenke [**13**] proposed a distributed diffusive heuristic for implicit modularity-based graph clustering. Classic work on parallel modular decompositions [**24**] finds a different kind of module, one where any two vertices in a module have identical neighbors and are somewhat indistinguishable. This could provide a scalable preprocessing step that collapses vertices that will end up in the same community, although removing the degree-1 fringe may have the same effect.

Work on sequential multilevel agglomerative algorithms like [**23**] focuses on edge scoring and local refinement. Our algorithm is agnostic towards edge scoring methods and can benefit from any problem-specific methods. The Cray XMT's word-level synchronization may help parallelize refinement methods, but we leave that to future work.

6. Observations

Our algorithm and implementation, the first published parallel algorithm for agglomerative community detection, extracts communities with apparently high modularity or low conductance in a reasonable length of time. Finding modularity-maximizing communities in a graph with 105 million vertices and over 3.3 billion

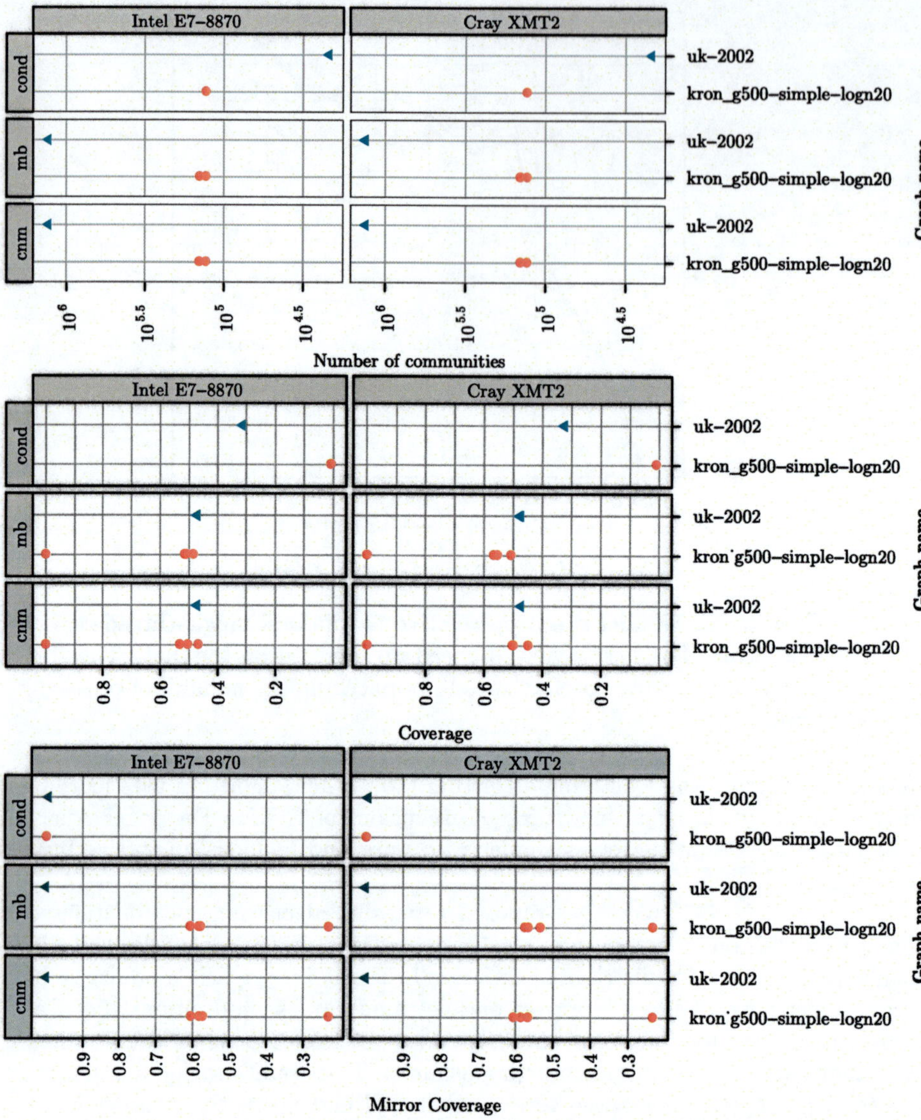

FIGURE 3. Coverage, modularity, and average conductance for the two graphs. The graphs are split vertically by platform and horizontally by scoring method. Here "cnm" and "mb" are the Clauset-Newman-Moore and McCloskey-Bader modularity maximizing heuristics, and "cond" minimizes the conductance.

edges requires a little over eight minutes on a four processor, Intel E7-8870-based server. Our implementation can optimize with respect to different local optimization criteria, and its modularity results are comparable to a state-of-the-art sequential implementation. By altering termination criteria, our implementation can examine some trade-offs between optimization quality and performance. As a twist to established sequential algorithms for agglomerative community detection, our parallel algorithm takes a novel and naturally parallel approach to agglomeration with maximum weighted matchings. That difference appears to reduce differences between the CNM and MB edge scoring methods. The algorithm is simpler than existing sequential algorithms and opens new directions for improvement. Separating scoring, choosing, and merging edges may lead to improved metrics and solutions. Our implementation is publicly available[1].

Outside of the edge scoring, our algorithm relies on well-known primitives that exist for many execution models. Much of the algorithm can be expressed through sparse matrix operations, which may lead to explicitly distributed memory implementations through the Combinatorial BLAS [**7**] or possibly cloud-based implementations through environments like Pregel [**19**]. The performance trade-offs for graph algorithms between these different environments and architectures remain poorly understood.

Besides experiments with massive real-world data sets, future work includes the extension of the algorithm to a streaming scenario. In such a scenario, the graph changes over time without an explicit start or end. This extension has immediate uses in many social network applications but requires algorithmic changes to avoid costly recomputations on large parts of the graph.

Acknowledgments

We thank PNNL and the Swiss National Supercomputing Centre for providing access to Cray XMT systems. We also thank reviews of previous work inside Oracle and anonymous reviewers of this work.

References

[1] R. Andersen and K. Lang, *Communities from seed sets*, Proc. of the 15th Int'l Conf. on World Wide Web, ACM, 2006, p. 232.

[2] D.A. Bader and J. McCloskey, *Modularity and graph algorithms*, Presented at UMBC, September 2009.

[3] D.A. Bader, H. Meyerhenke, P. Sanders, and D. Wagner, *Competition rules and objective functions for the 10th DIMACS Implementation Challenge on graph partitioning and graph clustering*, `http://www.cc.gatech.edu/dimacs10/data/dimacs10-rules.pdf`, September 2011.

[4] J.W. Berry., B. Hendrickson, R.A. LaViolette, and C.A. Phillips, *Tolerating the community detection resolution limit with edge weighting*, CoRR **abs/0903.1072** (2009).

[5] Béla Bollobás, *Modern graph theory*, Graduate Texts in Mathematics, vol. 184, Springer-Verlag, New York, 1998. MR1633290 (99h:05001)

[6] Ulrik Brandes, Daniel Delling, Marco Gaertler, Robert Görke, Martin Hoefer, Zoran Nikoloski, and Dorothea Wagner, *On modularity clustering*, IEEE Trans. Knowledge and Data Engineering **20** (2008), no. 2, 172–188.

[7] Aydın Buluç and John R Gilbert, *The Combinatorial BLAS: design, implementation, and applications*, International Journal of High Performance Computing Applications **25** (2011), no. 4, 496–509.

---

[1] `http://www.cc.gatech.edu/~jriedy/community-detection/`

[8] A. Clauset, M.E.J. Newman, and C. Moore, *Finding community structure in very large networks*, Physical Review E **70** (2004), no. 6, 66111.

[9] Facebook, *Fact sheet*, February 2012, `http://newsroom.fb.com/content/default.aspx?NewsAreaId=22`.

[10] B. O. Fagginger Auer and R. H. Bisseling, *Graph coarsening and clustering on the GPU*, Tech. report, 10th DIMACS Implementation Challenge - Graph Partitioning and Graph Clustering, Atlanta, GA, February 2012.

[11] S. Fortunato and M. Barthélemy, *Resolution limit in community detection*, Proc. of the National Academy of Sciences **104** (2007), no. 1, 36–41.

[12] Santo Fortunato, *Community detection in graphs*, Phys. Rep. **486** (2010), no. 3-5, 75–174, DOI 10.1016/j.physrep.2009.11.002. MR2580414 (2011d:05337)

[13] Joachim Gehweiler and Henning Meyerhenke, *A distributed diffusive heuristic for clustering a virtual P2P supercomputer*, Proc. 7th High-Performance Grid Computing Workshop (HGCW'10) in conjunction with 24th Intl. Parallel and Distributed Processing Symposium (IPDPS'10), IEEE Computer Society, 2010.

[14] Robert Görke, Andrea Schumm, and Dorothea Wagner, *Experiments on density-constrained graph clustering*, Proc. Algorithm Engineering and Experiments (ALENEX12), 2012.

[15] Jaap-Henk Hoepman, *Simple distributed weighted matchings*, CoRR **cs.DC/0410047** (2004).

[16] P. Konecny, *Introducing the Cray XMT*, Proc. Cray User Group meeting (CUG 2007) (Seattle, WA), CUG Proceedings, May 2007.

[17] Andrea Lancichinetti and Santo Fortunato, *Limits of modularity maximization in community detection*, Phys. Rev. E **84** (2011), 066122.

[18] S. Lozano, J. Duch, and A. Arenas, *Analysis of large social datasets by community detection*, The European Physical Journal - Special Topics **143** (2007), 257–259.

[19] Grzegorz Malewicz, Matthew H. Austern, Aart J.C Bik, James C. Dehnert, Ilan Horn, Naty Leiser, and Grzegorz Czajkowski, *Pregel: a system for large-scale graph processing*, Proceedings of the 2010 international conference on Management of data (New York, NY, USA), SIGMOD '10, ACM, 2010, pp. 135–146.

[20] Fredrik Manne and Rob Bisseling, *A parallel approximation algorithm for the weighted maximum matching problem*, Parallel Processing and Applied Mathematics (Roman Wyrzykowski, Jack Dongarra, Konrad Karczewski, and Jerzy Wasniewski, eds.), Lecture Notes in Computer Science, vol. 4967, Springer Berlin / Heidelberg, 2008, pp. 708–717.

[21] M.E.J. Newman, *Modularity and community structure in networks*, Proc. of the National Academy of Sciences **103** (2006), no. 23, 8577–8582.

[22] M.E.J. Newman and M. Girvan, *Finding and evaluating community structure in networks*, Phys. Rev. E **69** (2004), no. 2, 026113.

[23] Andreas Noack and Randolf Rotta, *Multi-level algorithms for modularity clustering*, Experimental Algorithms (Jan Vahrenhold, ed.), Lecture Notes in Computer Science, vol. 5526, Springer Berlin / Heidelberg, 2009, pp. 257–268.

[24] Mark B. Novick, *Fast parallel algorithms for the modular decomposition*, Tech. report, Cornell University, Ithaca, NY, USA, 1989.

[25] NYSE Euronext, *Consolidated volume in NYSE listed issues, 2010 – current*, March 2011, `http://www.nyxdata.com/nysedata/asp/factbook/viewer_edition.asp?mode=table&key=3139&category=3`.

[26] OpenMP Architecture Review Board, *OpenMP application program interface; version 3.0*, May 2008.

[27] *STACS 99*, Lecture Notes in Computer Science, vol. 1563, Springer-Verlag, Berlin, 1999. Edited by Christoph Meinel and Sophie Tison. MR1734032 (2000h:68028)

[28] F. Radicchi, C. Castellano, F. Cecconi, V. Loreto, and D. Parisi, *Defining and identifying communities in networks*, Proc. of the National Academy of Sciences **101** (2004), no. 9, 2658.

[29] E. Ravasz, A. L. Somera, D. A. Mongru, Z. N. Oltvai, and A.-L. Barabási, *Hierarchical organization of modularity in metabolic networks*, Science **297** (2002), no. 5586, 1551–1555.

[30] E. Jason Riedy, David A. Bader, and Henning Meyerhenke, *Scalable multi-threaded community detection in social networks*, Workshop on Multithreaded Architectures and Applications (MTAAP) (Shanghai, China), May 2012.

[31] E. Jason Riedy, Henning Meyerhenke, David Ediger, and David A. Bader, *Parallel community detection for massive graphs*, Proceedings of the 9th International Conference on Parallel Processing and Applied Mathematics (Torun, Poland), September 2011.

[32] ———, *Parallel community detection for massive graphs*, Tech. report, 10th DIMACS Implementation Challenge - Graph Partitioning and Graph Clustering, Atlanta, GA, February 2012.

[33] C. Seshadhri, Tamara G. Kolda, and Ali Pinar, *Community structure and scale-free collections of Erdös-Rényi graphs*, CoRR **abs/1112.3644** (2011).

[34] Twitter, Inc., *Happy birthday Twitter!*, March 2011, `http://blog.twitter.com/2011/03/happy-birthday-twitter.html`.

[35] Ken Wakita and Toshiyuki Tsurumi, *Finding community structure in mega-scale social networks*, CoRR **abs/cs/0702048** (2007).

[36] Dennis M. Wilkinson and Bernardo A. Huberman, *A method for finding communities of related genes*, Proceedings of the National Academy of Sciences of the United States of America **101** (2004), no. Suppl 1, 5241–5248.

[37] Yuzhou Zhang, Jianyong Wang, Yi Wang, and Lizhu Zhou, *Parallel community detection on large networks with propinquity dynamics*, Proceedings of the 15th ACM SIGKDD international conference on Knowledge discovery and data mining (New York, NY, USA), KDD '09, ACM, 2009, pp. 997–1006.

Georgia Institute of Technology, 266 Ferst Drive, Atlanta, Georgia, 30332
E-mail address: `jason.riedy@cc.gatech.edu`

Karlsruhe Institute of Technology (KIT), Am Fasanengarten 5, 76131 Karlsruhe, Germany
E-mail address: `meyerhenke@kit.edu`

Georgia Institute of Technology, 266 Ferst Drive, Atlanta, Georgia, 30332
E-mail address: `dediger@gatech.edu`

Georgia Institute of Technology, 266 Ferst Drive, Atlanta, Georgia, 30332
E-mail address: `bader@cc.gatech.edu`

Graph coarsening and clustering on the GPU

B. O. Fagginger Auer and R. H. Bisseling

ABSTRACT. Agglomerative clustering is an effective greedy way to generate graph clusterings of high modularity in a small amount of time. In an effort to use the power offered by multi-core CPU and GPU hardware to solve the clustering problem, we introduce a fine-grained shared-memory parallel graph coarsening algorithm and use this to implement a parallel agglomerative clustering heuristic on both the CPU and the GPU. This heuristic is able to generate clusterings in very little time: a modularity 0.996 clustering is obtained from a street network graph with 14 million vertices and 17 million edges in 4.6 seconds on the GPU.

1. Introduction

We present a fine-grained shared-memory parallel algorithm for graph coarsening and apply this algorithm in the context of graph clustering to obtain a fast greedy heuristic for maximising modularity in weighted undirected graphs. This is a follow-up to [**8**], which was concerned with generating weighted graph matchings on the GPU, in an effort to use the parallel processing power offered by multi-core CPUs and GPUs for discrete computing tasks, such as partitioning and clustering of graphs and hypergraphs. Just as generating graph matchings, graph coarsening is an essential aspect of both graph partitioning [**4, 9, 12**] and multi-level clustering [**22**] and therefore forms a logical continuation of the research done in [**8**].

Our contribution is a parallel greedy clustering algorithm, that scales well with the number of available processor cores, and generates clusterings of reasonable quality in very little time. We have tested this algorithm, see Section 5, against a large set of clustering problems, all part of the 10th DIMACS challenge on graph partitioning and clustering [**1**], such that the performance of our algorithm can directly be compared with the state-of-the-art clustering algorithms participating in this challenge.

An *undirected graph* G is a pair (V, E), with vertices V, and edges E that are of the form $\{u, v\}$ for $u, v \in V$ with possibly $u = v$. Edges can be provided with weights $\omega : E \to \mathbf{R}_{>0}$, in which case we call G a *weighted undirected graph*. For vertices $v \in V$, we denote the set of all of v's *neighbours* by

$$V_v := \{u \in V \mid \{u, v\} \in E\} \setminus \{v\}.$$

2010 *Mathematics Subject Classification.* Primary 68R10, 68W10; Secondary 91C20, 05C70.
Key words and phrases. Graphs, GPU, shared-memory parallel, clustering.
This research was performed on hardware from NWO project NWO-M 612.071.305.

A *matching* of $G = (V, E)$ is a subset $M \subseteq E$ of the edges of G, satisfying that any two edges in the matching are disjoint. We call a matching M *maximal* if there does not exist a matching M' of G with $M \subsetneq M'$ and we call it *perfect* if $2\,|M| = |V|$. If $G = (V, E, \omega)$ is weighted, then the *weight* of a matching M of G is defined as the sum of the weights of all edges in the matching: $\omega(M) := \sum_{e \in M} \omega(e)$. A matching M of G which satisfies $\omega(M) \geq \omega(M')$ for every matching M' of G is called a *maximum-weight matching*.

Clustering is concerned with partitioning the vertices of a given graph into sets consisting of vertices related to each other, e.g. to isolate communities in graphs representing large social networks [**2**, **14**]. Formally, a *clustering* of an undirected graph G is a collection $\mathcal{C}$ of subsets of V, where elements $C \in \mathcal{C}$ are called *clusters*, that forms a partition of G's vertices, i.e.

$$V = \bigcup_{C \in \mathcal{C}} C, \qquad \text{as a disjoint union.}$$

Note that the number of clusters is not fixed beforehand, and that there can be a single large cluster, or as many clusters as there are vertices, or any number of clusters in between. A quality measure for clusterings, *modularity*, was introduced in [**16**], which we will use to judge the quality of the generated clusterings.

Let $G = (V, E, \omega)$ be a weighted undirected graph. We define the weight $\zeta(v)$ of a vertex $v \in V$ in terms of the weights of the edges incident to this vertex as

$$\zeta(v) := \begin{cases} \sum\limits_{\{u,v\} \in E} \omega(\{u, v\}) & \text{if } \{v, v\} \notin E, \\ \sum\limits_{\substack{\{u,v\} \in E \\ u \neq v}} \omega(\{u, v\}) + 2\,\omega(\{v, v\}) & \text{if } \{v, v\} \in E. \end{cases} \tag{1.1}$$

Then, the modularity, cf. [**1**], of a clustering $\mathcal{C}$ of G is defined by

$$\operatorname{mod}(\mathcal{C}) := \frac{\sum\limits_{C \in \mathcal{C}} \sum\limits_{\substack{\{u,v\} \in E \\ u,v \in C}} \omega(\{u, v\})}{\sum\limits_{e \in E} \omega(e)} - \frac{\sum\limits_{C \in \mathcal{C}} \left(\sum\limits_{v \in C} \zeta(v) \right)^2}{4 \left(\sum\limits_{e \in E} \omega(e) \right)^2}, \tag{1.2}$$

which is bounded by $-\frac{1}{2} \leq \operatorname{mod}(\mathcal{C}) \leq 1$, as we show in the appendix.

Finding a clustering $\mathcal{C}$ which maximises $\operatorname{mod}(\mathcal{C})$ is an NP-complete problem, i.e. ascertaining whether there exists a clustering that has at least a fixed modularity is strongly NP-complete [**3**, Theorem 4.4]. Hence, to find clusterings that have maximum modularity in reasonable time, we need to resort to heuristic algorithms. Many different clustering heuristics have been developed, for which we would like to refer the reader to the overview in [**19**, Section 5] and the references contained therein: there are heuristics based on spectral methods, maximum flow, graph bisection, betweenness, Markov chains, and random walks. The clustering method we present belongs to the category of greedy agglomerative heuristics [**2**, **5**, **15**, **17**, **22**]. Our overall approach is similar to the parallel clustering algorithm discussed by Riedy et al. in [**18**] and a detailed comparison is included in Section 5.

2. Clustering

We will now rewrite (1.2) to a more convenient form. Let $C \in \mathcal{C}$ be a cluster and define the weight of a cluster as $\zeta(C) := \sum_{v \in C} \zeta(v)$, the set of all internal edges as $\text{int}(C) := \{\{u,v\} \in E \mid u,v \in C\}$, the set of all external edges as $\text{ext}(C) := \{\{u,v\} \in E \mid u \in C, v \notin C\}$, and for another cluster $C' \in \mathcal{C}$, the set of all cut edges between C and C' as $\text{cut}(C,C') := \{\{u,v\} \in E \mid u \in C, v \in C'\}$. Let furthermore $\Omega := \sum_{e\in E} \omega(e)$ be the sum of all edge weights.

With these definitions, we can reformulate (1.2) as (see the appendix):

$$\text{mod}(\mathcal{C}) = \frac{1}{4\,\Omega^2} \sum_{C\in\mathcal{C}} \left(\zeta(C)\,(2\,\Omega - \zeta(C)) - 2\,\Omega \sum_{\substack{C'\in\mathcal{C} \\ C'\neq C}} \omega(\text{cut}(C,C')) \right). \tag{2.1}$$

This way of looking at the modularity is useful for reformulating the agglomerative heuristic in terms of graph coarsening, as we will see in Section 2.1.

For this purpose, we also need to determine what effect the merging of two clusters has on the clustering's modularity. Let $\mathcal{C}$ be a clustering and $C, C' \in \mathcal{C}$. If we merge C and C' into one cluster $C \cup C'$, then the clustering $\mathcal{C}' := (\mathcal{C} \setminus \{C, C'\}) \cup \{C \cup C'\}$ we obtain, has modularity (see the appendix)

$$\text{mod}(\mathcal{C}') = \text{mod}(\mathcal{C}) + \frac{1}{2\,\Omega^2}\Big(2\,\Omega\,\omega(\text{cut}(C,C')) - \zeta(C)\,\zeta(C')\Big), \tag{2.2}$$

and the new cluster has weight

$$\zeta(C \cup C') = \sum_{v\in C} \zeta(v) + \sum_{v\in C'} \zeta(v) = \zeta(C) + \zeta(C'). \tag{2.3}$$

2.1. Agglomerative heuristic. Equations (2.1), (2.2), and (2.3) suggest an agglomerative heuristic to generate a clustering **[15, 18, 22]**. Let $G = (V, E, \omega, \zeta)$ be a weighted undirected graph, provided with edge weights ω and vertex weights ζ as defined by (1.1), for which we want to calculate a clustering $\mathcal{C}$ of high modularity.

We start out with a clustering where each vertex of the original graph is a separate cluster, and then progressively merge these clusters to increase the modularity of the clustering. This process is illustrated in Figure 1. The decision which pairs of clusters to merge is based on (2.2): we generate a weighted matching in the graph with all the current clusters as vertices and the sets $\{C, C'\}$ for which $\text{cut}(C, C') \neq \emptyset$ as edges. The weight of such an edge $\{C, C'\}$ is then given by (2.2), such that a maximum-weight matching will result in pairwise mergings of clusters for which the increase of the modularity is maximal.

We do this formally by, starting with G, constructing a sequence of weighted graphs $G^i = (V^i, E^i, \omega^i, \zeta^i)$ with surjective maps $\pi^i : V^i \to V^{i+1}$,

$$G \;=\; G^0 \;\overset{\pi^0}{\to}\; G^1 \;\overset{\pi^1}{\to}\; G^2 \;\overset{\pi^2}{\to}\; \cdots$$

These graphs G^i correspond to clusterings $\mathcal{C}^i$ of G in the following way:

$$\mathcal{C}^i := \{\{v \in V \mid (\pi^{i-1} \circ \cdots \circ \pi^0)(v) = u\} \mid u \in V^i\}, \qquad i = 0, 1, 2, \ldots$$

Each vertex of the graph G^i will correspond to precisely one cluster in $\mathcal{C}^i$: all vertices of G that were merged together into a single vertex in G^i via $\pi^0, \ldots, \pi^{i-1}$, are considered as a single cluster. (In particular for $G^0 = G$ each vertex of the original graph is a separate cluster.)

From (2.3) we know that weights $\zeta(\cdot)$ of merged clusters should be summed, while for calculating the modularity, (2.1), and the change in modularity due to merging, (2.2), we only need the total edge weight $\omega(\text{cut}(\cdot,\cdot))$ of the collection of edges between two clusters, not of individual edges. Hence, when merging two

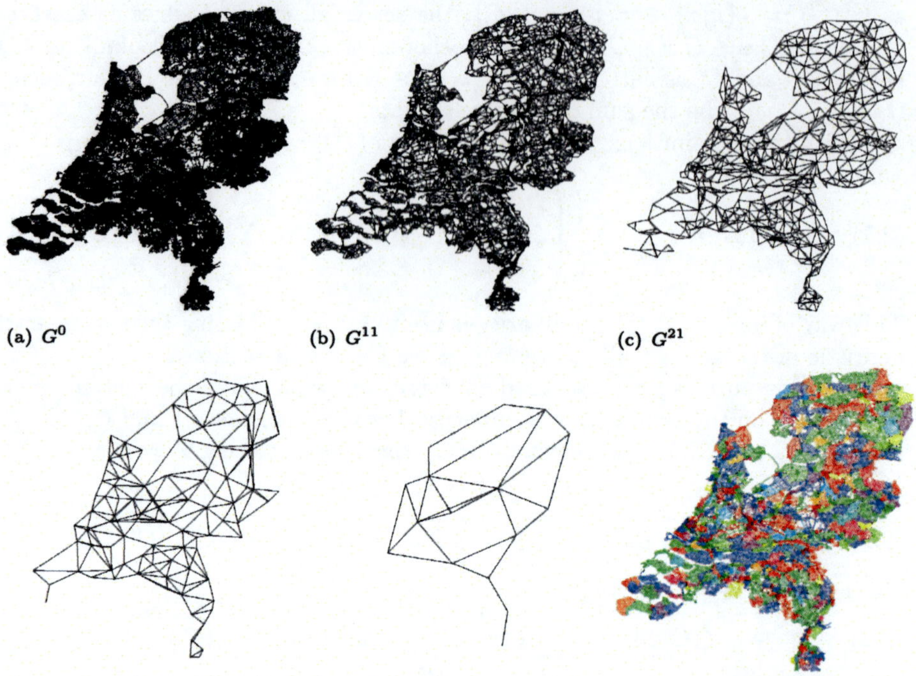

(a) G^0 (b) G^{11} (c) G^{21}

(d) G^{26} (e) G^{33} (f) Best clustering (G^{21}).

FIGURE 1. Clustering of `netherlands` into 506 clusters with modularity 0.995.

clusters, we can safely merge the edges in G^i that are mapped to a single edge in G^{i+1} by π^i, provided we sum their edge weights. This means that the merging of clusters in G^i to obtain G^{i+1} corresponds precisely to coarsening the graph G^i to G^{i+1}. Furthermore, weighted matching in the graph of all current clusters corresponds to a weighted matching in G^i where we consider edges $\{u^i, v^i\} \in E^i$ to have weight $2\,\Omega\,\omega^i(\{u^i, v^i\}) - \zeta^i(u^i)\,\zeta^i(v^i)$ during matching. This entire procedure is outlined in Algorithm 1, where we use a map $\mu : V \to \mathbf{N}$ to indicate matchings $M \subseteq E$ by letting $\mu(u) = \mu(v) \iff \{u, v\} \in M$ for vertices $u, v \in V$.

3. Coarsening

Graph coarsening is the merging of vertices in a graph to obtain a coarser version of the graph. Doing this recursively, we obtain a sequence of increasingly coarser approximations of the original graph. Such a multilevel view of the graph is useful for graph partitioning [**4**, **9**, **12**], but can also be used for clustering [**22**].

Let $G = (V, E, \omega, \zeta)$ be an undirected graph with edge weights ω and vertex weights ζ. A *coarsening* of G is a map $\pi : V \to V'$ together with a graph $G' = (V', E', \omega', \zeta')$ satisfying the following properties:

(1) $\pi(V) = V'$,
(2) $\pi(E) = \{\{\pi(u), \pi(v)\} \mid \{u, v\} \in E\} = E'$,

Algorithm 1 Agglomerative clustering heuristic for a weighted undirected graph $G = (V, E, \omega, \zeta)$ with ζ given by (1.1). Produces a clustering $\mathcal{C}$ of G.

$\text{mod}^{\text{best}} \leftarrow -\infty$
$G^0 = (V^0, E^0, \omega^0, \zeta^0) \leftarrow G$
$i \leftarrow 0$
$\mathcal{C}^0 \leftarrow \{\{v\} \mid v \in V\}$
while $|V^i| > 1$ **do**
    if $\text{mod}(G, \mathcal{C}^i) \geq \text{mod}^{\text{best}}$ **then**
        $\text{mod}^{\text{best}} \leftarrow \text{mod}(G, \mathcal{C}^i)$
        $\mathcal{C}^{\text{best}} \leftarrow \mathcal{C}^i$
    $\mu \leftarrow$ **match_clusters**(G^i)
    $(\pi^i, G^{i+1}) \leftarrow$ **coarsen**(G^i, μ)
    $\mathcal{C}^{i+1} \leftarrow \{\{v \in V \mid (\pi^i \circ \cdots \circ \pi^0)(v) = u\} \mid u \in V^{i+1}\}$
    $i \leftarrow i + 1$
return $\mathcal{C}^{\text{best}}$

(3) for $v' \in V'$,

$$\zeta'(v') = \sum_{\substack{v \in V \\ \pi(v) = v'}} \zeta(v), \tag{3.1}$$

(4) and for $e' \in E'$,

$$\omega'(e') = \sum_{\substack{\{u,v\} \in E \\ \{\pi(u), \pi(v)\} = e'}} \omega(\{u, v\}). \tag{3.2}$$

Let $\mu : V \to \mathbf{N}$ be a map indicating the desired coarsening, such that vertices u and v should be merged into a single vertex precisely when $\mu(u) = \mu(v)$. Then we call a coarsening π *compatible with* μ if for all $u, v \in V$ it holds that $\pi(u) = \pi(v)$ if and only if $\mu(u) = \mu(v)$. The task of the coarsening algorithm is, given G and μ, to generate a graph coarsening π, G' that is compatible with μ.

As noted at the end of Section 2.1, the map μ can correspond to a matching M, by letting $\mu(u) = \mu(v)$ if and only if the edge $\{u, v\} \in M$. This ensures that we do not coarsen the graph too aggressively, only permitting a vertex to be merged with at most one other vertex during coarsening. Such a coarsening approach is also used in hypergraph partitioning [**20**]. For our coarsening algorithm, however, it is not required that μ is derived from a matching: any map $\mu : V \to \mathbf{N}$ is permitted.

3.1. Star-like graphs. The reason for permitting a general μ (i.e. where more than two vertices are contracted to a single vertex during coarsening), instead of a map μ arising from graph matchings is that the recursive coarsening process can get stuck on star-like graphs [**6**, Section 4.3].

In Figure 2(a), we see a star graph in which a maximum matching is indicated. Coarsening this graph by merging the two matched vertices will yield a graph with only one vertex less. In general, with a k-pointed star, coarsening by matching will reduce the total number of vertices from $k + 1$ to k, requiring k coarsening steps to reduce the star to a single vertex. This is slow compared to a graph for which we can find a perfect matching at each step of the coarsening, where the total number of vertices is halved at each step and we require only $\log_2 k$ coarsening steps to reduce

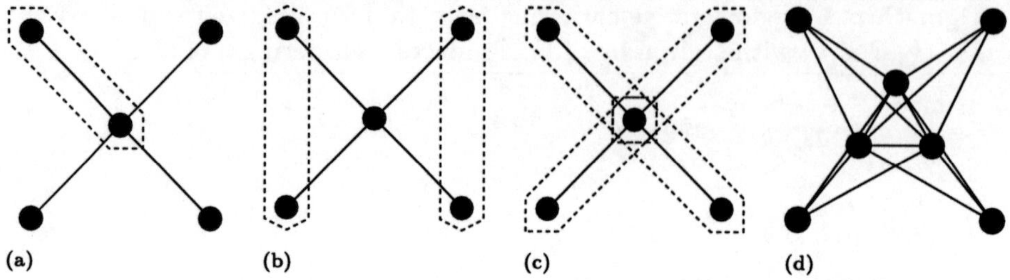

FIGURE 2. Merging vertices in star-like graphs: by matching in (a), by merging vertices with the same neighbours in (b), and by merging more than two vertices in (c). In (d) we see a star-like graph with a centre clique of 3 vertices and 4 satellites.

the graph to a single vertex. Hence, star graphs increase the number of coarsening iterations at line 5 of Algorithm 1 we need to perform, which increases running time and has an adverse effect on parallelisation, because of the few matches that can actually be made in each iteration.

A way to remedy this problem is to identify vertices with the same neighbours and match these pairwise, see Figure 2(b) [**7, 10**]. When maximising clustering modularity however, this is not a good idea: for clusters $C, C' \in \mathcal{C}$ without any edges between them, $\mathrm{cut}(C, C') = \emptyset$, merging C and C' will change the modularity by $\frac{-1}{2\,\Omega^2}\,\zeta(C)\,\zeta(C') \leq 0$.

Because of this, we will use the strategy from Figure 2(c), and merge multiple outlying vertices, referred to as *satellites* from now on, to the centre of the star simultaneously. To do so, however, we need to be able to identify star centres and satellites in the graph.

As the defining characteristic of the centre of a star is its high degree, we will use the vertex degrees to measure to what extent a vertex is a centre or a satellite. We propose, for vertices $v \in V$, to let

$$\mathrm{cp}(v) := \frac{\deg(v)^2}{\sum\limits_{u \in V_v} \deg(u)}, \tag{3.3}$$

be the *centre potential* of v. Here, the *degree* of a vertex $v \in V$ is defined as $\deg(v) := |V_v|$. Note that for satellites the centre potential will be small, because a satellite's degree is low, while the centre to which it is connected has a high degree. On the other hand, a star centre will have a high centre potential because of its high degree. Let us make this a little more precise.

For a regular graph where $\deg(v) = k$ for all $v \in V$, the centre potential will equal $\mathrm{cp}(v) = k^2/k^2 = 1$ for all vertices $v \in V$. Now consider a star-like graph, consisting of a clique of l vertices in the centre which are surrounded by k satellites that are connected to every vertex in the clique, but not to other satellites (Figure 2(d) has $l = 3$ and $k = 4$), with $0 < l < k$. In such a graph, $\deg(v) = l$ for satellites v and $\deg(u) = l - 1 + k$ for vertices u in the centre clique. Hence, for satellites v

$$\mathrm{cp}(v) = \frac{l^2}{l\,(l - 1 + k)} \leq \frac{l}{l - 1 + l + 1} = \frac{1}{2},$$

while for centre vertices u

$$\mathrm{cp}(u) = \frac{(l-1+k)^2}{(l-1)\,(l-1+k)+k\,l} = 1 + \left(\frac{k-1}{2\,l-1+\frac{(l-1)^2}{k}}\right) \geq \frac{4}{3}.$$

If we fix $l > 0$ and let the number of satellites $k \to \infty$, we see that

$$\mathrm{cp}(v) \to 0 \qquad \text{and} \qquad \mathrm{cp}(u) \to \infty.$$

Hence, the centre potential seems to be a good indicator for determining whether vertices v are satellites, $\mathrm{cp}(v) \leq \frac{1}{2}$, or centres, $\mathrm{cp}(v) \geq \frac{4}{3}$.

In Algorithm 1, we will therefore, after line 9, use $\mathrm{cp}(v)$ to identify all satellites in the graph and merge these with the neighbouring non-satellite vertex that will yield the highest increase of modularity as indicated by (2.2). This will both provide greedy modularity maximisation, and stop star-like graphs from slowing down the algorithm.

4. Parallel implementation

In this section, we will demonstrate how the different parts of the clustering algorithm can be implemented in a style that is suitable for the GPU.

To make the description of the algorithm more explicit, we will need to deviate from some of the graph definitions of the introduction. First of all, we consider arrays in memory as ordered lists, and suppose that the vertices of the graph $G = (V, E, \omega, \zeta)$ to be coarsened are given by $V = (1, 2, \ldots, |V|)$. We index such lists with parentheses, e.g. $V(2) = 2$, and denote their length by $|V|$. Instead of storing the edges E and edge weights ω of a graph explicitly, we will store for each vertex $v \in V$ the set of all its neighbours V_v, and include the edge weights ω in this list. We will refer to these sets as *extended neighbour lists* and denote them by V_v^ω for $v \in V$.

Let us consider a small example: a graph with 3 vertices and edges $\{1,2\}$ and $\{1,3\}$ with edge weights $\omega(\{1,2\}) = 4$ and $\omega(\{1,3\}) = 5$. Then, for the parallel coarsening algorithm we consider this graph as $V = (1,2,3)$, together with $V_1^\omega = ((2,4),(3,5))$ (since there are two edges originating from vertex 1, one going to vertex 2, and one going to vertex 3), $V_2^\omega = ((1,4))$ (as $\omega(\{1,2\}) = 4$), and $V_3^\omega = ((1,5))$ (as $\omega(\{1,3\}) = 5$).

In memory, such neighbour lists are stored as an array of indices and weights (in the small example, $((2,4),(3,5),(1,4),(1,5))$), with for each vertex a range in this array (in the small example range $(1,2)$ for vertex 1, $(3,3)$ for 2, and $(4,4)$ for 3). Note that we can extract all edges together with their weights ω directly from the extended neighbour lists. Hence, (V, E, ω, ζ) and $(V, \{V_v^\omega \mid v \in V\}, \zeta)$ are equivalent descriptions of G.

We will now discuss the parallel coarsening algorithm described by Algorithm 2, in which the **parallel_*** functions are slight adaptations of those available in the Thrust template library [11]. The **for ... parallel do** construct indicates a for-loop of which each iteration can be executed in parallel, independent of all other iterations.

We start with an undirected weighted graph G with vertices $V = (1, 2, \ldots, |V|)$, vertex weights ζ, and edges E with edge weights ω encoded in the extended neighbour lists as discussed above. A given map $\mu : V \to \mathbf{N}$ indicates which vertices should be merged to form the coarse graph.

Algorithm 2 Parallel coarsening algorithm on the GPU. Given a graph G with $V = (1, 2, \dots, |V|)$ and a map $\mu : V \to \mathbf{N}$, this algorithm creates a graph coarsening π, G' compatible with μ.

$\rho \leftarrow V$
$(\rho, \mu) \leftarrow$ **parallel_sort_by_key**(ρ, μ)
$\mu \leftarrow$ **parallel_adjacent_not_equal**(μ)
$\pi^{-1} \leftarrow$ **parallel_copy_index_if_nonzero**(μ)
$V' \leftarrow (1, 2, \dots, |\pi^{-1}|)$
append$(\pi^{-1}, |V| + 1)$
$\mu \leftarrow$ **parallel_inclusive_scan**(μ)
$\pi \leftarrow$ **parallel_scatter**(ρ, μ)
for $v' \in V'$ **parallel do** {Sum vertex weights.}
  $\zeta'(v') \leftarrow 0$
  for $i = \pi^{-1}(v')$ **to** $\pi^{-1}(v' + 1) - 1$ **do**
    $\zeta'(v') \leftarrow \zeta'(v') + \zeta(\rho(i))$
for $v' \in V'$ **parallel do** {Copy neighbours.}
  $V_{v'}^{\prime\omega'} \leftarrow \emptyset$
  for $i = \pi^{-1}(v')$ **to** $\pi^{-1}(v' + 1) - 1$ **do**
    for $(u, \omega) \in V_{\rho(i)}^{\omega}$ **do**
      append$(V_{v'}^{\prime\omega'}, (\pi(u), \omega))$
for $v' \in V'$ **parallel do** {Compress neighbours.}
  $V_{v'}^{\prime\omega'} \leftarrow$ **compress_neighbours**$(V_{v'}^{\prime\omega'})$

Algorithm 2 starts by creating an ordered list ρ of all the vertices V, and sorting ρ according to μ. The function **parallel_sort_by_key**(a, b) sorts b in increasing order and applies the same sorting permutation to a, and does so in parallel. Consider for example a graph with 12 vertices and a given μ:

ρ	1	2	3	4	5	6	7	8	9	10	11	12
μ	9	2	3	22	9	9	22	2	3	3	2	4

Then applying **parallel_sort_by_key** will yield

ρ	2	8	11	3	9	10	12	1	5	6	4	7
μ	2	2	2	3	3	3	4	9	9	9	22	22

We then apply the function **parallel_adjacent_not_equal**(a) which sets $a(1)$ to 1, and for $1 < i \leq |a|$ sets $a(i)$ to 1 if $a(i) \neq a(i-1)$ and to 0 otherwise. This yields

ρ	2	8	11	3	9	10	12	1	5	6	4	7
μ	1	0	0	1	0	0	1	1	0	0	1	0

Now we know where each group of vertices of G that needs to be merged together starts. We will store these numbers in the 'inverse' of the projection map π, such that we know, for each coarse vertex v', what vertices v in the original graph are coarsened to v'. The function **parallel_copy_index_if_nonzero**(a) picks out the indices $1 \leq i \leq |a|$ for which $a(i) \neq 0$ and stores these consecutively in a list, π^{-1} in this case, in parallel.

ρ	2	8	11	3	9	10	12	1	5	6	4	7
μ	1	0	0	1	0	0	1	1	0	0	1	0
π^{-1}	1	4	7	8	11							

This gives us the number of vertices in the coarse graph as $|\pi^{-1}| = 5$, so $V' = (1, 2, \ldots, |\pi^{-1}|)$. To make sure we get a valid range for the last vertex in G', at line 6 we append $|V|+1$ to π^{-1}. Now, we want to create the map $\pi : V \to V'$ relating the vertices of our original graph to the vertices of the coarse graph. We do this by re-enumerating μ using an inclusive scan. The function **parallel_inclusive_scan**(a) keeps a running sum s, initialised as 0, and updates for $1 \le i \le |a|$ the value $s \leftarrow s + a(i)$, storing $a(i) \leftarrow s$.

ρ	2	8	11	3	9	10	12	1	5	6	4	7
μ	1	1	1	2	2	2	3	4	4	4	5	5
π^{-1}	1	4	7	8	11	13						

From these lists, we can see that vertices $3, 9, 10 \in V$ are mapped to the vertex $2 \in V'$ (so, we should have $\pi(3) = \pi(9) = \pi(10) = 2$), and from $2 \in V'$ we can recover $3, 9, 10 \in V$ by looking at values of ρ in the range $\pi^{-1}(2), \ldots, \pi^{-1}(2+1)-1$. From the construction of ρ and μ we know that we should have that $\pi(\rho(i)) = \mu(i)$ for our map $\pi : V \to V'$. Note that $\rho(i)$ is the original vertex in V and $\mu(i)$ is the current vertex in V'. Hence, we use the $c =$ **parallel_scatter**(a, b) function, which sets $c(a(i)) \leftarrow b(i)$ for $1 \le i \le |a| = |b|$ in parallel, to obtain π. Now we know both how to go from the original to the coarse graph (π), and from the coarse to the original graph (π^{-1} and ρ). This permits us to construct the extended neighbour lists of the coarse graph.

Let us look at this from the perspective of a single vertex $v' \in V'$ in the coarse graph. All vertices v in the fine graph that are mapped to v' by π are given by $\rho(\pi^{-1}(v')), \ldots, \rho(\pi^{-1}(v'+1)-1)$. All vertex weights (line 9) $\zeta(v)$ of these v are summed to satisfy (3.1). By considering all extended neighbour lists V_v^ω (line 13), we can construct the extended neighbour list $V_{v'}'^{\omega'}$ of v'. Every element in the neighbour list is a pair $(u, \omega) \in V_v^\omega$. In the coarse graph, $\pi(u)$ will be a neighbour of v' in G', so we add $(\pi(u), \omega)$ to the extended neighbour list $V_{v'}'^{\omega'}$ of v'.

After copying all the neighbours, we compress the neighbour lists of each vertex in the coarse graph by first sorting elements $(u', \omega) \in V_{v'}'^{\omega'}$ of the extended neighbour list by u', and then merging ranges $((u', \omega_1), (u', \omega_2), \ldots, (u', \omega_k))$ in $V_{v'}'^{\omega'}$ to a single element $(u', \omega_1 + \omega_2 + \ldots + \omega_k)$ with **compress_neighbours**. This ensures that we satisfy (3.2).

Afterwards, we have V', $\{V_{v'}'^{\omega'} \mid v' \in V'\}$, and ζ', together with a map $\pi : V \to V'$ compatible with the given μ.

4.1. Parallelisation of the remainder of Algorithm 1. Now that we know how to coarsen the graph in parallel in Algorithm 1 by using Algorithm 2, we will also look at parallelising the other parts of the algorithm. We generate matchings μ on the GPU using the algorithm from [**8**], where we perform weighted matching with edge weight $2\,\Omega\,\omega(\{u, v\}) - \zeta(u)\,\zeta(v)$ (cf. (2.2)), for each edge $\{u, v\} \in E$.

Satellites can be marked and merged in parallel as described by Algorithm 3, where the matching algorithm indicates that a vertex has not been matched to any other vertex by using a special value for μ, such that the validity of $|\mu^{-1}(\{\mu(v)\})| = 1$ can be checked very quickly. Note that in this case the gain of merging a satellite with a non-satellite as described by (2.2) is only an approximation, since we can merge several satellites simultaneously in parallel.

In Algorithm 1 (line 11), we can also keep track of clusters in parallel. We create a clustering map $\kappa : V \to \mathbf{N}$ that indicates the cluster index of each

Algorithm 3 Algorithm for marking and merging unmatched satellites in a given graph $G = (V, E, \omega, \zeta)$, extending a map $\mu : V \to \mathbf{N}$.

for $v \in V$ **parallel do** {Mark unmatched satellites.}
  if $|\mu^{-1}(\{\mu(v)\})| = 1$ **and** $\text{cp}(v) \leq \frac{1}{2}$ **then**
    $\sigma(v) \leftarrow$ **true**
  else
    $\sigma(v) \leftarrow$ **false**
for $v \in V$ **parallel do** {Merge unmatched satellites.}
  if $\sigma(v)$ **then**
    $u^{\text{best}} \leftarrow \infty$
    $w^{\text{best}} \leftarrow -\infty$
    for $u \in V_v$ **do**
      $w \leftarrow 2\Omega\,\omega(\{u, v\}) - \zeta(u)\,\zeta(v)$
      if $w > w^{\text{best}}$ **and not** $\sigma(u)$ **then**
        $w^{\text{best}} \leftarrow w$
        $u^{\text{best}} \leftarrow u$
    if $u^{\text{best}} \neq \infty$ **then**
      $\mu(v) \leftarrow \mu(u^{\text{best}})$

vertex of the original graph, such that for $i = 0, 1, \ldots$, our clustering will be $\mathcal{C}^i = \{\{v \in V \mid \kappa^i(v) = k\} \mid k \in \mathbf{N}\}$ (i.e. vertices u and v belong to the same cluster precisely when $\kappa^i(u) = \kappa^i(v)$). Initially we assign all vertices to a different cluster by letting $\kappa^0(v) \leftarrow v$ for all $v \in V$. After coarsening, the clustering is updated at line 11 by setting $\kappa^{i+1}(v) \leftarrow \pi^i(\kappa^i(v))$. We do this in parallel using $c \leftarrow$ **parallel_gather**(a, b), which sets $c(i) \leftarrow b(a(i))$ for $1 \leq i \leq |a| = |c|$.

Note that unlike [**17, 22**], we do not employ a local refinement strategy such as Kernighan–Lin [**13**] to improve the quality of the obtained clustering from Algorithm 1, because such an algorithm does not lend itself well to parallelisation. This is primarily caused by the fact that exchanging a single vertex between two clusters changes the total weight of both clusters, leading to a change in the modularity gain of *all* vertices in both the clusters. A parallel implementation of the Kernighan–Lin algorithm for clustering is therefore even more difficult than for graph partitioning [**9, 12**], where exchanging vertices only affects the vertex's neighbours. Remedying this is an interesting avenue for further research.

To improve the performance of Algorithm 1 further, we make use of two additional observations. We found during our clustering experiments that the modularity would first increase as the coarsening progressed and then would decrease after a peak value was obtained, as is also visible in [**16**, Figures 6 and 9]. Hence, we stop Algorithm 1 after the current modularity drops below 95% (to permit small fluctuations) of the highest modularity encountered thus far.

The second optimisation makes use of the fact that we do not perform uncoarsening steps in Algorithm 1 (although with the data generated by Algorithm 2 this is certainly possible), which makes it unnecessary to store the entire hierarchy G^0, G^1, G^2, ... in memory. Therefore, we only store two graphs, G^0 and G^1, and coarsen G^0 to G^1 as before, but then we coarsen G^1 to G^0, instead of a new graph G^2, and alternate between G^0 and G^1 as we coarsen the graph further.

5. Results

Algorithm 1 was implemented using NVIDIA's Compute Unified Device Architecture (CUDA) language together with the Thrust template library [**11**] on the GPU and using Intel's Threading Building Blocks (TBB) library on the CPU. The experiments were performed on a computer equipped with two quad-core 2.4 GHz Intel Xeon E5620 processors with hyperthreading (we use 16 threads), 24 GiB RAM, and an NVIDIA Tesla C2075 with 5375 MiB global memory. All source code for the algorithms, together with the scripts required to generate the benchmark data, has been released under the GNU General Public Licence and are freely available from `https://github.com/BasFaggingerAuer/Multicore-Clustering`. It is important to note that the clustering times listed in Table 1, 2, and Figure 3 do include data transfer times from CPU to GPU, but not data transfer from hard disk to CPU memory. On average, 5.5% of the total running time is spent on CPU–GPU data transfer. The recorded time and modularity are averaged over 16 runs, because of the use of random numbers in the matching algorithm [**8**]. These are generated using the TEA-4 algorithm [**21**] to improve performance.

The modularity of the clusterings generated by the CPU implementation is generally a little higher (e.g. `eu-2005`) than those generated by the GPU. The difference between both algorithms is caused by the matching stage of Algorithm 1. For the GPU implementation, we always generate a maximal matching to coarsen the graph as much as possible, even if including some edges $\{u, v\} \in E$ for which $2\,\Omega\,\omega(\{u, v\}) - \zeta(u)\,\zeta(v) < 0$ will decrease the modularity. This yields a fast algorithm, but has an adverse effect on the obtained modularity. For the CPU implementation, we only include edges $\{u, v\} \in E$ which satisfy $2\,\Omega\,\omega(\{u, v\}) - \zeta(u)\,\zeta(v) \geq 0$ in the matching, such that the modularity can only be increased by each matching stage. This yields higher modularity clusterings, but will slow down the algorithm if only a few modularity-increasing edges are available (if there are none, we perform a single matching round where we consider all edges).

Comparing Table 1 with modularities from [**17**, Table 1] for `karate` (0.412), `jazz` (0.444), `email` (0.572), and `PGPgiantcompo` (0.880), we see that Algorithm 1 generates clusterings of lesser modularity. We attribute this to the absence of a local refinement strategy in Algorithm 1, as noted in Section 4.1. The modularity of the clusterings of irregular graphs from the `kronecker/` categories is an order of magnitude smaller than those of graphs from other categories. We are uncertain about what causes this behaviour.

Algorithm 1 is fast: for the `road_central` graph with 14 million vertices and 17 million edges, the GPU generates a clustering with modularity 0.996 in 4.6 seconds, while for `uk-2002`, with 19 million vertices and 262 million edges, the CPU generates a clustering with modularity 0.974 in 30 seconds. In particular, for clustering of nearly regular graphs (i.e. where the ratio $\left(\max_{v \in V} \deg(v)\right) / \left(\min_{v \in V} \deg(v)\right)$ is small) such as street networks, the high bandwidth of the GPU enables us to find high-quality clusterings in very little time (Table 2). Furthermore, Figure 3(a) suggests that in practice, Algorithm 1 scales linearly with the number of edges of the graph, while Figure 3(b) shows that the parallel performance of the algorithm scales reasonably with the number of available cores, increasingly so as the size of the graph increases. Note that with dual quad-core processors, we have eight physical cores available, which explains the smaller increase in performance when the number of threads is extended beyond eight via hyperthreading.

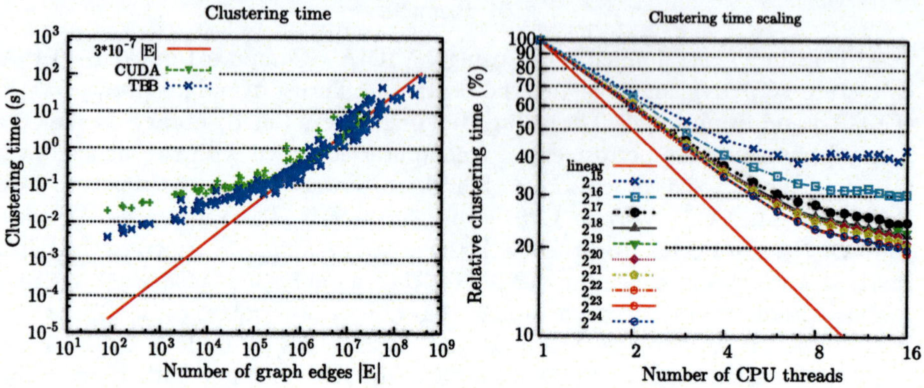

(a) (b)

FIGURE 3. In (a), we show the clustering time required by Algorithm 1 for graphs from the 10th DIMACS challenge [**1**] test set (categories `clustering/`, `streets/`, `coauthor/`, `kronecker/`, `matrix/`, `random/`, `delaunay/`, `walshaw/`, `dyn-frames/`, and `redistrict/`), for both the CUDA and TBB implementations. For large graphs, clustering time scales almost linearly with the number of edges. In (b), we show the parallel scaling of the TBB implementation of Algorithm 1 as a function of the number of threads, normalised to the time required by a single-threaded run for graphs `rgg_n_2_`k`_s0` with 2^k vertices, from the `random/` category. We compare this to ideal, linear, scaling. The test system has 8 cores and up to 16 threads with hyperthreading.

From Figure 3(a), we see that while the GPU performs well for large, $|E| \geq 10^6$, nearly regular graphs, the CPU handles small and irregular graphs better. This can be explained by the GPU setup time that becomes dominant for small graphs, and by the fact that for large irregular graphs, vertices with a higher-than-average degree keep one of the threads occupied, while the threads treating the other, low-degree, vertices are already done, leading to a low GPU occupancy (i.e. where only a single of the 32 threads in a warp is still doing actual work). On the CPU, varying vertex degrees are a much smaller problem because threads are not launched in warps: they can immediately start working on a new vertex, without having to wait for other threads to finish. This results in better performance for the CPU on irregular graphs.

The most costly per-vertex operation is **compress_neighbours**, used during coarsening. We therefore expect the GPU to spend more time, for irregular graphs, on coarsening than on matching. For the regular graph `asia` (GPU 3.4× faster), the GPU (CPU) spends 68% (52%) of the total time on matching and 16% (41%) on coarsening. For the irregular graph `eu-2005` (CPU 4.7× faster), the GPU (CPU) spends 29% (39%) on matching and 70% (57%) on coarsening, so coarsening indeed becomes the bottleneck for the GPU when the graph is irregular.

The effectiveness of merging unmatched satellites can also be illustrated using these graphs: for `asia` the number of coarsenings performed in Algorithm 1 is

TABLE 1. For graphs $G = (V, E)$, this table lists the average modularities $\text{mod}_{1,2}$, (1.2), of clusterings of G generated in an average time of $t_{1,2}$ seconds by the CUDA_1 and TBB_2 implementations of Algorithm 1. The '$\%_1$' column indicates the percentage of time spent on CPU–GPU data transfer. Results are averaged over 16 runs. A '-' indicates that the test system ran out of memory in one of the runs. This table lists graphs from the `clustering/` category of the 10th DIMACS challenge [**1**].

G	$\lvert V\rvert$	$\lvert E\rvert$	mod_1	t_1	$\%_1$	mod_2	t_2
`karate`	34	78	0.363	0.020	13	0.387	0.004
`dolphins`	62	159	0.453	0.027	7	0.485	0.007
`chesapeake`	39	170	0.186	0.024	7	0.220	0.005
`lesmis`	77	254	0.444	0.023	8	0.528	0.006
`adjnoun`	112	425	0.247	0.032	5	0.253	0.009
`polbooks`	105	441	0.437	0.034	6	0.472	0.008
`football`	115	613	0.412	0.033	5	0.455	0.009
`c...metabolic`	453	2,025	0.374	0.055	3	0.394	0.013
`celegansneural`	297	2,148	0.390	0.055	3	0.441	0.011
`jazz`	198	2,742	0.314	0.048	4	0.372	0.010
`netscience`	1,589	2,742	0.948	0.060	4	0.955	0.040
`email`	1,133	5,451	0.440	0.078	2	0.479	0.021
`power`	4,941	6,594	0.918	0.066	3	0.925	0.033
`hep-th`	8,361	15,751	0.795	0.093	2	0.809	0.070
`polblogs`	1,490	16,715	0.330	0.129	1	0.396	0.039
`PGPgiantcompo`	10,680	24,316	0.809	0.095	3	0.842	0.040
`cond-mat`	16,726	47,594	0.788	0.122	2	0.798	0.083
`as-22july06`	22,963	48,436	0.607	0.184	1	0.629	0.036
`cond-mat-2003`	31,163	120,029	0.674	0.195	2	0.690	0.103
`astro-ph`	16,706	121,251	0.588	0.219	1	0.611	0.085
`cond-mat-2005`	40,421	175,691	0.624	0.248	2	0.639	0.113
`pr...Attachment`	100,000	499,985	0.214	1.177	0	0.216	0.217
`smallworld`	100,000	499,998	0.636	0.468	2	0.663	0.175
`G_n_pin_pout`	100,000	501,198	0.241	0.851	1	0.246	0.231
`caida...Level`	192,244	609,066	0.768	0.506	2	0.791	0.198
`cnr-2000`	325,557	2,738,969	0.828	2.075	1	0.904	0.342
`in-2004`	1,382,908	13,591,473	0.946	4.403	3	0.974	1.722
`eu-2005`	862,664	16,138,468	0.816	8.874	1	0.890	1.854
`road_central`	14,081,816	16,933,413	0.996	4.562	11	0.996	13.058
`road_usa`	23,947,347	28,854,312	-	-.-	-	0.997	20.227
`uk-2002`	18,520,486	261,787,258	-	-.-	-	0.974	29.958

reduced from 47 to 37 (1.1× speedup), while for `eu-2005` it is reduced from 10,343 to 25 (55× speedup), with similar modularities. This explains the good speedup of our algorithm over [**18**] in Table 3 for `eu-2005`, while we do not obtain a speedup for `belgium`.

In the remainder of this section, we will compare our method to the existing clustering heuristic developed by Riedy et al. [**18**]. We use the same global greedy

TABLE 2. Continuation of Table 1: remaining graphs of the DIMACS clustering challenge instances. From top to bottom, we list graphs from the `streets/`, `coauthor/`, `kronecker/`, `numerical/`, `matrix/`, `walshaw/`, and `random/` categories.

G	$\lvert V\rvert$	$\lvert E\rvert$	mod_1	t_1	$\%_1$	mod_2	t_2
luxembourg	114,599	119,666	0.986	0.125	6	0.987	0.138
belgium	1,441,295	1,549,970	0.992	0.440	10	0.993	1.106
netherlands	2,216,688	2,441,238	0.994	0.615	13	0.995	1.716
italy	6,686,493	7,013,978	0.997	1.539	13	0.997	5.256
great-britain	7,733,822	8,156,517	0.997	1.793	13	0.997	5.995
germany	11,548,845	12,369,181	0.997	2.818	14	0.997	9.572
asia	11,950,757	12,711,603	0.998	2.693	15	0.998	9.325
europe	50,912,018	54,054,660	-	-.-	-	0.999	45.205
coA...Citeseer	227,320	814,134	0.837	0.420	3	0.848	0.225
coA...DBLP	299,067	977,676	0.748	0.592	3	0.761	0.279
cit...Citeseer	268,495	1,156,647	0.643	0.894	2	0.682	0.315
coP...DBLP	540,486	15,245,729	0.640	6.427	1	0.666	2.277
coP...Citeseer	434,102	16,036,720	0.746	6.490	2	0.774	2.272
kron...logn18	262,144	10,582,686	0.025	13.598	0	0.025	2.315
kron...logn19	524,288	21,780,787	0.023	28.752	0	0.023	5.007
kron...logn20	1,048,576	44,619,402	-	-.-	-	0.022	10.878
kron...logn21	2,097,152	91,040,932	-	-.-	-	0.020	23.792
333SP	3,712,815	11,108,633	0.983	2.712	7	0.984	4.117
ldoor	952,203	22,785,136	0.945	6.717	2	0.950	2.956
audikw1	943,695	38,354,076	-	-.-	-	0.857	4.878
cage15	5,154,859	47,022,346	-	-.-	-	0.682	13.758
memplus	17,758	54,196	0.635	0.160	1	0.652	0.043
rgg_n_2_20_s0	1,048,576	6,891,620	0.974	1.614	5	0.977	1.383
rgg_n_2_21_s0	2,097,152	14,487,995	0.978	3.346	4	0.980	2.760
rgg_n_2_22_s0	4,194,304	30,359,198	-	-.-	-	0.983	5.799
rgg_n_2_23_s0	8,388,608	63,501,393	-	-.-	-	0.986	12.035
rgg_n_2_24_s0	16,777,216	132,557,200	-	-.-	-	0.988	25.139

matching and coarsening scheme (Algorithm 1) to obtain clusters as [**18**]. However, our algorithm is different in the following respects. *Stopping criterion:* in [**18**] clusters are only merged if this results in an increase in modularity and if no such merges exist, the algorithm is terminated. We permit merges that decrease modularity to avoid getting stuck in a local maximum and continue coarsening as long as the modularity is within 95% of the highest encountered modularity so far. *Matching:* in [**18**] a $\frac{1}{2}$-approximation algorithm is used to generate matchings, while we use the randomised matching algorithm from [**8**]. *Coarsening:* in addition to merging matched edges, we propose a centre potential to treat star-like subgraphs efficiently, which is not done in [**18**]. *Data storage:* [**18**] uses a clever bucketing approach to only store each edge once as a triplet, while we use adjacency lists (Section 4), thus storing every edge twice. A direct comparison of the performance of the DIMACS versions of both algorithms is given in Table 3. We outperform the algorithm from [**18**] in terms of quality. A fair comparison of computation times is

TABLE 3. Comparison between Algorithm 1 and the algorithm from [18], using raw, single-run results for large graphs from the 10th DIMACS modularity Pareto benchmark, `http://www.cc.gatech.edu/dimacs10/results/`. Here, $\cdot_1$ and $\cdot_2$ refer to our CUDA and TBB implementations, while $\cdot_O$ and $\cdot_X$ refer to the OpenMP and Cray XMT implementations of the algorithm from [18]. Timings have been recorded on different test systems.

G	mod_1	t_1	mod_2	t_2	mod_O	t_O	mod_X	t_X
`caida...Level`	0.764	0.531	0.792	0.185	0.540	0.188	0.540	3.764
`in-2004`	0.955	4.554	0.976	1.887	0.475	55.986	0.488	294.420
`eu-2005`	0.829	9.072	0.886	1.981	0.420	90.012	0.425	1074.488
`uk-2002`	-	-.-	0.974	31.121	0.473	181.346	0.478	772.359
`uk-2007-05`	-	-.-	-	-.-	0.476	496.390	0.480	36229.531
`belgium.osm`	0.992	0.447	0.993	1.187	0.660	0.562	0.643	10.571
`coP...DBLP`	0.641	6.612	0.668	2.367	0.496	1.545	0.501	9.492
`kron...logn20`	0.021	59.144	0.022	13.897	0.001	538.060	0.001	8657.181
`333SP`	0.983	2.712	0.985	4.321	0.515	1.822	0.512	27.790
`ldoor`	0.944	6.799	0.950	3.071	0.542	1.348	0.611	10.510
`audikw1`	0.847	15.341	0.858	5.180	0.560	1.635	0.558	9.957
`cage15`	0.640	32.804	0.677	14.308	0.513	4.846	0.512	48.747
`memplus`	0.635	0.175	0.654	0.038	0.519	0.034	0.520	0.903
`rgg_n_2_17_s0`	0.958	0.247	0.963	0.174	0.619	0.102	0.619	1.949

hard because of the different test systems that have been used: we (t_1 and t_2) used two quad-core 2.4 GHz Intel Xeon E5620 processors with a Tesla C2050, while the algorithm from [18] used four ten-core 2.4 GHz Intel Xeon E7-8870 processors (t_O) and a Cray XMT2 (t_X).

6. Conclusion

In this paper we have presented a fine-grained shared-memory parallel algorithm for graph coarsening, Algorithm 2, suitable for both multi-core CPUs and GPUs. Through a greedy agglomerative clustering heuristic, Algorithm 1, we try to find graph clusterings of high modularity to measure the performance of this coarsening method. Our parallel clustering algorithm scales well for large graphs if the number of threads is increased, Figure 3(b), and can generate clusterings of reasonable quality in very little time, requiring 4.6 seconds to generate a modularity 0.996 clustering of a graph with 14 million vertices and 17 million edges.

An interesting direction for future research would be the development of a local refinement method for clustering that scales well with the number of available processing cores, and can be implemented efficiently on GPUs. This would greatly benefit the quality of the generated clusterings.

7. Appendix

7.1. Reformulating modularity. Our first observation is that for every cluster $C \in \mathcal{C}$, by (1.1):

$$\zeta(C) = 2\,\omega(\text{int}(C)) + \omega(\text{ext}(C)). \tag{7.1}$$

Now we rewrite (1.2) using the definitions we gave before:

$$\begin{aligned}\operatorname{mod}(\mathcal{C}) &= \frac{\sum\limits_{C\in\mathcal{C}} \omega(\operatorname{int}(C))}{\Omega} - \frac{\sum\limits_{C\in\mathcal{C}} \zeta(C)^2}{4\,\Omega^2}\\ &= \frac{1}{4\,\Omega^2}\sum_{C\in\mathcal{C}} \left(4\,\Omega\,\omega(\operatorname{int}(C)) - \zeta(C)^2\right)\\ &\overset{(7.1)}{=} \frac{1}{4\,\Omega^2}\sum_{C\in\mathcal{C}} \left(4\,\Omega\left[\frac{1}{2}\zeta(C) - \frac{1}{2}\omega(\operatorname{ext}(C))\right] - \zeta(C)^2\right).\end{aligned}$$

Therefore, we arrive at the following expression,

$$\operatorname{mod}(\mathcal{C}) = \frac{1}{4\,\Omega^2}\sum_{C\in\mathcal{C}} \Big(\zeta(C)\,(2\,\Omega - \zeta(C)) - 2\,\Omega\,\omega(\operatorname{ext}(C))\Big). \tag{7.2}$$

As

$$\operatorname{ext}(C) = \{\{u,v\} \in E \mid u \in C, v \notin C\} = \bigcup_{\substack{C'\in\mathcal{C}\\ C'\neq C}} \operatorname{cut}(C,C'),$$

as a disjoint union, we find (2.1).

7.2. Merging clusters. Let $C, C' \in \mathcal{C}$ be a pair of different clusters, set $C'' = C \cup C'$ and let $\mathcal{C}' := (\mathcal{C} \setminus \{C, C'\}) \cup \{C''\}$ be the clustering obtained by merging C and C'.

Then $\zeta(C'') = \zeta(C) + \zeta(C')$ by (2.3). Furthermore, as $\operatorname{cut}(C, C') = \operatorname{ext}(C) \cap \operatorname{ext}(C')$, we have that

$$\omega(\operatorname{ext}(C'')) = \omega(\operatorname{ext}(C)) + \omega(\operatorname{ext}(C')) - 2\,\omega(\operatorname{cut}(C,C')). \tag{7.3}$$

Using this, together with (7.2), we find that

$$\begin{aligned}4\,\Omega^2(\operatorname{mod}(\mathcal{C}') - \operatorname{mod}(\mathcal{C})) &= -\zeta(C)\,(2\,\Omega - \zeta(C)) + 2\,\Omega\,\omega(\operatorname{ext}(C))\\ &\quad - \zeta(C')\,(2\,\Omega - \zeta(C')) + 2\,\Omega\,\omega(\operatorname{ext}(C'))\\ &\quad + \zeta(C'')\,(2\,\Omega - \zeta(C'')) - 2\,\Omega\,\omega(\operatorname{ext}(C''))\\ &\overset{(7.3)}{=} -\zeta(C)\,(2\,\Omega - \zeta(C)) + 2\,\Omega\,\omega(\operatorname{ext}(C))\\ &\quad - \zeta(C')\,(2\,\Omega - \zeta(C')) + 2\,\Omega\,\omega(\operatorname{ext}(C'))\\ &\quad + (\zeta(C) + \zeta(C'))\,(2\,\Omega - (\zeta(C) + \zeta(C')))\\ &\quad - 2\,\Omega\left[\omega(\operatorname{ext}(C)) + \omega(\operatorname{ext}(C')) - 2\,\omega(\operatorname{cut}(C,C'))\right]\\ &= 4\,\Omega\,\omega(\operatorname{cut}(C,C')) - 2\,\zeta(C)\,\zeta(C').\end{aligned}$$

So merging clusters C and C' from $\mathcal{C}$ to obtain a clustering $\mathcal{C}'$, leads to a change in modularity given by (2.2).

7.3. Proof of the modularity bounds. Here, we contribute a generalisation of [**3**, Lemma 3.1] (where the bounds are established for unweighted graphs) to the weighted case. Let $G = (V, E, \omega)$ be a weighted graph and $\mathcal{C}$ a clustering of G, we will show that

$$-\frac{1}{2} \leq \operatorname{mod}(\mathcal{C}) \leq 1.$$

From (1.2),

$$\operatorname{mod}(\mathcal{C}) \leq \frac{\sum\limits_{C \in \mathcal{C}} \sum\limits_{\substack{\{u,v\} \in E \\ u,v \in C}} \omega(\{u,v\})}{\sum\limits_{e \in E} \omega(e)} \quad - \quad 0 \leq \frac{\sum\limits_{\substack{\{u,v\} \in E \\ u,v \in V}} \omega(\{u,v\})}{\sum\limits_{e \in E} \omega(e)} = 1,$$

which shows one of the inequalities. For the other inequality, note that for every $C \in \mathcal{C}$ we have $0 \leq \omega(\operatorname{int}(C)) \leq \Omega - \omega(\operatorname{ext}(C))$, and therefore

$$\begin{aligned}
\operatorname{mod}(\mathcal{C}) &= \frac{1}{4\,\Omega^2} \sum_{C \in \mathcal{C}} \left(4\,\Omega\,\omega(\operatorname{int}(C)) - \zeta(C)^2\right) \\
&\overset{(7.1)}{=} \frac{1}{4\,\Omega^2} \sum_{C \in \mathcal{C}} \Big(4\,\Omega\,\omega(\operatorname{int}(C)) - 4\,\omega(\operatorname{int}(C))^2 - 4\,\omega(\operatorname{int}(C))\,\omega(\operatorname{ext}(C)) \\
&\qquad\qquad - \omega(\operatorname{ext}(C))^2\Big) \\
&= \frac{1}{4\,\Omega^2} \sum_{C \in \mathcal{C}} \left(4\,\omega(\operatorname{int}(C))\,[\Omega - \omega(\operatorname{ext}(C)) - \omega(\operatorname{int}(C))] - \omega(\operatorname{ext}(C))^2\right) \\
&\geq \frac{1}{4\,\Omega^2} \sum_{C \in \mathcal{C}} \left(0 - \omega(\operatorname{ext}(C))^2\right) \quad = -\sum_{C \in \mathcal{C}} \left(\frac{\omega(\operatorname{ext}(C))}{2\,\Omega}\right)^2.
\end{aligned}$$

Enumerate $\mathcal{C} = \{C_1, \ldots, C_k\}$ and define $x_i := \frac{\omega(\operatorname{ext}(C_i))}{2\,\Omega}$ for $1 \leq i \leq k$ to obtain a vector $x \in \mathbf{R}^k$. Note that $0 \leq x_i \leq \frac{1}{2}$ (as $0 \leq \omega(\operatorname{ext}(C_i)) \leq \Omega$) for $1 \leq i \leq k$, and because every external edge connects precisely two clusters, we have $\sum_{i=1}^k \omega(\operatorname{ext}(C_i)) \leq 2\,\Omega$, so $\sum_{i=1}^k x_i \leq 1$. By the above, we know that

$$\operatorname{mod}(\mathcal{C}) \geq -\|x\|_2^2,$$

hence we need to find an upper bound on $\|x\|_2^2$, for $x \in [0, \frac{1}{2}]^k$ satisfying $\sum_{i=1}^k x_i \leq 1$. For all $k \geq 2$, this upper bound equals $\|(\frac{1}{2}, \frac{1}{2}, 0, \ldots, 0)\|_2^2 = \frac{1}{2}$, so $\operatorname{mod}(\mathcal{C}) \geq -\frac{1}{2}$. The proof is completed by noting that for a single cluster, $\operatorname{mod}(\{V\}) = 0 \geq -\frac{1}{2}$.

Acknowledgements

We would like to thank Fredrik Manne for his insights in parallel matching and coarsening, and the Little Green Machine project, `http://littlegreenmachine.org/`, for permitting us to use their hardware under project NWO-M 612.071.305.

References

[1] D. A. Bader, P. Sanders, D. Wagner, H. Meyerhenke, B. Hendrickson, D. S. Johnson, C. Walshaw, and T. G. Mattson, *10th DIMACS implementation challenge - graph partitioning and graph clustering*, 2012. `http://www.cc.gatech.edu/dimacs10`

[2] H. Bisgin, N. Agarwal, and X. Xu, *Does similarity breed connection? - an investigation in Blogcatalog and Last.fm communities.*, Proc of. SocialCom/PASSAT'10, 2010, pp. 570–575. DOI 10.1109/SocialCom.2010.90.

[3] U. Brandes, D. Delling, M. Gaertler, R. Gorke, M. Hoefer, Z. Nikoloski, and D. Wagner, *On modularity clustering*, IEEE Trans. Knowledge and Data Engineering **20** (2008), no. 2, 172–188. DOI 10.1109/TKDE.2007.190689.

[4] T. Bui and C. Jones, *A heuristic for reducing fill-in in sparse matrix factorization*, Proc. Sixth SIAM Conference on Parallel Processing for Scientific Computing (Philadelphia, PA, USA), SIAM, 1993, pp. 445–452.

[5] A. Clauset, M. E. J. Newman, and C. Moore, *Finding community structure in very large networks*, Phys. Rev. E **70** (2004), 066111. DOI 10.1103/PhysRevE.70.066111.

[6] T. A. Davis and Y. Hu, *The University of Florida sparse matrix collection*, ACM Trans. Math. Software **38** (2011), no. 1, Art. 1, 25pp, DOI 10.1145/2049662.2049663. MR2865011 (2012k:65051)

[7] I. S. Duff and J. K. Reid, *Exploiting zeros on the diagonal in the direct solution of indefinite sparse symmetric linear systems*, ACM Trans. Math. Software **22** (1996), no. 2, 227–257, DOI 10.1145/229473.229480. MR1408491 (97c:65085)

[8] B. O. Fagginger Auer and R. H. Bisseling, *A GPU algorithm for greedy graph matching*, Proc. FMC II, LNCS, vol. 7174, Springer Berlin / Heidelberg, 2012, pp. 108–119. DOI 10.1007/978-3-642-30397-5_10.

[9] B. Hendrickson and R. Leland, *A multilevel algorithm for partitioning graphs*, Proc. Supercomputing '95 (New York, NY, USA), ACM, 1995. DOI 10.1145/224170.224228.

[10] B. Hendrickson and E. Rothberg, *Improving the run time and quality of nested dissection ordering*, SIAM J. Sci. Comput. **20** (1998), no. 2, 468–489, DOI 10.1137/S1064827596300656. MR1642639 (99d:65142)

[11] J. Hoberock and N. Bell, *Thrust: A parallel template library*, 2010, Version 1.3.0.

[12] G. Karypis and V. Kumar, *Analysis of multilevel graph partitioning*, Proc. Supercomputing '95 (New York, NY, USA), ACM, 1995, p. 29. DOI 10.1145/224170.224229.

[13] B. W. Kernighan and S. Lin, *An efficient heuristic procedure for partitioning graphs*, Bell System Technical Journal **49** (1970), 291–307.

[14] J. Leskovec, K. J. Lang, A. Dasgupta, and M. W. Mahoney, *Statistical properties of community structure in large social and information networks*, Proc. WWW '08 (New York, NY, USA), ACM, 2008, pp. 695–704. DOI 10.1145/1367497.1367591.

[15] M. E. J. Newman, *Fast algorithm for detecting community structure in networks*, Phys. Rev. E **69** (2004), 066133. DOI 10.1103/PhysRevE.69.066133.

[16] M. E. J. Newman and M. Girvan, *Finding and evaluating community structure in networks*, Phys. Rev. E **69** (2004), 026113. DOI 10.1103/PhysRevE.69.026113.

[17] M. Ovelgönne, A. Geyer-Schulz, and M. Stein, *Randomized greedy modularity optimization for group detection in huge social networks*, Proc. SNA-KDD '10 (Washington, DC, USA), ACM, 2010.

[18] E. J. Riedy, H. Meyerhenke, D. Ediger, and D. A. Bader, *Parallel community detection for massive graphs*, Proc. PPAM11 (Torun, Poland), LNCS, vol. 7203, Springer, 2012, pp. 286–296. DOI 10.1007/978-3-642-31464-3_29.

[19] S. E. Schaeffer, *Graph clustering*, Computer Science Review **1** (2007), no. 1, 27–64. DOI 10.1016/j.cosrev.2007.05.001.

[20] B. Vastenhouw and R. H. Bisseling, *A two-dimensional data distribution method for parallel sparse matrix-vector multiplication*, SIAM Rev. **47** (2005), no. 1, 67–95 (electronic), DOI 10.1137/S0036144502409019. MR2149102 (2006a:65070)

[21] F. Zafar, M. Olano, and A. Curtis, *GPU random numbers via the tiny encryption algorithm*, Proc. HPG10 (Saarbrucken, Germany), Eurographics Association, 2010, pp. 133–141.

[22] Z. Zhu, C. Wang, L. Ma, Y. Pan, and Z. Ding, *Scalable community discovery of large networks*, Proc. WAIM '08, 2008, pp. 381–388. DOI 10.1109/WAIM.2008.13.

Mathematics Institute, Utrecht University, Budapestlaan 6, 3584 CD, Utrecht, the Netherlands

E-mail address: `B.O.FaggingerAuer@uu.nl`

Mathematics Institute, Utrecht University, Budapestlaan 6, 3584 CD, Utrecht, the Netherlands

E-mail address: `R.H.Bisseling@uu.nl`